Experimental Design

From User Studies to Psychophysics

Experimental Design

From User Studies to Psychophysics

Douglas W. Cunningham
Christian Wallraven

CRC Press is an imprint of the
Taylor & Francis Group, an **informa** business
AN A K PETERS BOOK

CRC Press
Taylor & Francis Group
6000 Broken Sound Parkway NW, Suite 300
Boca Raton, FL 33487-2742

First issued in paperback 2019

ISBN-13: 978-1-56881-468-1 (hbk)
ISBN-13: 978-0-367-38215-5 (pbk)

Library of Congress Cataloging-in-Publication Data

Cunningham, Douglas W. (Douglas William)
Experimental design : from user studies to psychophysics / Douglas W. Cunningham, Christian Wallraven.
p. cm.
Includes bibliographical references and index.
ISBN 978-1-56881-468-1 (alk. paper)
1. Computer science--Experiments. 2. Human-computer interaction--Experiments. 3. Experimental design. 4. Psychophysics I. Wallraven, Christian. II. Title.

QA76.27.C82 2011
519.5'7--dc23 2011031078

Visit the Taylor & Francis Web site at
http://www.taylorandfrancis.com

and the CRC Press Web site at
http://www.crcpress.com

Contents

Preface

For over 150 years, physiologists and psychologists have been performing experiments to determine what signals in the world humans and animals can extract, how those signals are converted into information, and how the information is then represented and processed. Recently, there has been an increasing trend of computer scientists performing similar experiments, although often with quite different goals. In computer science, the experiments are sometimes generically collected under the term *user studies* since they often study how people use (or react to) newly designed techniques, images, or interfaces. Properly defined, however, a user study—which is also called usability testing—refers to a class of human-factors experimentation that examines whether a finished product meets its design goals. While such studies are invaluable and should certainly be performed, they are rather inflexible since the experimenter has little control over what the users sees or does—that is, over the stimuli and task—because the system already exists. Moreover, most of the human experiments performed in computer science are *not* user studies. They do not test an existing, end-to-end system. Instead, these experiments seek to discover more general knowledge about the underlying parameters of the system and its influence on the user. For example, rather than performing a user study to see if the bathroom tile adviser expert system that we just finished creating meets its design goals,[1] we might wish to know whether our facial animation system is "better" than the others, or whether our dialog manager or text generator produced more natural conversations than existing state of the art techniques. We might even focus on more specific elements of our new technique: which aspects of the new technique are good enough (i.e., can people use the technique with sufficient skill, accuracy, or ease)? Which range of parameters is optimal? Naturally, we might even wish to focus on more general questions: how much can we compress these images or audio files before

[1] In the EU project Conversational Multimedia Interaction with Computers, we designed and tested a natural language interface (including an avatar) for an expert system, and the testbed did indeed specialize in bathroom tiles. For more, see EU project COMIC, IST-2001-32311.

people begin to notice? What sort of information must be present in a facial animation for it to look realistic and believable? To answer these and other questions, we need to perform something that is more akin to a perceptual experiment. Perceptual experiments have considerably more leeway in terms of choosing what to show, how and when to show it, whom to show it to, and what the users should do. In short, rather than conducting a rigid user study on an existing system, what most computer scientists would like to do is to explore the parameter space of their system in order to optimize it towards a certain goal.

Computer scientists are not only increasingly conducting human experiments, but are in fact increasingly being *required* to do so. While at first glance such experiments seem to be easy to design, carry out, and analyze, a century and a half of experience has taught psychologists that there are numerous hidden traps which can waylay the unaware. This book will focus on basic experimental methodology: how do we go about designing a valid perceptual experiment? What are the advantages and disadvantages of the different methodologies? Without knowledge of experimental design in general and the hidden traps of various tasks specifically, seemingly valid conclusions that have been derived logically from solid-looking results will most likely be completely inaccurate. Additionally, there are many tips and tricks that can make conducting experiments easier, more accurate, subtler, or more efficient. Unfortunately, most of these traps and tips are not obvious. In short, the increased flexibility that perceptual experiments offer comes at the cost of a need for increased vigilance, rigor, and expertise. Depending on the exact question something as strict as a proper *psychophysical* design might be needed. Psychophysics is a term used to describe the field or the set of experimental methodology that was first created by Gustav Fechner in 1860. The goal of psychophysics is generally considered to be one of providing mathematical descriptions of the functional relationship between variations in the physical world and the resulting variations in the psychological (or perceptual) world.

Given the intricacies involved in experimental design, attaining an advanced degree in experimental or cognitive psychology requires the successful completion of many years worth of classes—both theoretical and practical—on experimental design. Most computer science curricula do not—currently—include such courses. While there are a few reference works on experimental design, they are all explicitly written for an audience with the background experience, knowledge, and mindset of experimental psychologists. Thus, they are rarely accessible to computer scientists. Moreover, existing works focus exclusively on very, very simple stimuli or tasks (for very important reasons that will hopefully become clearer in Chapter 2). The use of complex, real-world images and situations—as is required in most of computer science—carries an additional set of pitfalls and even violates some of the central assumptions behind many of the traditional experimental designs.

The goal of this book is to provide a basic background on the design, execution, and analysis of perceptual experiments for the practicing computer scientist. This book is written for a general audience. Beyond basic math and logic skills, together with some knowledge of basic calculus, no specialized knowledge is required. The information in this book provides an aid to the design and execution of effective experiments as well as to understanding and judging existing experiments. Moreover, it offers an introduction to the basic jargon and concepts used in experimental psychology, opening up access to existing experimental design texts if more information on a specific technique or analysis is needed.

Acknowledgments

We would like to thank the many people who helped in creating this book. We would like to thank Erik Reinhard, Tania Pouli, Ben Long, and Monica De Filippis for their insightful comments and constructive criticisms of the manuscript. We would like to thank Alice Peters for her guidance, understanding, and support throughout the project. Thanks also go to the reviewers and editorial team, including Elaine Coveney. Finally, a special thanks goes out to Sarah Cutler for her care and consideration throughout the project. It was a true pleasure working with you.

I would like to thank my Beloved for being infinitely supportive and understanding during the writing process, as well as for being inquisitive about the associated deadlines ("Wait...*this* February?"). Also, thanks to Kiran and Finley for being the two most wonderful and inspiring children one can possibly imagine—both in 2D (via Skype) and, most importantly, in 3D. –CW

I would like to thank my loving wife, Monica, for her support both emotional and technical. I would also like to thank my son Samuel for being a great source of warmth, amusement, and distraction. As always, my parents deserve medals. Without them, I would not be a fraction of what I am, and I doubt this book would ever have been finished. I would also like to thank my brother Thomas, who is an inspiration to me in so many ways. A special thanks goes to my doctoral adviser Thomas F. Shipley who fostered in me a deep appreciation for the complexities inherent in proper experimentation (regardless of the research question) as well as a love for the mathematical elegance that can be found in a well-designed experiment. –DWC

Acknowledgments

1

Introduction

Chapter 1

What Is an Experiment?

Most people today have an intuitive understanding of what *experimentation* is. This can be formulated along the lines of: perform an action—or a series of actions that are variants of each other—in order to answer a question. The answers provided by experimentation may tell us about small and simple things (such as answering the question, "What does this switch do?") or it might tell us about something large and complex (such as answering the question, "How do people recognize objects?"). What kind of answer we get depends on many factors, including what the question was and what we did during the experimentation.

At one extreme, the question itself may be something formal, such as a *hypothesis* (or formal statement of what one thinks should happen in a given circumstance) that has been derived through logical principles from basic assumptions. At the other extreme, the question can be something much more informal. It can even be as vague as, "What happens if . . . ?" In some respects, then, experimenting with something is often thought of as controlled playing with the aim of understanding something better (whether it is an idea, situation, toy, contraption, technique, etc.). Most questions fall somewhere between these two extremes.

The types of things one can study also fall somewhere along a continuum. On the one side, there are *observational* studies, which examine events or situations that arise spontaneously, without intervention by the experimenter. For example, Jane Goodall became quite famous in the 1970s and 1980s for studying the naturally occurring behavior of chimpanzees in the wild. This form of experiment is the norm in many areas of science, from anthropology to astronomy to zoology. In some cases, such as with the natural behavior of wild chimpanzees, native tribes in the Amazon, or supernovas, it is impossible—even in principle—to cause the events that interest us. In such cases, observational experimentation is the only option available. In other cases, such as in studying epidemic outbreaks, it is possible to generate the events of interest, but unethical to do so. Regardless of the reason, in an observational experiment we must observe the events or actions in the exact form in which they have arisen and wait for the specific events or variants that interest us to occur naturally.

At the other extreme lies *controlled* research (also called experimental research). This form of experimentation relies on events or actions that are caused by the experimenter intentionally, and is the method of choice in most natural sciences, including psychology. Indeed, the ability to repeatedly and reliably produce a specific event, image, situation, etc. (referred to as the *stimulus*), as well as finely controlled variants of it, is at the core of most perceptual research methodology. The reasons for this will become much clearer in Chapter 2.

Increasingly, computer scientists are being required to conduct controlled, human experiments, generally using variants of established perceptual research methodology. Although experimental psychologists undergo several years worth of training in basic research methodology in addition to several years worth of supervised practical application of that methodology, computer scientists generally do not. Indeed, very few computer science programs offer even a single course on designing controlled, human experiments. This book will focus exclusively on the basic issues involved in designing and analyzing controlled research, with particular emphasis on the need for rigor, the various tasks that are available, and the types of data analyses that are appropriate. These topics provide the fundamental knowledge necessary to more easily understand advanced or specialized texts on experimental design, such as Falmagne (1985); Gescheider (1997); Maxwell and Delaney (1990); Oppenheim (1992).

1.1 The Research Question

Given that most people have a clear—and rather accurate—conception of experimentation, it is surprising that most people do not have a clear idea of what constitutes a *single* experiment. If we want to know what a switch does, we can certainly experiment to find out: we can perform different actions to gain information about the switch and its functionality. We may have a specific hypothesis in mind—such as, if I move the switch to this position, then that light will turn on—or we may have no clear idea before starting the experiment what will happen. The presence or absence of a hypothesis does not make what we do any less of an experiment; in both cases, we are searching for knowledge by following specific procedures. It *does*, however, make a difference for the statistical analysis of the results. Performing a series of actions (e.g., performing an experiment) just to see what happens and then running statistical tests on all differences to see if one might be significant is strongly discouraged. The reasons for this will become clearer in Chapters 2 and 12, but essentially have to do with false positives: if we perform enough analyses we will eventually find some result to be significant purely by chance. It should be said that questions like, "What happens if I do X?" (where X is some action) represents a situation where one has no a priori idea what will happen and therefore the lack of a hypothesis, and that would preclude statistical tests. The statement, "Doing X will affect Y" (where Y is some measured value), however, is a clear hypothesis and a minor rephrasing of the question, "Does performing X affect Y?" In practice—since every experiment requires us to vary something and measure something—the difference between the two questions is usually just a matter of wording.

What are the specific procedures? What are the precise elements of a single experiment? If we flip the switch once, is that an experiment? The answer is yes and no. Experimentation, as we saw above, is about answering a question. To determine if a single flip of the switch represents a single, complete experiment, we need to know what the research question is. For example, imagine that we know what the switch does (e.g., it turns a specific light on) and that the switch is in the off position. If we want to know if the bulb is broken, then we can flip the switch once and get an answer. In this case, a single flip would represent a single experiment. Notice that since we have a specific hypothesis, we can engage in confirmatory experimentation. If, on the other hand, we do not know what the switch does and want to find out—i.e., we lack a specific hypothesis, and thus must engage in exploratory research—then flipping the switch once is only part of what we need to do to answer that question. As such, it would not constitute a *complete* experiment.

If we turn the previous observation around, we can see three critical aspects of experimental design. The first is that the research question is the core of an experiment: if we do not know what the question is, we cannot go about finding an answer to it. While this may seem obvious, many, many experiments fail to specify explicitly the central question. The value of a clearly formulated research question cannot be overstated.

The second observation is derived from the first: the more precisely the question is formulated, the clearer it will be *what* we must do to answer the question. As we saw, the research question, "What does this switch do?" is too vague either to provide a unique answer or to be answered quickly and easily—what precisely does "do" mean here? It can be interpreted to mean either that we want to know if the switch controls a specific light bulb, if it flips up and down (i.e., a toggle switch), or if it rotates (i.e., a dimmer switch), or any of a number of other possibilities. The set of actions that we need to perform depends on the precise meaning of the question. If we want to determine how the switch moves, then the set of actions is clear (e.g., we will try to flip or rotate the switch). Likewise, the type of answer that is valid is also clear: the answers we are looking for will come in the form of "the switch toggles between two alternate positions (up and down) when force in the appropriate direction is applied, but does not rotate." If, on the other hand, we are not interested in the switch's *type* but instead are interested in its *function*, then we will perform a very different set of actions and expect a very different form of answer.

The third observation is that creating a precise, explicit research question not only involves using specific terms, but also involves listing known assumptions. This can be seen in the first interpretation of our question (whether the switch controlled a specific lamp). Here, we have a very specific question and several assumptions (including that the bulb is not broken, that the switch is a toggle switch, and that the switch is in the off position). The more specific the question and assumptions are, the clearer it will be what needs to be done to get an answer.

Given the need for specificity, we might rephrase the interpretation of our research question, "What does this switch do?" to be, "What elements (if any) of this room does this switch control?" While this version of the question is clearer, it still needs some improvements. For example, what does the term "elements" mean? What does "control" mean? What time frame between the flip of the switch and the change in an "element" will we define as constituting control? By specifying precisely what we mean for the purposes of *this* experiment by certain terms, we are providing *operational definitions*. We might also wish to list any additional assumptions (for example, that we know that the switch affects things in this room and not the next one; that we know what type of switch we have, how many positions it has, and what its current position is, etc.). Normally, we are not explicitly aware of *all* of the assumptions that we have, but it can be very useful to try to determine as many of them as possible. Knowing all of our assumptions can be particularly helpful in figuring out what occurred in our experiment when we obtain a result that is either unexpected or contrary to existing (i.e., published) results.

1.2 The Relationship Between Hypothesis and Task

Once we have formulated our research question, the next step is often to list the actions that are possible in the experimental situation. This step constrains the types of experiments that are either possible or necessary. To help illustrate how to go about this, let us take as an example the research question, "What elements in this room (if any) does this switch control?" and make the assumptions that the switch is a toggle switch with two positions (up and down), that it is currently in the down position, and that the current state (off versus on) is unknown. In this situation, the list of actions that might plausibly answer the research question is very small: we can only change the switch to the up position. Other actions that are physically possible (e.g., trying to set the switch exactly halfway between up and down, dancing in front of the switch) are ruled out since prior experience with toggle switches suggests that such a position is highly unlikely to lead to an answer to our question. This prior experience might come from previous experiments or even from everyday-life experience. In the context of systematic experimentation, everyday-life experiences are often referred to as *anecdotal* evidence, and is considered to be important, but is not given as much weight as scientific evidence. Beyond the single, simple action of moving the switch the up position, we might consider extended or combined actions. These include the *back-and-forth flip*, where we change the switch from down to up and back to down one time; the *rapid flick*, where we change the switch from down to up and back to down one time as fast as possible; and the *repeated rapid flick*, where we change the switch from down to up and back to down as fast as possible many times.

For some questions (like this one), it will be easy to list all the possible actions. For others, such as, "Does my facial animation system produce better results than other systems?" it is not. In such cases, this may be an indication that the question is too vague. For example, what does "better" mean, operationally? We might decide that we want people to think that our animations involve a real face, in which case "better" means "more realistic." On the other hand, we might decide that "better" means that the animations are "better able to convey meaning" or "seen as being more sincere." Refinements of the question might be, "Does my animation system lead to higher recognition of facial expressions than other systems?" or "Are the facial expressions generated by my animation system seen as being more sincere than those of other systems?"

A large list of possible actions may also merely be an indication that the question involves a large or complex domain where a nearly infinite number of actions are possible. In such a case, we must find some other way of constraining the possible actions. For example, how many other animation systems are we referring to when we say "than other systems"? Do we want to test *all* systems that have ever existed? Or can we make the (reasonable) assumption that state-of-the-art animation systems are better (however we defined that term) than those created in—and not altered since—the 1980s or earlier? If we can make this further assumption (and change "than other systems" to "than state-of-the-art systems"), we can vastly reduce the number of animation systems that need to be tested. We might further restrict the testing we need to do by wondering if we need to test *all* state-of-the-art systems or if it is sufficient to test only the animations systems that fall into the same general category as ours.

Once we have refined our question and made a few additional assumptions, we may still find that the number of actions is too large. For example, the actions that people might take when confronted with a particular facial expression is rather large. Just as we used prior experience to rule out the halfway position on a toggle switch as unlikely to produce a useful answer, we might be able to shorten the list of possible actions for testing our facial animation system by determining which potential actions reflect the conveyance of meaning. We might also choose to initially test the two conditions that are—based on casual observation or anecdotal evidence—the most different. If such a *pilot* experiment does not find a difference between the extremes, then the smaller differences are unlikely to be different. Finally, additional, "quick-and-dirty" pilot experiments can be run using a subset of the actions or stimuli to rule out options or support new assumptions.

This process of defining a research question, noting assumptions, and listing and prioritizing potential actions has two primary results. First—based on the question we are asking, the actions that can occur, the type of real-world situations that are important, the resources available, and a host of other factors—we select a task. The task should be as close to the real-world behavior that we are interested in as possible, but still be amenable to study in a highly controlled environment. Note that all of the tasks listed in Part II have been used to help answer different aspects of the question, "How do people recognize facial expressions?" The issue of tasks is discussed in more detail in Chapter 3.

Second—once we decide what can, in principle, be done—we need to decide what we expect to happen (our hypotheses). While these can be specific (e.g., based on the exact task we have chosen), they are usually more general (i.e., using terms defined in our research question). The hypotheses can be very simple (e.g., that people will recognize expressions more often with our system than with any other), or very elaborate (e.g., a specific list of the cases and conditions where the different systems will prove to be superior can be constructed). Hypotheses can also be theory-driven, or more exploratory.

1.3 The Experiment

Once it is clear what we want to know, what actions are possible, and what we expect to happen, we can run a set of tests. That is, we try specific variations, one at a time, under specific circumstances. The variations and circumstances chosen depend on other factors in the experimental design, and are discussed in Chapter 2 and Part III. Each single execution of a variation (whether a simple action like flipping the switch or a complex action like a rapid flip) is called a *trial.* The full collection of trials that addresses the current research question is referred to as an *experiment* (since only the full set of trials can provide us with an answer to the question).

Except under certain special conditions, it is a very bad idea to change the experimental design in the middle of an experiment, since this will almost always lead to uninterpretable data. Plan the experiment well, then run it and analyze the full dataset. After running an experiment it will often be clear that some possible actions were overlooked, or that some new variations would be worth looking at, or that we have new expectations, or even that some of the operational definitions need to be changed. Testing these new hypotheses, actions, and variations will require new experiments.

A final word of warning on interpreting the results of experiment is warranted. A considerable amount has been written in both philosophy and psychology on causality and correlation. If you enter a room, and precisely at that instant the lights turn on, the temporal co-occurrence of the two events tends to encourage us to wonder if our entering the room (i.e., our action) caused the lights to go on or whether it was just a coincidence. When relying on anecdotal or observational evidence, it is very difficult—if not impossible—to be sure. If we experiment with the situation—such as by repeatedly entering and leaving the room, etc.—we increase the certainty with which we can say that the action caused the effect, rather then merely being temporally coincidental with it. That is, a properly designed experience can be a great help in determining causality. Care needs to be taken nonetheless when interpreting the results to be sure that the experiments allows one to make causal attributions. After reading through this book, you should have a much clearer understanding not only of why this is so, but also of how to design an experiment so that you can be sure of your interpretation. For more on causality itself, we recommend that the reader consult the extensive philosophical literature, most notably David Hume, Auguste Comte, Rudolf Carnap, and Karl Popper.

1.4 How to Use This Book

Chapter 2 examines the general, overarching issues behind experimental design. It describes some of the guiding design criteria that underlie the issues included in the rest of the book, and examines in detail the need for control and specificity on the one hand, and the need for generality on the other. Since the unique interplay of these mutually exclusive needs lies at the core of nearly every decision in experimental design, we recommend that most readers look at this chapter. If you are new to experimental design, it will provide a framework for understanding the rest of the book. Experienced experimenters are also encouraged to read the chapter, as many of the terms and concepts that reappear throughout the book are introduced there.

The next three parts of the book are designed as a reference, and provide an overview of the elements of an experiment. They will provide descriptions and discussions of the different options and variants for each aspect of an experiment. Part II examines the many different classes of task, with each chapter covering a different one. Each chapter starts with a brief description of the task, including the typical questions that can be asked with it and what it generally measures. This should help in deciding if a particular task might help answer the research question. Following a more detailed description of the task and its issues, the chapters end with one or more concrete examples. At least one detailed example for each task will be drawn from research on the recognition of facial expressions. This allows not only deeper insight into the elements of an experiment when using complex stimuli, it also provides the opportunity to directly compare the different techniques. Part III discusses issues related to stimulus selection and presentation, including specific software for these issues. Part IV looks at analysis methods that are appropriate for the tasks discussed in Part II.

It is worth noting that the more you know about what is and is not possible for humans, the easier it will be for you to formulate concrete, plausible hypotheses and avoid hidden mistakes. Describing what is currently known about human perception is beyond the scope of this book, but we highly recommend the very thorough recent book Thompson et al. (2011).

Chapter 2

Designing an Experiment

This chapter examines the basics of experimental design and lays the foundation for the rest of the book. Section 2.1 starts with a discussion of why precision must lie at the core of any experiment—particularly experiments that involve human participants—and how this need for control affects all aspects of an experiment. In Section 2.2 the individual elements of an experiment are introduced and briefly discussed.

2.1 Specificity versus Generality

Experimental design is a balancing act between *specificity* and *generality*. On the one hand, experimental design demands extreme precision and control to ensure that usable, uniquely interpretable results can *in principle* be obtained. One fundamental rule in experimental design is that we can make *definitive* claims only about the specific situations we actually have measured. If, for example, we measured how well people point to a target, then we can draw conclusions about pointing behavior but not about how well people *walk* to the target. Any discussion of walking would not be based on actual data, and would therefore be speculation. Likewise, if we measure performance in only one specific situation or *condition*, then since any of a nearly infinite number of factors may have influenced the specific behavior found in that specific condition—we cannot make any definitive claims about *why* our participants behaved as they did. That is, we cannot *generalize* our conclusions beyond the condition we measured. Similarly, if we measure performance for only one person, then we can make detailed claims only about that one person's behavior. As an extension of this specificity, if we measure two conditions that differ along many dimensions—any of which might have caused or affected the behavior—then we again cannot draw any definitive conclusions about the causes or influences of the differences in behavior between the conditions. We can say only that the conditions differ, but not why they differ. In short, the more conditions we measure and the more tightly controlled the variation between conditions is, the more reliably we can talk about situations in general and about the causes and influences of behavior.

If this is taken to its logical extreme—that we can discuss only the exact items or situations in our experiment—then in order to be able to talk about large, broad categories we would have to measure every member of that category. For example, in order to talk about how people *in general* point to a target we would have to measure *all* people. Since that is neither possible nor desirable, we have to find a compromise. Fortunately, there are conditions under which we can generalize beyond what we actually measured, as long as a few simple rules are obeyed. These rules become obvious if experimentation is viewed as sampling an unknown function to determine its shape.

2.1.1 Exploring an Unknown Function

Section 2.1 contains the core of what one needs to know to understand experimental design, but it is only an overview. Here, we will discuss the issues in more detail. Any function has a range of possible values that can be given to it as input, and a range of possible values that it returns. For example, in the equation $f(x) = mx + b$, the term x represents the input to the function and $mx + b$ represents the output that will be returned for a given x.

An unknown function is shown in Figure 2.1. Imagine that we have a method for obtaining the results for specific input; that is, when we input the value of x_1 into a black box, we receive the value y_1 (we measure the output of the function at that point; see Figure 2.2). We now know precisely what the value of the function is at that point. We do not know, however, what the function's value might be elsewhere, such as at the point x_2. To know the function's value at x_2, we would have to actually measure it. If we measured all possible points (x_i), then we would know the exact shape of the full function. This is what we mean by specificity: we can only talk with certainty about points we have actually measured.

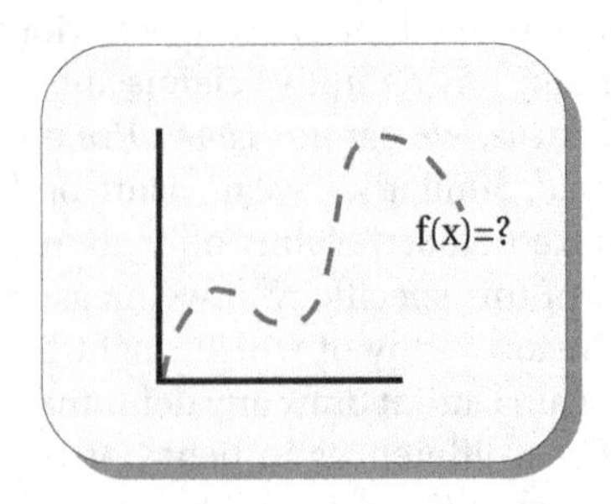

Figure 2.1. An example of an unknown function shown in dotted lines.

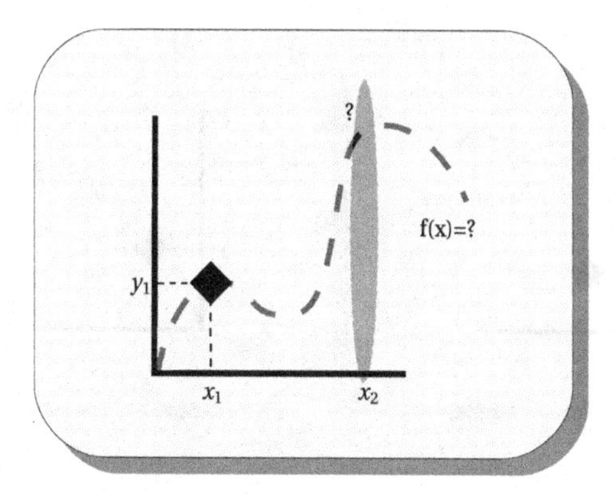

Figure 2.2. Sampling the unknown function. The function has been *sampled* (i.e., the value of the function has been measured) at point x_1, so we know with certainty what the function looks like at that point. Since we have not sampled at point x_2, we have no idea what the value of the function is at the point: this uncertainty as to the value of the function at x_2 is represented by a gray ellipse.

If we measure a number of points near each other along a single dimension (i.e., sample from a narrow range in the distribution of possible values on the dimension), then we can be reasonably certain of what the measurements at the intervening points would yield. For example, a linear interpolation between known values would produce an reliable estimate, assuming that the function is analytic and at least locally close to being linear (see Figure 2.3(a)). These two assumptions are very common in perception perception research and usually hold for most perceptual phenomena. Thus, by varying our stimuli systematically along a dimension, we can make clear, reliable claims about both the measured points and the intervening points (due to generalization). In fact, if enough points are sampled, and if the sampled points are well chosen, we can make claims about the dimension in general. Sampling is an important topic not only in experimental psychology but also in a vast number of other fields. The information provided here about sampling does not even begin to scratch the surface of what is known. For more information on sampling within a perception context, see Thompson et al. (2011). For more information on sampling within a mathematical or statistical context, see Bracewell (2000); Hayes (1994). For more on information sampling within a signal processing context, see Orfanidis (1995). For more information on sampling within a image processing context, see Castleman (1996); Gonzalez and Woods (2008). For more information on sampling within a computer graphics context, see Nguyen (2007).

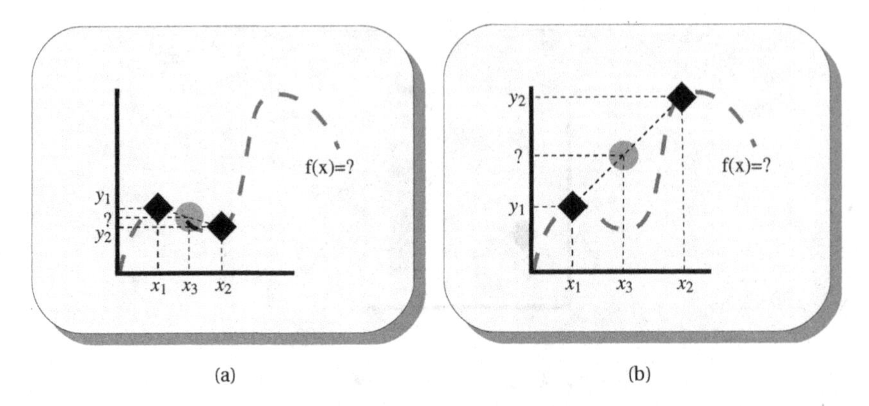

Figure 2.3. We know the value of the function at points x_1 and x_2, and can interpolate between them. (a) Closely sampled points (b) Points that are far apart.

Thus, we see that specificity and generality can be seen as belonging along a continuum. With enough specificity, we can be quite general. In addition to gaining enough specific information to make reliable generalizations by sampling more points, we can gain the needed information from other sources—such as samplings that other people have performed (i.e., previous studies) or general knowledge about the type of function we are sampling (i.e., general laws or principles based upon theory and previous research).

Naturally, the interpolation or generalization involves some uncertainty. If the sampled points are far apart from each other, then there is a high chance that the interpolated value might vary from the true value (see Figure 2.3(b)). The closer the sampled points are to each other, the more reliable the interpolated values will be. The downside of closely sampled points is that we have no idea what values the function will take outside the sampled region (that is, we will need many more sample points to adequately cover the function; see Figure 2.3(a)). Thus, there is a fundamental tension between measurement and interpolation—between specificity and generality—and this lies at the heart of experimental design.

2.1.2 Modeling Experimental Design

To make the balance between specificity and generality even clearer, let us take a simple example and try to express these intuitions mathematically.[1] Imagine

[1]The following equations are designed more to illustrate the concepts of experimental design than they are to actually mathematically model experimental design. For more mathematical models of experimental design, see Chapter 12 and Hayes (1994).

that we wanted to measure pointing accuracy. We will start with a very simple and well known target—a bullseye. We will take one participant, and have that participant reach and touch the target once (with his or her right hand). The distance between the center of the target and the point the participant touched will be our measurement. Since this value is the result of the experiment, it can be said to be dependent upon it, and is therefore referred to as the *dependent measure* or a *dependent variable.* We will refer to this measure as M. Naturally, if pointing accuracy is perfect, M will be zero (i.e., no offset). This experiment is sketched in Figure 2.4. If we let x be a vector representing a complete description of the specific situation (i.e., this includes absolutely everything, including the fact that there was a high-contrast bullseye target, there was one participant, the participant used his or her right hand, there was one one trial, the participant was paid for his or her participation, etc.) then $M(x)$ is the measured response for that particular situation. Furthermore, we define B as the internal process we are interested in. Since we cannot directly measure the participant's mental representation of the target or his or her internal motor planning, B represents the entire chain of internal events, from converting the signal into electrochemical signals (*transduction*) through the actual execution of the action (in this case, pointing). This chain of actions is called a *perception-action loop.* It is exceedingly difficult, if not impossible, to separate the *perception* side of this chain (all of the processes involved in extracting and representing the stimuli) from the *action* side (those processes involved in planning, preparing, and executing the motor behavior). The reasons why this is so will become apparent. First, however, we will summarize our conclusions so far in Equation (2.1), where e_w is an error term for that participant:

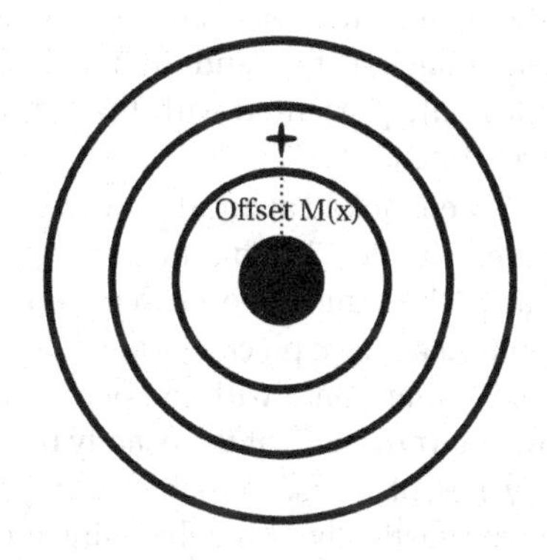

Figure 2.4. Pointing to a simple bullseye. The participant's response is represented with a cross, and the dotted line is the measured response M, for the specific situation x.

$$M(x) = B(x) + e_w. \tag{2.1}$$

At its simplest level, the error term (e_w) means that the action the participant produces is usually not exactly the action he or she meant to perform. Many factors are involved in this deviation. For example, the human retina has a finite resolution in space and time, so there is a degree of approximation when the stimulus is converted into electrochemical signals in the retina. In other words, there is an error in locating the target. There is also *physiological noise*—such as the random background firing of neurons or when a neuron is repeatedly presented with the same stimulus (Bradley et al., 1987; Newsome et al., 1989).

When looked at another way, this inaccuracy of the human system actually represents a critical but often forgotten aspect of humans: we *cannot* produce exactly the same action twice, no matter how hard we try. For example, take a pen and a blank piece of paper. Swing the pen towards the paper quickly and make a mark on it with the pen. Now, try to hit precisely the same spot again, moving your hand at the exactly the same speed, along exactly the same trajectory. It is, of course, possible to hit precisely the same spot—but only by slowing down drastically or maybe using a different trajectory. In fact, the pressure of the pen on the paper will probably also be different, as might be the amount of time the pen remains in contact with the paper. This natural variance probably lies at the core of our ability to learn new (and unexpected) things, as well as the ability to adapt existing perception-action loops to a changing environment (for more on this, see Baum, 1994).

Regardless of the source or meaning of the error term, it is clear that there is inherent, unintended variation in human behavior. That is, the actual behavior we measure is not solely a function of the desired perception-action loop, but is also a function of some inherent (and usually unintentional) noise. This needs to be represented in our equation. This error term e_w, then, is critical in understanding humans and for understanding experimental design. If e_w were always the same, then it would be a form of constant *bias* in that it would introduce a constant offset from the true value of $B(x)$. This means that we would be able to exactly produce the same behavior every time we tried, and by extension the measured behavior $M(x)$ would be solely a function of the underlying perception-action loop $B(x)$. This would mean that a single measurement is fully sufficient to determined $B(x)$.

Likewise, if the variance in e_w is small, or even negligible, then any individual, random sample of the function $M(x_i)$ will be close to $B(x_i)$. Thus, any given measurement will not truly reflect the underlying perception-action loop, but will be really close to it (see Figure 2.5(a)). If, on the other hand, the variance in e_w is large, any single measurement might diverge very strongly from the true value of $B(x)$ (see Figure 2.5(b)). Since the value of e_w (the amount of divergence of behavior from intent) is not constant (it varies from trial to trial), and we almost never know a priori how big the variance in e_w is, we need to find some way to estimate it. To put this another way, since Equation (2.1) has two unknowns ($B(x)$ and e_w), a single measurement (i.e., a single sampling of the function $M(x)$) *cannot* tell us *precisely* what $B(x)$ is. A single trial of a single condition can tell us what the behavior in a very specific situation is, but not why it is that way or even how close repetitions will be to that measured value.

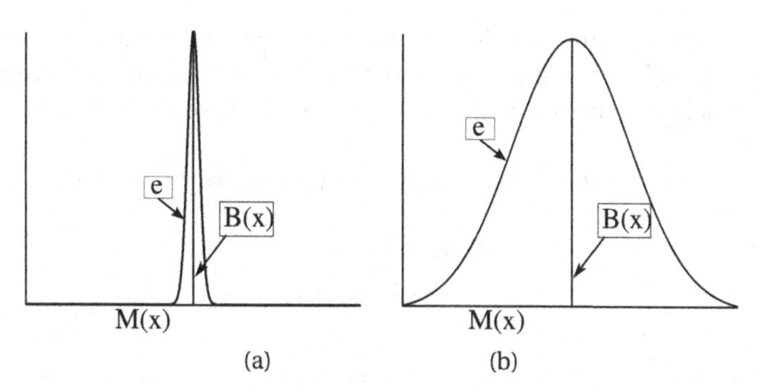

Figure 2.5. (a) If the variance of the error term e_w is small, the variance of the measurement $M(x)$ will be small. Thus, any random point on the function $M(x)$ will be close to $B(x)$. (b) A larger error term leads to a wider spread in the measurement.

2.1.3 Repeated Measures

The first method for isolating $B(x)$ derives from the observation that while e_w is different every time we measure $M(x)$ (i.e., for each trial), our underlying perception-action loop $B(x)$ is constant. This is one of the fundamental assumptions in perception research. Of course, there is some debate about whether $B(x)$ is truly constant. Clearly, behavior changes over time. For example, we can and do get better at pointing. Most evidence suggests, however, that $B(x)$ will remain constant (or nearly so) *for short periods of time.* This is one of the central reasons why experiments should generally be conducted over a limited period of time (usually two hours): to ensure that the perception-action loop we are trying to define—i.e., $B(x)$—does not change. Another reason for keeping a single experiment short is to reduce fatigue in our participants.

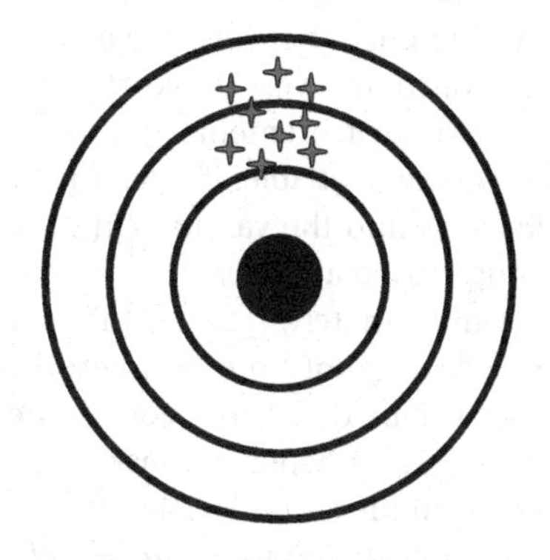

Figure 2.6. Pointing to a bullseye several times. The average of pointing errors will be a good approximation of $B(x)$.

Notice that the error in pointing will sometimes be to the left of the intended point, and sometimes to the right (and sometimes above, sometimes below). On average, the deviation from the intention is zero: there is no constant bias.[2] Thus, averaging over several measurements will yield a good

[2]Since in addition to having no offset, the frequency of occurrence of the different errors—left, right, above, or below—is equal, this is equivalent to saying that the values for the error function

estimate of $B(x)$. In other words, if we simply repeat the experiment exactly (x must be identical) over a short period of time, then the average pointing accuracy will be a good estimate of the underlying pointing ability (see, for example, Figure 2.6).

We can state this more formally, for multiple measures M_i and their error terms e_{wi}, as

$$\begin{aligned} M_1(x) &= B(x) + e_{w1}, \\ M_2(x) &= B(x) + e_{w2}, \\ &\dots \\ M_n(x) &= B(x) + e_{wn}, \end{aligned} \tag{2.2}$$

$$\bar{M}(x) = B(x). \tag{2.3}$$

As can be seen in Equation (2.2), we have multiple measurements M_i of a given situation x. While x will remain constant across the different measurements, the measurements themselves will not all be identical. We have previously defined the output of B to be constant for a constant situation, and thus any differences between the measurements M_i are due to variations in e_{wi}. As mentioned, it is very unlikely that e_w will be zero for any single measurement. With two measurements, it is more likely that the average will be zero, but it is still very improbable. If $B(x)$ is truly constant (and the situation x is absolutely identical) and the variance in e_w is small, then two measurements might be enough to isolate $B(x)$.

Unfortunately, it is hard to claim that x is *absolutely* identical. First, the mere passage of time between the two trials mean that many physiological states will be different. Second, the first trial can essentially be seen as an opportunity to practice this task before performing the second trial. Thus, performance on the second trial reflects more practice or learning than the first trial. This is referred to as a *learning effect* or *order effect*, and will be addressed in more detail below.

In general, more measurements mean a greater likelihood that e_w will average out of the equation. This is equivalent to the observation made above about functions (i.e., the more samples of a function we have, the easier it is to map it) applied to the distribution of e_w. In classical psychophysics, it is not uncommon to measure a given situation tens of thousands of times—for each participant! Since the effects in most perception experiments are rather large, a very rough estimate of e_w is sufficient to be able to get a decent idea of $B(x)$. Thus, five to twenty repetitions are usually sufficient. Fewer than five repetitions do not usually provide enough information to define e_w. Using more than twenty repetitions in a short period of time without specific precautions

are normally distributed, which is usually true in perception research. For more on Gaussian and other distributions see Chapter 12.

is likely to lead to fatigue, overlearning, or other masking effects (for more on what those precautions are, please refer to a text on advanced experimental design such as Falmagne, 1985; Gescheider, 1997; Maxwell and Delaney, 1990). If one is interested in a more subtle or a smaller effect (as is often the case, for example, in the study of language), then more measurements are needed. What precisely is meant by "effect" will be discussed more below. It should be mentioned that there are methods for determining the number of repetitions that one needs, and some of these methods are discussed in Chapter 12. The use of multiple measurements or repetitions of a given situation in an experiment is called a *repeated measures design.*

2.1.4 Multiple Participants

As already stated, we can reliably make definitive claims only about the specific situation x that we measured, and a natural part of the measured situation is the participant. In other words, if we only measure one person, we can talk only about that one person's results (and consequently, only his or her perception-action loop). We have no information whatsoever about how other people would perform. Maybe they will perform similarly, but we do not know this and we certainly do not know enough to define what "similarly" really means. So, in order to make claims about more than one person, we need to measure more than one person.

Since measuring more than one person will yield more than one measurement, it should be possible to use these multiple measurements—rather than repeated measurements of an identical situation—to isolate $B(x)$. If $B(x)$ is *identical* for all people, then measuring one person many times or many people one time become identical methods (e.g., the multiple crosses in Figure 2.6 might have come from one person measured many times, many people measured once, or some combination).

Unfortunately, in almost no real-world case is $B(x)$ perfectly identical across people. We might, however, assume that $B(x)$ consists of some element that *is* constant across all people and some element that varies between people. In other words, we can assume that there is a constant *population* perception-action loop $B(x)$ and that each person's *individual* perception-action loop is a (minor) deviation from it. This is very similar to the assumption made for the measurement $M(x)$ (i.e., where each person has an underlying perception-action loop and each measurement is a minor deviation from it). To model this, then, we can say that there is an underlying perception-action loop $B(x)$ and *two* error terms. The first error term is our original error term e_w: variations within a given participant or the *within-participant* error. The second error term reflects the deviations of each person from the population $B(x)$, which we will call the *between-participant* error, or e_b. We explicitly make the assumption that e_w and e_b are independent of each other and that they are linearly combined in determining $M(x)$.[3]

Notice that this new error term will affect the number of measurements (or samples of $B(x)$) that we will need, just as the within-participant term did. More specifically, the larger the between-participant noise e_b is, the more samples (or participants, in this case) we will need to be sure we have a good approximation of $B(x)$. How many participants will be enough? Just as with determining the proper number of repetitions, there are formulas for calculating this, which can be found in any standard statistics text (see, for example Hayes, 1994; Maxwell and Delaney, 1990), or in Chapter 12. History and tradition have shown that common-sense reasoning about sampling distributions in general can be a good guide here. In many low-level perceptual processes (e.g., the processing of line orientation, the detection of motion), the between-participant variance is very, very small. Thus, as long as we are sure that we have a good estimate for each specific person (which means running many trials for each person so that e_w is well known), then four people are sufficient to yield a reliable estimate of the entire population $B(x)$. In higher-level processes (such as the ability to solve math problems), e_b can be quite large and many more participants are required. Therefore, as more variance is expected, more participants will be needed. This is another reason that many cognitive psychology experiments have hundreds of participants, while perceptual experiments can be fully valid with ten. If we make the further assumption that the within-participant variance e_w is similar for all people, then we can use a large number of people with one trial per person with in order to estimate the underlying perception-action loop of all people $B(x)$. For more detail on how to decide how many participants to choose, see Chapter 12 or (Hayes, 1994; Maxwell and Delaney, 1990). For a multiple-participant version of our pointing experiment, with one measurement per person, we have

$$\begin{aligned} M(x_1) &= B(x_1) + e_{w1}, \\ M(x_2) &= B(x_2) + e_{w2}, \\ &\dots \\ M(x_n) &= B(x_n) + e_{wn}, \end{aligned} \tag{2.4}$$

$$\begin{aligned} M_1(x) &= B(x) + e_{w1} + e_{b1}, \\ M_2(x) &= B(x) + e_{w2} + e_{b2}, \\ &\dots \\ M_n(x) &= B(x) + e_{wn} + e_{bn}, \end{aligned} \tag{2.5}$$

$$\bar{M}(x) = B(x). \tag{2.6}$$

[3]These assumptions are fundamental to most of the statistical procedures that are used in psychology. For more information about this, see Chapter 12 and Hayes (1994).

In Equation (2.4) we see that we are measuring different situations x_i, and thus the function B is also given different inputs. Since the difference between the situations is (hopefully) solely a change in participants, we can parcel this out of x, yielding multiple measures of a single situation (defined as x). This allows a constant input to both M and B, with any variations between the values of M being again captured in the error terms, as seen in Equation (2.5).

It is important to note that in some cases the variance between participants (e_b) is extremely large. This not only tends to make it harder to precisely measure the effect we are interested in (i.e., it *masks* the effect), but it might lead us to wonder why our participants deviate so much from each other. A *bimodal distribution* (i.e., a distribution with two main peaks) might, for example, suggest that there are two distinct sub-populations—one of which is very accurate and one of which is not. Once we have identified a source of variance we can control for it. The resulting reduction in variance will reduce the error term e_b, and thus make it easier to detect the effect we were initially interested in. For example, imagine that in our multiple-participant pointing experiment we find a bimodal distribution, which might encourage us to classify the participants as belonging to one of two groups (those who were accurate and those who were not). Closer examination of the people in the two groups might reveal that all the people in the inaccurate group were left-handed. Since we had everyone point with his or her right hand (in an attempt to keep things as consistent as possible), all of the left-handed people were pointing with their nondominant hands. It is possible that this is biasing (adding a constant offset from the true pointing accuracy function $B(x)$) the results for these people. This is an example of what is referred to as a *confound* in psychology—some variable that we did not control for has such a large effect on the pattern of results that it is possible that this uncontrolled variable is producing our effects and not the variable that we manipulated. To test whether handedness had indeed affected the results of our hypothetical experiment, we can rerun the experiment with everyone pointing with his or her dominant hand. If this was the source of the bimodal distribution, then the new data should be unimodal (i.e., one main peak) and cleaner (i.e., less variance).

2.1.5 Two Conditions

In order for the average of the measurements M_i to approximate the perception-action loop $B(x)$, every aspect of the situation x must be identical across all samples. *Any* change from one trial to the next might add noise (random variations from $B(x)$). As we mentioned above, changing the participant from one measurement to the next alters the situation x. Adding the variance caused by this change of participant as an explicit term e_b allows us to more easily isolate $B(x)$. If we assume that all other possible changes are also

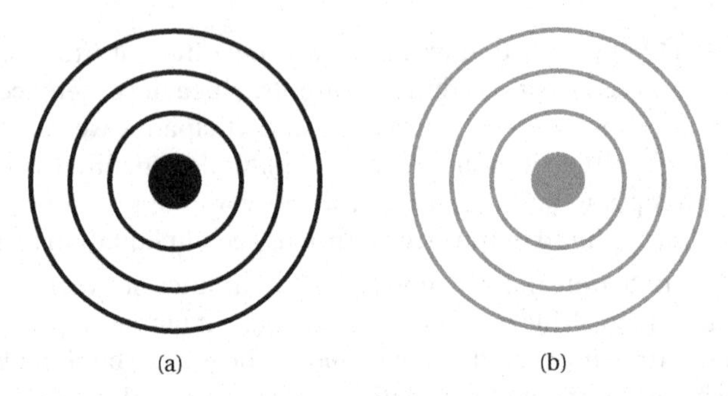

Figure 2.7. Two targets. (a) 100% contrast. (b) 50% contrast. Note that the contrast was defined and measured in stimulus space and preparation, and printing in this book may have reduced or otherwise altered it. If the targets were used as stimuli in an actual experiment, we would need to verify that the contrast presented is in fact the contrast we intended.

independent of each other and that their effects can be modeled as a weighted linear sum, then we can represent them as additional terms as well.

In our pointing example, most people would probably say that it is unlikely that reducing the contrast by a small amount will effect pointing behavior (see Figure 2.7). It is, however, possible. In fact, if the contrast change is large enough, it almost certainly will. If the contrast changes randomly between trials (both within-participant as well as between-participant), an additional noise term (e_c) can be added so that

$$M(x) = B(x) + e_c + e_w + e_b. \tag{2.7}$$

If we know that a small contrast change does not affect performance (i.e., causes no additional error) and the changes are all small, then contrast need not show up as a term in the equation (since the term will be zero or close to it). In order to determine that it has no effect, however, we need to first *explicitly control* for it. To do this, we will need (at least) two measurements; one with a high-contrast target—$M(x)$—and one with a lower one—$M(x+\Delta c)$. Here, we are explicitly varying our stimuli along a given dimension (contrast). The dimension along which the stimuli vary is referred to as a *factor*. Factors are the items that we explicitly alter and whose effects we wish to study. The specific fixed values that we use are *levels*. Since the only thing that determines the levels of a factor is the experiment, they are independent of the actual experiment (i.e., their value is known before the experiment). For this reason, they are called *independent variables*. This terminology also subtly highlights some of the assumptions in psychology: statistical psychology and experimental design explicitly assume that the independent variables are independent of each other and can be modeled as a weighted linear sum (see, for example, Hayes, 1994).

The complete combination of factors and levels for any single trial is a *condition.* If there were two factors (call them size and color) with two levels each (big and small for size and red and blue for color), then we would have four conditions (i.e., big and red, big and blue, small and red, and small and blue).

Obviously, some changes will be trivial and will not affect our measurement. The question of which changes will be trivial and which are important is one of the central challenges in scientific experimentation. Some changes will be unexpectedly important. Our a priori hunches about which changes will affect our experiment are not always accurate. If they were, we would not need to do experiments in the first place! Thus, without knowing beforehand which factor will play a role and which will not, it is best to try to make all trials as identical as possible along all dimensions (even down to the room the experiment is conducted in, the color of the walls in that room, the number and placement of objects in that room, etc.). This point really cannot be overemphasized.

In our hypothetical experiment on pointing to a high- and a low-contrast target, we have one factor with two levels. Notice that the two conditions are identical, with the sole exception that the contrast of the target is different. This difference is represented by defining one condition as the *baseline* situation (x, see Equation (2.8)), and the other as the variation on it ($x+\Delta c$, where Δc represents the change in contrast; see Equation (2.9)).[4]

$$M(x) = B(x) + e_w + e_b, \tag{2.8}$$

$$\begin{aligned} M(x+\Delta c) &= B(x+\Delta c) + e_w + e_b, \\ &= B(x) + B(\Delta c) + e_w + e_b. \end{aligned} \tag{2.9}$$

By taking the difference between Equations (2.8) and (2.9), something amazing happens: we can determine the precise effect of contrast change on pointing behavior:

$$\begin{aligned} M(x+\Delta c) - M(x) &= [B(x) + B(\Delta c) + e_w + e_b] - [B(x) + e_w + e_b] \\ &= B(\Delta c) \end{aligned} \tag{2.10}$$

The size of the difference between the two conditions—$B(\Delta c)$—is one meaning of the term *effect.* A large effect, then, would refer to a large difference in performance between the two conditions. The careful reader will notice that if $B(\Delta c)$ is zero, then the difference in contrast had no effect on performance and we do not need to control for it in future experiments. Likewise, if $B(\Delta c)$ is large, then it will be easier to detect that it is not zero (i.e., that there is an

[4]Note that splitting the function $B(x+\Delta c)$ into its component parts ($B(x)$ and $B(\Delta c)$) requires that the function B be homomorphic. Linear functions satisfy this property. Since we have assumed that the elements of x are independent of each other and can be modeled as a linear, weighted sum, B is homomorphic. For a detailed mathematical approach to experimental design, see Hayes (1994).

effect), which is usually the goal of an experiment. Thus, a large value for $B(\Delta c)$ can be detected even with an imprecise estimate of the error terms e_w and e_b. Contrariwise, small effect sizes will be masked by large error terms and need a more precise estimate of e_w and e_b. Thus, the number of participants and repetitions that are necessary are correlated strongly with the size of effect that we are searching for.

2.1.6 Change versus Transformation

The above conclusions are worth stating again, more generally: if we have two conditions that differ along one dimension, then we can determine the precise effect of the change between them on our measured behavior, regardless of how complex the stimuli and situation are. A note on interpretation and conclusions is important here. Although we designed the stimuli by taking one stimulus (the high-contrast target, which had a 100% contrast) and altering something to produce the other (the low contrast stimulus, which had a contrast of 50%), the experiment does not allow us to talk about the effect of *changing* the contrast. Why? First, we have only two conditions, so we can only talk about those two levels: we cannot talk about contrast in general. We cannot say whether other 50% differences in contrast (such as between a 10% and a 60% contrast target) are. Nor can we draw conclusions about larger or smaller contrast differences. We can only reliably discuss the difference in pointing to a 100% contrast target and a 50% contrast target (for more on this point, see Section 2.1.9). Second, and perhaps more critically, there is an important distinction between a *difference* between two conditions and a change or *transformation* between them. Talking about the perceptual effect of changing some variable requires that the participants actually see the temporal transformation from one condition to the other. Since the participants saw only two different levels of a contrast at two different points in time, we can talk about the effects of one condition in terms of the its difference to the other: the results of our hypothetical experiments inform us about $B(\Delta c)$, the effect of a specific 50% contrast difference. Since the participants do not see the transformation from high contrast to low (or from low to high) happen, we cannot say what the perceptual properties of that transformation are.

To make this a bit clearer, imagine that we perform two experiments. Both experiments consist of two stimuli: a picture of a vase and a picture of some ceramic shards. In both experiments we ask people to indicate if objects or events seen in the pictures are natural or unnatural. In the first experiment, we will show the participants the two pictures one at a time (i.e., on different trials). Prior experience would lead us to expect that most people will report that both stimuli are natural. Thus, we can say that there is no difference in naturalness between an intact vase and a broken vase. We cannot say anything about the transition from intact to broken. In the second experiment we show the participants the transformation of an intact vase into the ceramic shards in one condition and the transformation from a group of shards to an intact vase (without any obvious external influence) in the other condition. Most people would say that destruction of an intact vase is natural, but the sudden self-assembly of some shards into a vase is unnatural. That is, the change in one direction between conditions is natural, but the change in the other is unnatural—even though the conditions themselves do not differ in their naturalness.[5] Thus, returning to the pointing experiment, we can say that the high-contrast target is easier or more difficult to point to than the low-contrast target, but cannot say that increasing contrast by 50% decreases performance: no participant ever actually saw the contrast be transformed.

2.1.7 Symmetrical Experiments

In traditional experimental terminology, the 100% contrast condition is our baseline or *control condition* for the 50% condition. In this particular experiment, both conditions have all the same elements, and only one factor is changed between them. No dimension is added or subtracted, only altered. That means that we can use either condition as a baseline for the other. Such symmetry is not common, but it is very desirable. It is far more common for one condition to be a basic situation—and thus the baseline or control condition—and the other condition(s) to add new information or qualities (such as going from black and white to color) as opposed to changing a quality that is already present.

2.1.8 Measuring Perception

At this point, it might be a bit clearer why it is exceedingly difficult to isolate the perception side of a perception-action loop. We cannot—even in principle—directly measure perception. Perception, by definition, is not directly observable or measurable. All we can do is measure behavior (either overt behavior such as button pushing or covert behavior such neural activity). Thus, the $B(x)$ was defined as the entire perception-action loop. A closer examination of the implications of Equation (2.10) suggests that the only way to measure the perception side of the perception-action loop is to partition $B(x)$ into two parts: perception ($P(x)$) and the rest ($R(x)$)—which includes strategies, memory, decisions, motor activity, etc. This, of course, makes the strong assumptions that perception is in fact independent of everything else and its effects are linearly additive to it. Even if these assumptions were true, we would still need to construct two conditions or situations that are absolutely identical in terms of literally everything except the perception part, which is even harder to do in practice than in theory.

[5]For more on the importance of time in perception and on the directional nature of temporal perception, see Gibson (1979).

2.1.9 Multiple Levels of a Factor

The previous experiment provides information about the differences in pointing at a 100% contrast target and at a 50% contrast target. What would pointing at a 90% contrast target look like? We might be able to guess by linearly interpolating between our two known points (100% and 50%). What about pointing at a 30% contrast target? That is less clear. We might *extrapolate* (extend the line generated in a linear interpolation between the two known values). While extrapolation might provide reliable estimates near the known value of 50%, it will be less reliable the farther out we get. To be able to generalize along the entire dimension, then, we would need to choose several contrasts along the entire range of possible contrasts. These points should be close to one another to allow good interpolation, but far enough apart that they cover as much of the dimension as possible. In other words, we need to ensure that we have a *representative sample* of the whole dimension. For example, if we wanted to be able to talk about the entire dimension of contrast, then we might choose 11 levels ranging from no target (0% contrast) to full contrast (100%) in steps of 10%. Since people will not be able to point to a target they can't see (and such a condition will confuse and annoy them), it would probably be best not to use the 0% contrast condition, leaving us with only ten levels.[6] Since there is only one factor (contrast), the number of conditions is the same as the number of levels. In creating the stimuli, care should be taken so that each of the ten conditions is identical to all the others, with the sole exception of the change in contrast level. In our thought experiment, we will measure each of several participant's behavior several times. Since this design is symmetric, the designation of one of the conditions as baseline is arbitrary. For Equation (2.11), we choose to set 100% as the baseline. Since the basic change along our factor is a 10% drop in contrast compared to baseline, we can represent all the changes in the stimulus with a single variable p, and the different conditions as multiples of this. By comparing performance in any of the conditions to the baseline, we can precisely determine the effect of that specific contrast decrease. As mentioned, since this is a symmetric design, we can compare performance in any one condition to any other to see the effect of that specific change. By treating the results for ten conditions as samples from the underlying function, we can estimate the effect of any contrast decrease on pointing behavior as follows:

$$\begin{aligned} M(x_1) &= B(x) + e_w + e_b, \\ M(x_1 - \Delta p) &= B(x - \Delta p) + e_w + e_b, \\ M(x_1 - 2\cdot\Delta p) &= B(x - 2\cdot\Delta p) + e_w + e_b, \\ &\dots \\ M(x_1 - 9\cdot\Delta p) &= B(x - 9\cdot\Delta p) + e_w + e_b. \end{aligned} \tag{2.11}$$

[6]There are valid reasons for including blank trials, but none of them are applicable in the present case. For more on the inclusion of catch trials, see Chapter 3 or Gescheider (1997)

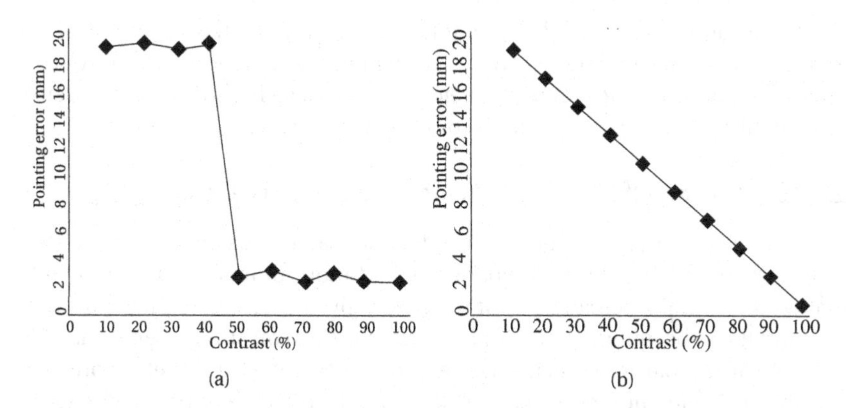

Figure 2.8. Hypothetical results for pointing to many targets of different contrast. (a) The pointing offset is more or less constant for values between 10% and 40%, as well as for values between 50% and 100%, suggesting that a threshold lies somewhere between 40% and 50%. (b) Pointing performance is a constant function of target contrast: equal changes in contrast produce identical changes in performance, regardless of the base contrast level.

Note that the input to both the measurement M and the perception-action loop B are represented in terms of a base condition x and multiples of 10% contrast change. While the input to the perception-action is indeed periodic, there is no guarantee that its output is. For example, performance may not differ much between 100% and 50% contrast (a contrast difference of 50%), but it certainly would between 50% and 0% (which is also a contrast difference of 50%). In short, a constant change on the physical side does not necessarily reflect a constant change in the perception-action loop (in fact, it rarely does).

If we compare any two conditions, as we have done so far, we learn only about those two conditions. Logically, then, if we compare multiple conditions, we might begin to learn the nature of the underlying function. The quickest and easiest way to do this is to visualize the function. Since we are measuring one value (pointing performance) at sampled locations along a single dimension, it would be reasonable to plot the results in a standard two-dimensional graph. Figure 2.8 shows two sets of hypothetical results. In Figure 2.8(a), pointing performance is more or less constant for values between 10% and 40% and for values between 50% and 100%, suggesting that a *threshold* lies somewhere between 40% and 50%. In Figure 2.8(b), pointing performance is a constant function of target contrast (within the range measured): equal changes in contrast produce identical changes in performance regardless of the base contrast level.

While some aspects of the function may be immediately apparent in the graph, others may be more elusive. A wide range of statistical tests have been developed, both to find descriptive elements of these functions and to determine if the differences are reliable. For more on statistics, see Part IV.

2.1.10 Order Effects and Between-Participants Design

As mentioned above, the second trial in a repeated-measures design differs from the first in that the participant has had more practice. In a more general sense, what participants see or do on any given trial might affect how they perceive the subsequent stimuli or how they perform on subsequent trials. For example, if the participants always saw the 100% contrast target before they saw the 50% one, then performance with a 50% target will reflect a change in contrast as well as a change in practice level. Thus, any difference between the two conditions might be due to contrast or to order of presentation, and we can make no definitive claims about which really was the cause. Updating Equation (2.9) to reflect the effects of order (Δo) gives us

$$\begin{aligned} M(x+\Delta c+\Delta o) &= B(x+\Delta c+\Delta o)+e_w+e_b, \\ &= B(x)+B(\Delta c)+B(\Delta o)+e_w+e_b. \end{aligned} \tag{2.12}$$

Updating Equation (2.10) to show how this affects the difference between two conditions yields

$$\begin{aligned} M(x+\Delta c+\Delta o)-M(x) &= [B(x)+B(\Delta c)+B(\Delta o)+e_w+e_b]-[B(x)+e_w+e_b], \\ &= B(\Delta c)+B(\Delta o). \end{aligned} \tag{2.13}$$

In other words, not controlling for order has introduced a confound. There are three general approaches to avoiding this confound. Just as we decided to explicitly control for contrast change to see if it had an effect, the first approach requires us to explicitly control for the change in order to see if it has an influence on our measurements. The first way to do this would be to use a pure *between-participants design.* In such a design, each person sees one—and only one—level of one factor (i.e., one condition, although he or she might see multiple repetitions of that condition). By ensuring that each person sees only one condition, there is no order of the conditions, and therefore there is no order effect. Notice, however, that differences between the two conditions now reflect changes in contrast *and* participants. If, for example, we have only two participants and one saw the 100% target while the other saw the 50% target, then differences between their performance might be due to the contrast change or to some intrinsic difference between the two people. Another way of phrasing this is that such a design relies very heavily on the assumption that the between-participant error e_b is small. The higher e_b is, the more participants one will need *for each condition* to ensure that any difference between the two conditions is not due to potential participant differences. Thus, one group of participants would see only the 100% target, while a second group would see only the 50% target.

As we saw in Equation (2.2), the between-participant error does not show up if we measure only one person. Thus, if each person sees all conditions, then any differences between conditions cannot be due to between-participant error.Each person can be seen as acting as his or her own control or baseline. An experiment where each participant sees every condition (i.e., every level of every factor) is, naturally enough, referred to as a *within-participant design.* The second approach to avoiding order effect difficulties has us treat contrast as a within-participant independent variable and order as a between-participant variable—each person will see all contrasts, but different groups will see different orders. For example, one group will see the 100% target first and the 50% target second, while a second group will see the two targets in the opposite order. This allows us to remove one source of noise (namely, e_b) in determining the effect of contrast per person. Between-participant noise will, however, still be present for the order effect: if the average within-participant effect of contrast is different between the two groups, it might be due to either differences between the groups or to differences in the order. To help control for this, many participants are required in each group (so that we can estimate e_b).

While this approach provides a cleaner measure of the effect of contrast differences, it comes at the cost of a potentially large number of groups. To completely control for order all possible orders need to be examined. For two conditions, there are two possible orders: [1,2] and [2,1]. For three conditions (e.g., red, blue, and green targets), there are six possible orders: [1,2,3]; [1,3,2]; [2,3,1]; [2,1,3]; [3,1,2]; and [3,2,1]. For four there are 24 possible orders. For our ten levels of contrast experiment, there would be 3,628,800 groups. In general, there will be $X!$ possible orders, with X being the number of conditions. Since we will need to have many participants in each group, this rapidly becomes prohibitively expensive. With advanced experimental designs, it is sometimes possible to use a subset of the total number of orders (for more on this kind of design, please see Creswell, 2009; Falmagne, 1985; Gescheider, 1997; Maxwell and Delaney, 1990).

The third approach is to randomize order: each participant sees every contrast in a different, randomly chosen order. In Equations (2.12) and (2.13), where every participant had the same order, the effect of order might have systematically biased the results, preventing us from knowing whether the difference between the two targets was due to the systematic change in contrast or the systematic change in order. Randomizing the order of trials converts the systematic-order term into another noise term. As with other noise terms, collective measurements for many participants eventually averages the effect of order out of the equation. Another way of looking at this is that we are sampling from the many order groups that are possible in a fully controlled order experiment. Again, because of the additional noise term, such a design will require more participants.

2.1.11 Two or More Factors

As we saw in the previous section, where we varied both order and contrast, it is possible to manipulate more than one factor in a single experiment. Let us imagine, for example, that we want to vary target size. To do this, we will need (at least) two measurements: one with a small target $M(x)$ and one with a large target $M(x+\Delta s)$. Here, we are explicitly varying our stimuli along a given dimension (size). Thus, we have one factor with two levels (big and small), which gives us the following equations:

$$\begin{aligned} M(x+\Delta s) &= B(x+\Delta s) + e_w + e_b \\ &= B(x) + B(\Delta s) + e_w + e_b, \end{aligned} \tag{2.14}$$

$$\begin{aligned} M(x+\Delta s) - M(x) &= [B(x) + B(\Delta s) + e_w + e_b] - [B(x) + e_w + e_b] \\ &= B(\Delta s). \end{aligned} \tag{2.15}$$

If we want to measure both size and contrast, then we need to manipulate both simultaneously. That is, every possible combination of each of the factors must be examined. For example, if we wish to vary size (with two levels) and contrast (also with two levels), we will have a 2 × 2 design (this is a standard representation of an experiment, and refers to the number of levels in factor 1 crossed with the number of levels in factor 2). A full factorial combination of the two factors gives us four conditions (two contrast levels for large targets and two contrast levels for small targets), each of which needs to be measured. This gives us

$$\begin{aligned} M(x+\Delta s) &= B(x+\Delta s) + e_w + e_b \\ &= B(x) + B(\Delta s) + e_w + e_b, \end{aligned} \tag{2.16}$$

$$\begin{aligned} M(x+\Delta p) &= B(x+\Delta p) + e_w + e_b \\ &= B(x) + B(\Delta p) + e_w + e_b, \end{aligned} \tag{2.17}$$

$$\begin{aligned} M(x+(\Delta s,\Delta p)) &= B(x+(\Delta s,\Delta p)) + e_w + e_b \\ &= B(x) + B(\Delta s,\Delta p) + e_w + e_b. \end{aligned} \tag{2.18}$$

Note that Equations (2.16), (2.17), and (2.18) explicitly include terms for size differences Δs and contrast change Δp, as well as a term representing the *interaction* between the two variables $(\Delta s, \Delta p)$. This allows us to examine factors that are not linearly separable or orthogonal, and strongly resembles the general linear model that underlies one of the most common statistical tests in perception research: *The Analysis of Variance* (ANOVA). For more on ANOVA, see Part IV.

2.1.12 Validity

The definition of generality used in this chapter is more properly referred to as *external validity* (Maxwell and Delaney, 1990). In fact, when discussing experiments, there are four levels of validity. The first and most simple is *statistical conclusion validity*, which addresses the issue of whether the statistical statement can be considered true. Factors that affect statistical conclusion validity include whether the assumptions and requirements for the statistical analyses have been met and whether there is enough data to be able to make a statement (this later point is referred to as *power* and is discussed in more detail in Chapter 12).

Building on that is *internal validity*. Essentially, this deals with the issue of causality. The statistical tests merely told us if two (or more) measurements were reliably different. Once we have established that they are in fact different, we can begin to ask what caused the difference. If the experiment is constructed properly, we can start to make statements about the independent variable causing the dependent variable. The third level of validity is *construct validity*, which asks whether the interpretations placed on the independent and dependent variables are correct (e.g., did we really measure what we thought we measured?). The final level, of course, is *external validity*. For more on the issue of validity and how it relates to experimental design, we recommend Maxwell and Delaney (1990).

2.1.13 Real-World Example

The examples so far in this chapter may seem to be rather simple. Regardless of whether someone would actually want to run such experiments or not, it is important to know that the concepts are easy to carry over to experiments using more complex stimuli (or tasks). Imagine, for example, that we are designing an expert system on a mobile computing platform. Space on the screen is very limited. Nonetheless, we want an avatar of a virtual expert to be present. How much of the screen space must we dedicate to the avatar and how much can we use for other items? To examine this, we need to decide what our avatar will be doing. In this case, the avatar will mostly be talking to a customer. Therefore, we can restrict ourselves to facial (or face and hand) animation and do not need to present the whole body. Likewise, it means that we are interested in conversational behaviors. Previous research and common experience have shown that a spoken statement of "surprised" that is accompanied by a neutral expression or a look of boredom has a very different meaning than one accompanied by a surprised expression. In fact, whenever spoken and facial statements conflict, the facial meaning tends to dominate (Carrera-Levillain and Fernandez-Dols, 1994; Fernandez-Dols et al., 1991; Mehrabian and Ferris, 1967). Before we can determine how much screen space our avatar needs in

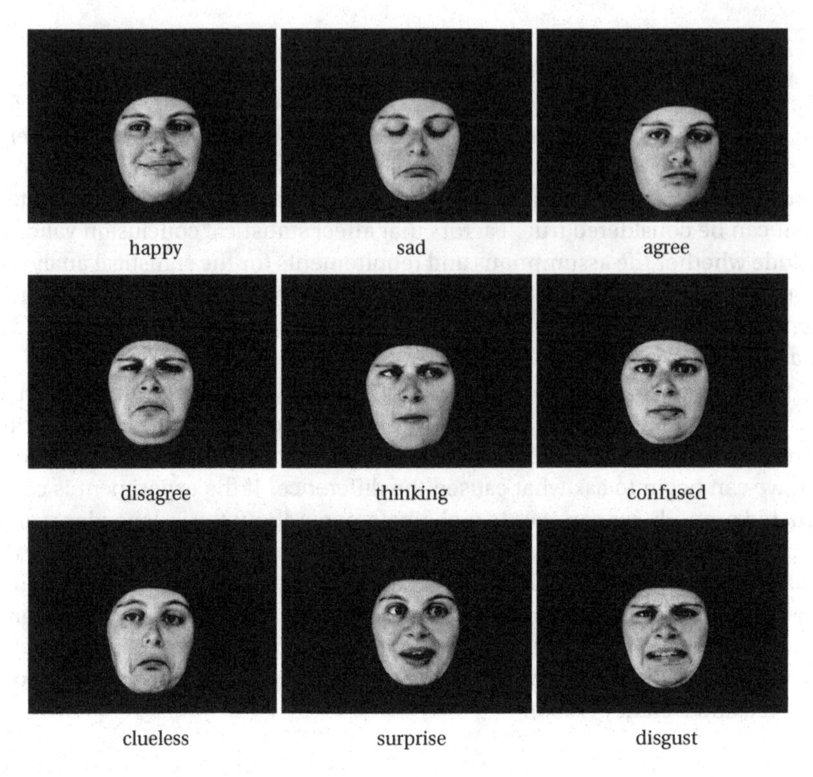

Figure 2.9. Static snapshots of nine conversational expressions from one actress.

order for the facial expressions to be properly perceived and understood, we need to know how people in general recognize expressions.

If we show the happy expression picture in Figure 2.9 to one person and ask that person to identify the expression (using a *non-forced-choice task*, see Chapter 6), we can say how well that one person is able to identify that one expression on that one trial. We would not be able to say anything about how well other people might identify the expression, let alone how well people can identify expressions in general. If we show that expression to a representative sample of the population, then we can begin to talk about how well people on average can identify that one expression. We still can say nothing about expressions in general. To do that, we would need to measure several expressions. To do this, we constructed a short list of expressions that were likely to show up in an expert system and recorded them from real people (note that videos and not static photographs were used—not only because dynamic expressions will be used in the final expert system, but because dynamic expressions are the norm for people in the real world). This initial list had nine expressions (see Figure 2.9). This is near the lower limit of expressions necessary to be able to say something about expressions in general. Of course, if we were to use expressions only from one person, we would not be able to say anything about expressions in general. Thus, expressions were recorded from six individuals (for more on the database of expressions, see Section 5.3.3.6 or Cunningham et al., 2005).

To determine how much screen space our avatar will need, we presented each of the nine expressions from each of the six actors at each of several image sizes (Cunningham et al., 2004a). We chose six image sizes (512 × 384 pixels—approximately 20 × 15 degrees of visual angle—256 × 192, 128 × 96, 64 × 48, 32 × 24, and 16 × 12 pixels). As can be seen, the face does not cover the full image. At the smallest size, the face in the recording is a mere 8 × 6 pixels! This represents the lowest level possible where a face still might be recognized without context. The largest size was determined by the size of the original recordings, as well as the computer monitor. This gives us three factors (expression, actor, and image size), with nine, six, and six levels, respectively, for a total of 324 conditions (A 9 × 6 × 6 fully factorial design). Fully controlling for order is not logistically possible, so we randomized order and each participant received a different order. A within-participant design was used, with each participant seeing each condition once. Repetitions of each condition were not used. There were two reasons for this. First, there were already a large number of trials, and using only one repetition of each condition helps to reduce participant fatigue as well as potential practice effects. Second, previous work using this task with the video sequences from this and similar databases had shown that experiments with a single repetition provided nearly the same

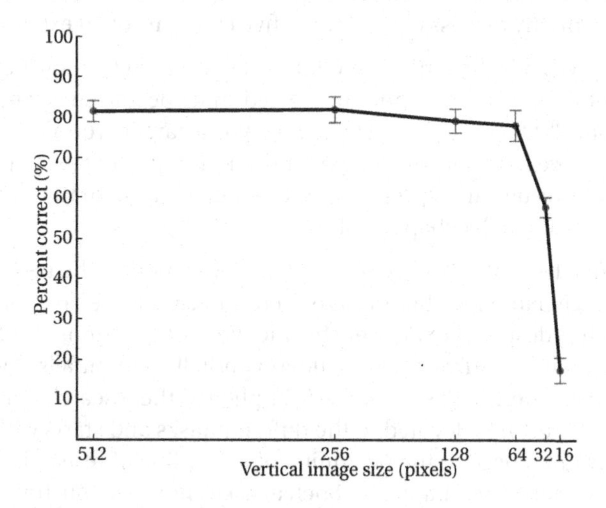

Figure 2.10. The results of the size experiment averaged across participants, expressions, actors/actresses.

results as experiments with five repetitions. A ten alternative, non-forced-choice task (see Chapter 6) was used, allowing us to measure the number of trials on which the expressions were correctly identified. Ten people participated in the study in exchange for financial compensation. The number of participants was chosen based on previous experience with this and similar phenomena; since we were searching for medium or larger effect sizes, ten participants should be sufficient.

The results are depicted in Figure 2.10. As can be seen, expression identification is largely independent of image size until the images become quite small. At 64 × 48 pixels, where identification performance begins to deteriorate, the face subtended approximately two degrees of visual angle, which is about the size of the *fovea* (i.e., the area of the retina that has the highest spatial resolution).[7]

2.2 The Elements of an Experiment

In Chapter 1, we saw that an experiment is a collection of trials designed to answer a specific research question. In general, a research question in the behavioral sciences asks how the members of some specific group perform some specific task. To examine this, experiments generally present some specific items (stimuli) to a subset of the group of interest (the participants), and ask the participants to perform some specific task. The way the stimuli are presented and the manner in which the collected data are subsequently analyzed also need to be planned, and count as elements of a full experimental design.

Below, we briefly discuss each of these five elements of an experiment.

Participants. What is the target population? Do we wish to know how all people do something, or are we more interested in a specific subgroup, such as expert surgeons? Once we have decided what our target group is, we need to make sure that we have a large enough random sampling of that population to ensure that our results are representative of (or generalize to) the whole group. This was discussed in the chapters of Part I.

Task. Perhaps the most obvious aspect of an experiment is the task. There are an astonishingly large number of tasks that are used in behavioral research, each of which is designed to answer specific types of questions. Each task has its advantages and disadvantages—hidden windfalls and pitfalls. Quite often the task that an experiment should use is implicit in the research question that is being asked. Part II is devoted to the different tasks and starts with the type of task that is often used at the beginning of a new line of research (namely, a *free-description* task, see Chapter 4). Such a task is not very constrained, which

[7] Interestingly, the results graph is not too dissimilar from the hypothetical results for the target contrast experiment above; see Figure 2.8.

means that it is very useful if the boundaries of a research area or if the form that the answers might take are unknown. This lack of constraint also makes it extremely tricky to use and means that the answers are almost always ambiguous. Free-description tasks are referred to as *qualitative tasks* since they provide text-based answers (such as a verbal description of something). The tasks described in Chapters 5 through 9, in contrast, all directly result a numerical value and are called *quantitative tasks.* They are increasingly targeted and constrained. The more constrained a task is, the more unambiguously interpretable and unequivocal the answer is.

Stimulus selection. In any perception-action loop, there will be some environmental information that forms the input to the participants. This information might be visual, auditory, gustatory, etc. Regardless of the sensory modality, we need to make sure that information we present is as highly controlled as possible and that it is relevant to the task and to the research question. The information or stimuli in different conditions must be as identical as possible, differing solely along the factors that we either know play no role in our task (e.g., from previous experiments) or that we are explicitly manipulating for the present experiment. For more on deciding what information needs to be presented, how many stimuli are needed, and other issues relating to stimulus selection, see Part III.

Stimulus presentation. Once we have decided who our participants are and what they will see, we need to actually show it to them. Merely throwing images on a computer screen is extremely uncontrolled, and will most certainly add noise. Order effects are only one small issue that arises when considering which participants see which stimuli and how. These issues are discussed in Part III.

Analysis. Once we have collected our measurements, we need to examine the data. The type of task chosen will strongly influence the form the data takes. Free description tasks often provide long, verbose text passages for each trial, while *forced-choice tasks* yield a single number for each trial. Thus, the analysis used is to some degree influenced by the task. It is also influenced by the question we want to ask, and by certain assumptions about the data (is the variance in the different conditions the same? do the data follow a Gaussian distribution?). The chapters in Part IV discuss some of the central issues in data analysis.

II

Response Measures

Chapter 3

The Task

One of most difficult aspects of actually designing an experiment is to decide precisely what the participants should do. Once the participants are sitting in the experimental chamber staring at the stimuli, they have to actually perform some task, but which one? The possibilities are nearly endless. We might ask them to describe what they see in their own words. We might ask them to rate some specific aspect of what they see. We might even ask them to interact with the stimuli (e.g., driving a virtual car or flying a virtual plane). We might even forgo overt actions and measure how their brains, hearts, or sweat-glands respond to what they see.

On the one hand, this variety is a very good thing, since it means that there is almost certainly a task that is perfect for any research question. It is also unfortunate, since it makes it very, very difficult even for experienced experimenters to decide which task is the best one for the current experiment. Indeed, many experimenters avoid the issue altogether by simply using one task and asking essentially the same research question (or variants of it) over the course of their entire careers.

The goal of Part II is to provide a brief overview of the types of tasks that exist. This chapter will present a brief discussion of tasks in general, including some of the issues involved in choosing the right task for the experiment. The remaining chapters in this part of the book provide a look at the each of the major types of task. Each of the task-specific chapters will begin with a brief overview of that task type, including the types of questions that one can ask with it. Then the task and its major variants will be introduced and discussed in more detail, including the task's advantages and disadvantages. Where possible, solutions or workarounds to the difficulties and disadvantages will be presented. Each chapter will include one or more sample experiments. To help facilitate comparison of the tasks—and to allow a more honest and direct criticism of the shortcomings of the examples presented—most of the examples will be based on our published work on the recognition of facial expressions. Even if you do not plan on using some of the tasks in the chapters of Part II, it is recommended that you read them as they contain additional clues and insights into experimental design.

3.1 Task Taxonomy

One of the things that makes choosing a task more difficult than is necessary is that it is often the first thing an experimenter thinks about explicitly. As was discussed in the last two chapters, the goal of an experiment is to answer a research question. Therefore, the first thing an experimenter *should* be explicit about is, what is the research question? As trivial as this seems, it really solves many difficulties. The more clearly and explicitly the research question is defined, the more obvious it is which tasks will be appropriate.

Once it is clear what we want to know (i.e., what our research question is) the next step is usually to decide what would serve as an answer. If we want to know which value for a specific parameter in a new visualization technique provides the fastest and most accurate identification of brain tumors, then we will want to use a task that can measure response speed as well as identification accuracy (and maybe localization accuracy). Therefore, simply asking participants to describe what they see will not provide us with the information we want. On the other hand, if we want to know whether people see some form of facial expression in the motion of a collection of dots—and if so which expression—then simply asking people what they see, which allows the participants as much leeway as possible without influencing the nature of the answers, may indeed provide us with the information we want.

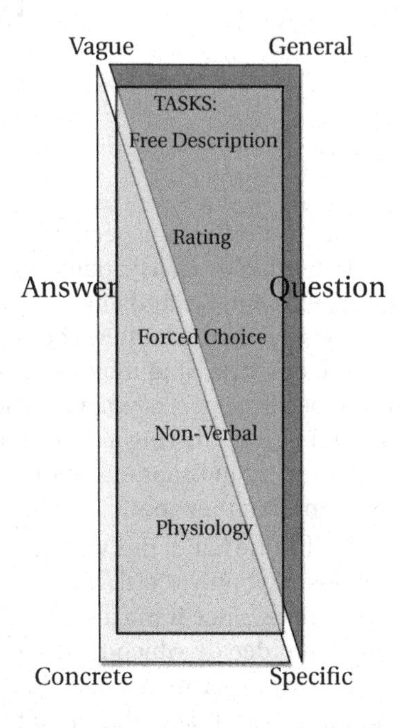

Figure 3.1. Experimental tasks can be seen as lying along a continuum. Tasks at one end are very flexible and are good for answering open-ended questions. These questions can provide a substantial amount of information and insight but are very difficult to interpret uniquely. Tasks at the other end of the continuum are much more constrained and provide very focused answers that are generally unique and easy to interpret.

This distinction between the specific and the general represents the endpoints of a continuum, as can be seen in Figure 3.1. At one end of the continuum are tasks that can answer broad, vague questions but which are very difficult to interpret uniquely. At the other end are tasks that easily support unique interpretations, but focus on very specific questions (and thus provide very specific answers).

Not coincidentally, the tasks on this continuum can be grouped roughly into *meta-tasks, direct tasks,* and *physiological tasks.* The tasks at the broad-question end of the continuum—such as *free description* (see Chapter 4) and some forms of *rating* and *forced-choice* tasks (see Chapters 5 and 6, respectively)—tend to be meta-tasks. In a meta-task, the participants are essentially asked how they *think* or *believe* they would act in a given situation. For example, we could present a number of images of facial expressions rendered in different manners (for example, rendered in different styles such as medical illustrated, cartoon, pointillism, etc.) and ask the participant which style they think is best at communicating the expressions. This task is a meta-task because it asks for a form of prediction or introspection: if a bunch of people were required to identify the expressions in these images, how well do the participants *think* those people would recognize the expressions as rendered in the different styles? A meta-task does not measure how well someone actually *can* recognize expressions. Meta-tasks are very useful, especially if we are interested in what participants believe to be true (and how those beliefs are influenced by what they see or experience). Such tasks are particularly useful at the onset of a new line of experimentation, as they can help us to obtain unbiased insights into the natural boundaries of a phenomenon. They are also useful for cases where participants cannot be placed in a situation for ethical or physical reasons. Meta-tasks can be awkward, however, since how people think they would act and how they actually act are not always the same. Moreover, it is not always easy at first glance to determine if the question is a meta-task or not. This is especially true for forced-choice tasks.

A direct task, on the other hand, asks the participants to actually perform an act. Some forms of rating and forced-choice tasks (see Chapters 5 and 6, respectively), the specialized forced-choice tasks (Chapter 7) and the *nonverbal tasks* or *real-world tasks* (Chapter 8) are all direct tasks. If we wanted to see which of the rendering styles mentioned above was *actually* better at communicating expressions, we could ask the participants to identify the expressions, and the style that gave the best scores would be—by definition—the best at conveying expressions. Direct tasks are very useful since they provide direct evidence of how people will respond in certain situations. Unfortunately, direct tasks are difficult to use for precisely the same reason: in order to be useful in answering a real-world question, the situation surrounding a direct task must be as close to the real world as possible. This is never easy, and is sometimes not physically or ethically possible.

At the most precise end of the scale are physiological tasks. They measure the body's reactions such as heart rate, body temperature, neural firings, etc.

These are very useful since they can provide a very direct, unbiased view of what elements of the stimulus the participants really saw or how they really felt about a stimulus. Physiological tasks are exceedingly difficult to use because most research questions involve real-world behavior or subjective experiences, and making solid, definitive connections between physiology and real-world behavior or subjective experiences is an unfinished task, to say the least.

Given that the tasks at the broadest end of the spectrum (e.g., free description) deliver text-based answers, they are referred to as *qualitative tasks.* Tasks further down the continuum, on the other hand, deliver one or more numbers as data, and as such are referred to as *quantitative tasks.* Since the majority of experiments within perception and computer science are quantitative, this book will focus on them. It should be noted that a mixed design, using some qualitative and some quantitative elements is possible. For more on hybrid designs, the reader is directed to Creswell (2009).

3.2 Which Method Should I Use to Answer My Question?

This cannot be said too often: there is no "best" method.

Yes, most researchers have their preferred methods. Often these researchers will insist that their methods are objectively better than all others. The reality of the situation is quite the contrary: every task has advantages and disadvantages. The preference for a given task is not normally due to the superiority of that particular task, but to the wealth of experience that the researcher has collected with it (and the concomitant lack of experience with other methods).

It is true, however, that some methods are more appropriate than others for certain types of questions. The tasks that are called for to answer a particular research question involving high-level cognitive processes (such as would be the case if we were studying risk taking) are very different from the tasks that are called for if the research question involves mostly low-level processes (such as color perception).

Given that experiments are performed in order to answer questions, it is tempting to try to get to the answer as quickly as possible. This is a fundamental—and often fatal—mistake. As was pointed out in Chapter 2, in order to be certain that the interpretation of the results is correct, we need to be absolutely sure that we have controlled for all possible sources of external noise or variance. If we have not controlled for everything, then there will be ways of explaining the data other than the one we wish to believe in. There are at least two practical consequences of this need to control for every possible source variance. The first is that answering any given research question will require that many experiments be performed, often using different tasks.

No single experiment is perfect. No single experiment can control for all possible variance along all possible dimensions. No single experiment can have an infinite sampling of the dimension that interests us or control for all possible variables.

The second consequence of the need to control for noise is that experiments need to be thought through fully and planned out *before* they are performed. Some methods are seductively compelling since they seem to promise quick answers. In just a few minutes, we can go from a vague question to data! The problem with many of these quick methods (such as the free-description task described in Chapter 4) is that although they do allow us to get data very quickly, the data is much, much harder to analyze. In some cases, there may not even be an objective way to analyze the data. Other tasks require considerably more investment before the participants show up, but have a greatly reduced analysis phase. In fact, these tasks also tend to be cleaner or more objective and less open to misinterpretation. These tasks, however, tend to focus on detail questions, and are less suited for addressing broad, sweeping issues. As noted before, many researchers tend to ask the same type of question over and over, and this leads to a preference for a particular method. Regardless, each task has its niche and its uses.

In addition to being dependent upon how specific the question is and what kind of answer we need, the choice of method is of course dependent on the nature of the phenomenon itself. If we are studying a phenomenon that has typical behaviors associated with it—such as driving a car—then performing that task directly is obviously very appropriate. Since most of the questions that will interest the practicing computer scientist involve real-world applications, most of the research questions will be amenable to such nonverbal or real-world tasks. If, on the other hand, we wish to understand a phenomenon whose relevant behavior is less clearly defined (e.g., "How do people represent faces?" or "Which virtual reality system establishes a greater sense of presence?"), then a less constrained task (such as a questionnaire or a rating task; see Chapters 4 and 5, respectively) might be more appropriate.

3.3 Naive Participants

One final general issue that needs to be considered when choosing a task is how much the participants will be allowed to know before and during the experiment. Within an information processing system, it is possible to categorize a process by how late in the information processing pipeline it occurs. Sensation processes convert external physical changes into internal nervous system signals (this is also called *transduction*) and are the very first step in the information processing pipeline. At the other end of the pipeline are such high-level processes as aesthetic judgment and decision making, which are generally considered to require intention (i.e., they are done on purpose) and in humans usually involve awareness (i.e., you know that you are doing them and have control over the processes while they are occurring). For our present purposes, one aspect of this description is very critical and relatively uncontroversial: can the participants intentionally influence how the task is performed?

For low-level processes, the answer is clearly no. People have little or no conscious impact on when or how sensory process occur. This feature is sometimes referred to as *cognitive impenetrability*, since such processes are separate from—and not directly affected by—higher level (or cognitive) processes. Imagine, for example, that we present a bright square on a black background to a participant. As long as his or her eyes are open, the cells in the retinal will convert the light into electrochemical signals. What the participant wants, thinks, or believes has no affect on this, since the only way the participant could affect the transduction would be to close his or her eyes. The participant cannot choose to transduce the signal differently than he or she normally would, nor can he or she transduce only part of the signal. The participant cannot affect how the process occurs.

If we wish to study a low-level process, then, we generally do not need to worry whether the participant knows what the research question is. In such cases, we can use an *overt* task. In an overt task, participants know precisely what they are doing and why.

3.3.1 Response Bias

Even for low-level processes, however, participants can still try to give us the pattern of answers that they think we want. That is, while high-level processes may not influence the process we want to study, they can (and probably will) influence the responses that participants give. For example, imagine that we are studying brightness thresholds; we want to determine the minimum amount of light necessary to just barely detect that a square is present (see Chapter 6). If the participants have jobs which require perfect vision, then they may not want to admit that they do not see something that is there. They will want to be seen as being very sensitive. That is, they will have a strong *motivation* to have their results indicate that they have perfect vision. Therefore they will most likely say that they see something if there is even the slightest chance that there might be something there. On the other hand, if the participants have jobs where accuracy is critical, they will have a high motivation to indicate that they see something only when they are really sure that they see it. As a result, motivation and *strategies* will influence the *response criteria*, and therefore the pattern of results, even in the study of low-level processes. In other words, the task that participants are performing might not always be the task that they were assigned. In the example above, participants were supposed to try to detect a square but instead tried to produce a specific response profile. This decision might have been conscious and might not have. Everyone has response criteria, and they are rarely the same across all people. Thus, even in the best of cases, we need to worry about the task that the participants were actually performing and what (if any) *response biases* were there.

The problem of response biases becomes worse when participants know both what we are studying and what sort of answer we expect (or hope for). They may choose (consciously or unconsciously) to provide a pattern of results that matches our expectations. That is, we will never know whether the results we obtained reflect underlying perceptual processes or merely response biases.

If we are studying processes that require participants to intentionally choose to perform the task, and that provide considerable leeway in how participants proceed, the problem becomes even trickier. If, for example, we wish to know if words referring to things that are normally found up high (such as the sky and the clouds) are processed faster than words referring to things that are generally found low (such as the ground or a dog), then the very fact that participants know what we are looking for might contaminate their results.

3.3.2 Addressing Response Bias

There are at least six solutions for dealing with response biases, which are described below.

3.3.2.1 Conceal Expectations

Participants should be unaware of the purposes of the experiment (the psychological terminology for this is *naive*). That is, the participants should *never* know what answers we expect. Admittedly, this is sometimes very hard to achieve. Nonetheless, unless the research question explicitly deals with prior knowledge—such as is the case in an *implicit learning* experiment—participants should always be naive.

3.3.2.2 Preserve Anonymity

Participants should always be convinced that data will be collected anonymously and that no one (not even the experimenter) will be able to connect their data with them. Note that the mere presence of an experimenter or other participants in the room means that the experimenter or the other participants can connect a given participant with that participant's data. As such, it is advisable to have each participant perform an experiment alone, without other people present. Furthermore, it is critical for experimental as well as ethical and legal reasons, not only that participants should *believe* that the data are recorded and maintained in an anonymous fashion, but that the data *are* in fact anonymous. Finally, we should mention that unconscious response biases that people have (e.g., some people are more likely to take a risk while others are more conservative) will not be mitigated by this solution.

3.3.2.3 Use Statistics

Statistics such as those derived from signal detection theory (see Chapter 12) can be used to figure out the response bias. Note that the use of these statistics might require specific experimental designs.

3.3.2.4 Alter the Response Criteria

Altering the response criteria might be done by adding and systematically varying rewards for correct answers and punishments for incorrect answers. While reward structures will work, adding them does tend to lead to longer experiments and requires that participants be provided *feedback* on accuracy for each trial. Providing feedback, however, has its own problems, the most critical of which is that we will probably change the participants' answering strategy from giving the response they think is best, to giving the response they think that we think is best.

3.3.2.5 Add Catch Trials

Catch trials are trials that are designed explicitly to catch specific response biases. To return to the brightness detection example, if we wanted to see whether people were simply saying yes all the time, we could add trials where no square is present. If participants are still answering yes on these trials, we will know that there is a response bias (and that we will need to use complicated statistics to get at the real thresholds). Interestingly, telling the participants that there will be catch trials (and letting them know what that means) can increase the chance that they will actually perform the task we chose. That is, we can influence the participant's motivation without providing accuracy feedback. For most low-level processes, however, such measures are not required.

3.3.2.6 Use a Covert Task

In addition to being sure that the participants do not know the answer we expect, use of a *covert* task will ensure that they do not even know the question! If we are examining whether "up" words are processed faster, as mentioned above, we would need to get the participants to read the words and process them without them knowing the real reason why we want them to process the words. For example, we could ask participants if the string of letters on the screen is a word or not. To perform this task (a lexical decision task), participants will process the words and we will examine the reaction time (to see if the "up" words are processed faster). Of course, if every single trial included a word, then people would rapidly stop paying attention (see the comments on balance in a forced-choice task in Chapter 6). Thus, we will need to add *filler trials.* These are trials that are not relevant to the real research question but are necessary for the cover task to be believable. In the present example, we would need to add a number of trials that have strings of letters that are not words.

Note that filler trials increase the length of an experiment considerably but do not contribute to answering the real research question.

Note that all covert tasks require us to use *deception*; that is, we lie to the participants. In the simple example above, we told the participants that we are studying words versus non-words when we are not. Deception is always tricky to use, for many reasons. In addition to ensuring that the participants believe the cover story, there are numerous ethical issues. For this reason, we recommend not using deception unless you absolutely have to. Even then, we strongly recommend collaborating with an expert on the specific form of deception.

3.4 Ethics Review

In nearly every country, performing experiments on humans requires that approval be obtained from an ethics review board before the experiment takes place. Over the course of the last 150 years, a number of tasks have been shown to have unexpected, long-term, negative consequences for the participants. Ethics-review boards help to ensure that participants will not be hurt. Moreover, many journals will not publish experimental results unless proper ethics guidelines have been followed. An excellent start for learning about these is the Declaration of Helsinki created by the World Medical Association. The current version of the declaration can be found on their website. Likewise, the American Psychological Association has a detailed ethics code, which can be found on their website. Furthermore, nearly every research institute that regularly performs research has an internal review board (also called an institutional review board) established precisely for this purpose, and they will certainly have more information on ethical guidelines. Failure to follow established ethical guidelines or to obtain proper approval for an experiment can be considered scientific misconduct and can lead to serious—possibly even legal—repercussions.

3.5 General Guidelines

For ease of reference, a list of a few of the general issues involved in choosing a task follows. Many of these have been addressed earlier in this chapter.

3.5.1 Response Bias

Do people have a preferred answer? Do they tend to say yes when they are unsure, or do they tend to say no unless they are 100% sure? Special statistics (and sometimes special experimental designs) may be necessary to determine what portion of the variance in our data is due to response biases and what portion to the phenomenon we are studying.

3.5.2 Motivation

Why is a participant answering the way he or she is? Is he or she trying to impress us or trying to provide us with the answer we want? Did the participant have a bad day and want to mess up our data? Does the participant just want to get through the experiment as quickly as possible so that he or she can get paid and go? Does the participant want the experiment to last as long as possible so that he or she can collect as much money as possible? Different motivations will lead to different response biases.

3.5.3 Instructions

The instructions tell the participants what to do. The instructions need to be as clear and consistent as possible. All participants should get identical instructions to ensure that they have the same understanding of the task. It is a good idea to write the instructions down and have the participants read them.

3.5.4 Relevance

The task needs to be as similar to the real-world behavior that is of interest as possible. All tasks will yield data, but not all tasks will help to answer our research question. Since the real world does not allow us the rigor, control, and repeatability we need to perform an experiment, we need to use an abstracted or simplified version of the real-world task in a controlled environment. The more abstracted from its real-world counterpart the task is, however, the less likely it is that the pattern of results found will relate to the real world.

3.5.5 Catch Trials

Catch trials are added to see if a particular response bias is in fact present. Informing the participants that there are catch trials may also help to reduce the likelihood of that response bias occurring in the first place.

3.5.6 Feedback

Does the task have an objectively correct answer? If so, are participants allowed to learn how well they did? In some cases providing feedback is unavoidable (such as in a driving simulator); in others it is easy to avoid. Providing feedback is very likely to affect how people act, especially since it gives them a concrete, immediate indication of how we want them to act, and thus increases the possibility of response bias.

3.5.7 Practice Trials

Sometimes it is good to let participants get familiar with the setup and task before the real experiment begins. This is particularly true of virtual reality experiments. In some cases, practice trials provide the participant with feedback on accuracy, but in other cases feedback is not provided. To determine if practice trials are needed and, if so, whether feedback is appropriate, please read the relevant psychological research literature—that is, if we are planning an experiment on the role of optic flow in route learning in a virtual environment, then it would be advisable to read the reports on existing experiments that address that research question.

3.5.8 Anonymity

No one should be able to connect a specific set of responses with a specific person. Participants should know that their responses will be anonymous.

3.5.9 Overt versus Covert Tasks

Do participants know what we are studying? Can they figure out from the task they are performing what the real research question is? Does it matter? In some cases, it is critical that the participant be kept in the dark about what it is we are trying to study (until after the experiment). In such a case, a covert task with its relevant cover story is required.

3.5.10 Filler Trials

Filler trials are added to an experiment so that the covert task and its cover story seem realistic and believable. They increase the length of the experiment and provide no answers to the real research question, but are required to keep up the masquerade if we are using a covert task.

3.5.11 Deception

Sometimes the participants must be lied to. This is generally a tricky thing to do and has many serious consequences. Do not use deception unless it is absolutely unavoidable. Even then, special permission from the appropriate ethics review boards will be necessary.

3.5.12 Informed Consent

Prior to starting the experiment, the participants must be informed of the risks posed by the experiment, and (at least roughly) of what will be required of them. This information must be presented in written form and the participants must sign it. Failure to obtain written informed consent can be considered scientific misconduct and can lead to serious repercussions.

3.5.13 The Experimental Chamber

Experiments are usually performed in a small, enclosed, dark experimental chamber. The walls of the chamber are usually black, and the room is usually empty of everything except the experimental apparatus. The reason for this is that it has been shown repeatedly that participants' responses on a wide variety of tasks will change depending on subtle aspects of the experimental setup (such as wall color, the presence of an experimenter, the gender of the experimenter, etc). This is often referred to as *context-* or *state-dependent memory,* and the extraneous features are sometimes referred to as *occasion setters.* For more on these topics, see Grahame et al. (1990); Holland (1983); Overton (1964, 1991). Presenting the stimuli to participants when they are alone in a dark room is an attempt to eliminate all these extraneous sources of noise or bias in the results.

3.6 Summary

In this chapter, we examined some of the broad issues related to to task selection. Some research questions call for a less constrained task, which provides a vague answer. Other questions demand considerably more rigor and as such get a much more quantitative answer. After 150 years of psychological experimentation, the odds are very good that someone has already performed an experiment to address a research question that is similar to your question. It is strongly advised that you search for and read published scientific articles on similar experiments. Even if the questions addressed in the published literature are not similar enough to help you find an answer, they can help provide insights into how you might better formulate your question, whether you need to provide feedback, what factors you need to control for (these vary considerably from one research area to another!), or what type of statistics you might need to run. Reading the relevant psychological literature can also help find the key phrases that you might use when you write up your experiment (for more on how to write up a paper in a standard psychology format, see the publication manual of the American Psychological Association (2010)).

Chapter 4

Free Description

The free-description task asks people to describe their beliefs and opinions. It is a meta-task. Free-description tasks are also qualitative tasks, which means that participants are asked to provide an explicit, word-based answer (as opposed to a numerical answer) to a question. Since qualitative tasks are not common in perceptual research, and are indeed not typically viewed positively, free descriptions will be the only qualitative task in this book. More information on qualitative tasks can be found in Creswell (2009).

In a free-description task, the questions are usually presented in written form. The participants normally perform the experiment with the experimenter not present, to help increase the feeling of anonymity. The responses are usually written, although sometimes audio or video tapes of the responses are made. If tapes are used, it is critical to obtain the participant's permission to be recorded prior to the recording. In all cases, one must ensure that all the proper steps are taken to ensure that the anonymity of the answers is maintained. Likewise, for ethical and legal reasons, permission is needed before any of the recordings can be shown or played for anyone not directly involved in the experiment. The participants' answers will generally vary considerably in content, length, and coherence.

4.1 Overview of Free-Description Tasks

Here, we provide a brief overview of the highlights of a free-description task.

4.1.1 Typical Questions

Free-description tasks are best at answering broad, general questions that seek broad, general, and vague answers. They are also useful at the beginning of a new line of experiments, where it is unclear what type of responses participants might give. Some examples of typical questions follow.

- What words would people use to describe this facial expression (or painting, etc.)?
- What is the most dominant aspect of this display?
- What do people notice first in this display?
- What kind of observations would people make about this situation, concept, or scene and how often does each observation occur?

4.1.2 Data Analysis

The answers are generally independently examined by two or more individuals, who use their own judgment to either rate each answer or place each answer in specific categories. The rating scales or categories are devised (preferably before the begin of the experiment) by the experimenter based on the specifics of the research question. For more information on the analysis of free-description tasks, questionnaires, and rating scales, see Chapter 13.

4.1.3 Overview of the Variants

The techniques that are either derived from or related to free-description are listed briefly here, and discussed in more detail later in the chapter.

4.1.3.1 Interview

In an *interview*, questions and answers are in the form of a spoken interaction between the experimenter and the participant. While the questions are generally formulated in advance, additional questions may arise during the interview to extract additional or more specific information from the participant. One special form of interview is the exit interview, or debriefing, where the goal is to find out what the participant felt or thought about the experiment.

4.1.3.2 Questionnaire

As with an interview, a *questionnaire* presents a series of specific questions, but this time in written form. Data analysis should include examining for *repeatability*: the same answer should be given to a question each time the participant is asked that question. Care also needs to be taken to ensure that participants' understanding of the questions is the same as that of the experimenter.

4.1.3.3 Long Answer

The *long answer* is supposed to be a logical, complete, well-thought-out statement or series of statements. The answer should be as explicit as possible and provide as much information as possible. Long answers can be anywhere from a paragraph to several pages. A special kind of the long answer is the essay, which places additional emphasis on the form and structure of the answer.

4.1.3.4 Short Answer

As with the long answer, the *short answer* is still supposed to be well-thought-out, but considerably less emphasis is placed on length or completeness. The answer can be a single sentence or word.

4.1.3.5 Partial Report

With a *partial report*, a complex stimulus (usually an array of simple stimuli, such as a grid of letters) is briefly shown. Different portions of the stimulus (such as rows in the grid) are given different identifying codes (such as different tones). After the stimulus is removed, the participant is asked to describe a specific portion of it (the code for the relevant portion is given). Thus, the task can be seen as a combination of multiple-choice and free description. Since there is an objectively correct answer for this task, the responses can be easily converted to a number (i.e., how many stimuli were correctly identified). Therefore, it is sometimes classified as a quantitative task.

4.2 Task Description

Often the most direct way to get an answer to a question is to simply—and explicitly—ask. If, for example, we want to know which painting people generally prefer, we could ask a number of people what their favorite painting is. This is the core to a free-description task. Not only can it meet with considerable success in everyday life, but it also plays a valuable role in scientific investigations. It should, however, be said that despite some advantages it can be one of the trickiest tasks to design and run properly, and its results are definitely the most difficult to cleanly and uniquely interpret.

4.2.1 Advantages and Disadvantages of Unconstrained Answers

The greatest advantage of a free-description task is that it has the potential to deliver a considerable amount of information. The information obtained, however, will probably vary considerably in terms of depth, quality, sincerity, and appropriateness. When asking someone which painting he or she prefers, for example, that person may simply name a painting. We would learn his or her preference—clearly and unambiguously—but little else. The person may, on the other hand, also describe why he or she likes that painting and how those reasons relate to the history of the artist and to the time period in which the painting was created. In such a case, we can learn a tremendous amount beyond the mere preference. Such verbose answers contain clues about the respondents' level of education, their history with art, the degree to which they have previously thought about art, how much they enjoy talking, their familiarity with conversational etiquette and language, their degree of extroversion, and a wide variety of other things. Regardless of what additional information can be gleaned in both of these answers, they provide a clear indication of preference.

A clear, unambiguous answer is more the exception than the rule. For example, it is possible—or even likely—that the person will not know the name of the painting or the painter. In this situation, some people might try to give a description of the painting. For the experimenter, there are three possible results. Some of the time the experimenter will recognize the painting correctly, and thus an unambiguous answer is possible. Some of the time the experimenter will not recognize the painting, in which case an unambiguous answer is not possible. The third case is the most dangerous: some of the time the experimenter will *think* that he or she recognizes the painting but in fact be incorrect.

Instead of naming or describing a painting, participants might state that they once owned a copy of a particular painting. How should this answer be interpreted? Does this mean they liked the painting? Or did they have it for some other reason (e.g., it was a gift, or they felt that one should own that sort of painting, or they purchased it as a financial investment)? Since they said they *had* owned a copy, do they *still* own it? If not, why not? If they *did* like it, do they still like it? Even if one can ascertain that they did—and still do—like the painting, this does not necessarily mean that they *prefer* it to all other paintings (which was the original question). Thus, while an answer has been given, it is by no means unambiguous.

Likewise, respondents might choose to discuss the merits and flaws of various paintings without ever implying a preference explicitly. While it is sometimes possible to extract a preference from such an answer, the answer will most certainly *not* be clear or unambiguous. That is, some people will interpret the long-winded oration as meaning one thing, while others will come to a different conclusion—which is the *correct* interpretation? While being certain that our interpretation of someone's answer may or may not be central to everyday life questions, it is absolutely critical for science.

As the above hypothetical example suggests, there are few direct restrictions on the type, range, or depth of the information that one receives with a free-description task. In other words, nearly any answer is possible. Sometimes, the answers can be very exotic. For example, in Cunningham et al. (1998) we showed a series of very simple displays to participants and asked them "to describe the display as carefully as you can—as if you are trying to describe what you see to someone who is not in the room, and has never seen these displays." Each display consisted of a black screen with a random array of small white dots on it. A black triangle moved over the white dots and a second field of dots was superimposed over this display, reducing the overall contrast between the figure (the triangle) and the background (the dot fields). While many of the responses referred to dots and moving figures, a few did not. One participant (who later indicated that he was convinced that the experiment was a personality test) answered, "I see Picasso's *Nude Descending a Staircase.*" Another participant indicated that he saw ". . . a happy face, no wait now it is sad." While informative, these responses had little to do with the research question (which was how often people would spontaneously report seeing a solid black surface in these displays).

The greatest advantage of a free-description task is simultaneously its greatest problem. The great wealth of information that one obtains may or may not be relevant to the research question. Being absolutely certain that one has correctly interpreted the results is exceedingly difficult. For this and other reasons, it is very difficult to get studies that use free-description tasks published in traditional perception journals.

A corollary of the great latitude that participants have in forming their answers is that the more specific a research question is, the more specific the free-description question can be. In fact, the more specific the research question is, the less reason there is to use a free-description task. Since one understands a given phenomenon more clearly the longer one studies it, free-description tasks generally show up at the beginning of a new line of research. When one starts on a new line of questioning, it is quite possible that one will not know the natural boundaries of the phenomenon, or how people respond to it, or what aspects of the phenomenon are most critical, and so on. Gathering unconstrained, unbiased information in a free-description task can be of considerable importance in determining how to proceed with further experiments. For example, the specific words or actions that people use are likely to be the most natural ones for the situation, and thus are reasonable choices for responses in later, more constrained tasks. That is, if a lot of people tend to describe a given facial expression as "scrunchy" or as "a maniacal grin" when they are free to call it anything they want, then these may be acceptable terms to use in a forced-choice task in a later experiment (see, for example, Section 4.3.3.2 or Kaulard et al., 2010).

4.2.2 Interviews and Social Factors

For any given experiment that we design, we know what our research question is and that we want a clear and unambiguous answer to it. If the experiment is structured as a dialogue in which we sit down with participants and get answers to a series of spoken questions, then the dialogue will quickly become something like an interview: regardless of the initial answer, we will most likely ask follow-up questions. Naturally, these questions will try to steer the participant towards making some definitive statement that is relevant to our research question. The more follow-up questions we ask, the less constrained and unbiased the results will be. To illustrate this process, let us imagine we wish to know if people can detect a square in a given display. We begin by sitting down with a participant in a small enclosed room, showing him or her the display, and asking, "Do you see anything here?" Given the very vague nature of the task question, it is quite plausible that some participants will just answer: yes. Since we cannot conclude uniquely from this answer that they saw a square (the participants might have meant the experimenter, or the monitor, or the experimental chamber, etc.) we will be tempted to ask them to elaborate. Other participants, in a desire to be thorough, might describe the room in general terms and include the presence of a computer monitor. We would then further direct the questioning to see if they saw anything *on* the monitor, and so on. Depending on many factors, we may at some point end up asking directly if they see a white square on a black background on the computer monitor. Obviously at this point the task has ceased to be free description (the last question is more characteristic of a forced-choice task). The answers will cease to be unconstrained and, more critically, will certainly not be unbiased.

Naturally, since the experiment has been arranged as a face-to-face interview the whole range of social factors that are involved in a conversation will come into play. There are two core problems here: the experimenter and the participant. The experimenter usually has a specific hypothesis or answer to the research question in mind. Thus, the experimenter might—consciously or unconsciously—lead the conversation in a direction that is most likely to provide us the expected or desired answer. This can be done very directly (such as by asking, "You do see a large, very obvious white square right here in the middle of the monitor, don't you?"). It can also be done more subtly (e.g., via a host of nonverbal cues). The degree to which very subtle, indirect aspects of an interview situation can impact the answer has been demonstrated elegantly in the field of eyewitness reports. For example, participants in Loftus and Palmer (1974) were asked questions about a film of a car accident. The use of the verb "smash" in the question, "About how fast were the cars going when they smashed into each other?" yielded higher speed estimates than questions that used the verbs "collided," "bumped," "contacted," or "hit." Moreover, when retested a week later participants who had received the question with the word "smashed" were more likely to answer that they had seen broken glass—even though there was no glass in the film. For a recent review of eyewitness reports and leading questions, see Wells and Olson (2003).

Even if we experimenters do our very best to remain neutral in our directed questioning, there is still the participant. Even though participants may know that their answers are anonymous, they also know that there is at least one person (the experimenter) who can connect them with their answers. For example, when we ask people about their painting preference, they will consider not only the content of the question and their answers, but the social implications as well. For instance, if the painting they really prefer is considered unacceptable in society, then some people will be unwilling to admit openly that such a painting is their favorite. In an interview, that person may instead try to guess what we wish to hear and give that as an answer instead. Alternately, perhaps the person is in a bad mood and decides to give us the name of a painting that he or she does not like but thinks will mess up our statistics. In other words, even for clearly stated answers we are not always certain about whether the answer *really* reflects the person's opinions. Moreover, most participants know or assume that the person conducting the interview is directly related to the experiment and may well have a vested interest in getting particular results. Some participants might think that if they are "good" participants that provide "good" data, then they will be invited back for additional experiments (and the concomitant financial compensation).

Finally, since the free-description task is not only unconstrained, but also intimately verbal, the results will naturally be couched in words, terms, metaphors, and idioms that are specific to the participant. Therefore, an outgoing person might use the same terms as an introverted person but mean completely different things by them, which makes direct comparison of answers problematic. For more on this issue, see Asch (1946) and Kelley (2003).

Due to the overwhelmingly complex social factors, it is strongly advisable to conduct any free-description task in a one-directional, non-interactive manner. We might choose to record (either audio or video) the answer—which also provide paralinguistic information such as tone of voice or body gestures—but issues of anonymity arise in this approach as well. Written responses provide the most—and most transparent—anonymity, and thus are recommended.

4.2.3 General Guidelines

In general, one must take great care in both the construction and the analysis of a free-description task to ensure that clear, unambiguous, reliable, and valid results are obtained. As many potential sources of bias as possible should be removed. Before a free-description task is used, it should be mentioned that—no matter how much care has been taken or how many precautions have been made—the interpretation of the results will not be very stable, solid, or reliable. That is, it is not possible to be very certain about the meaning of the results.

As a general rule, the following steps are recommended.

4.2.3.1 Anonymity

The more participants are convinced that no one can connect their answers to them, the more likely they are to be completely honest. It is critical not only that participants *believe* that the data are recorded and maintained in an anonymous fashion, but that (for both ethical and legal reasons) the data *are* in fact anonymous.

4.2.3.2 Clarity

All ambiguity should be avoided. This is particularly true of the free-description question itself. Each and every participant should know precisely what is being asked of them. Ambiguity must also be avoided when creating the categories used in the data analysis.

4.2.3.3 Relevance

The task should not only address the research question, but should also reflect the real-world situation as much as possible. For a free-description task, the particular question or questions given to the participants should be specific enough to address the research question but general enough to avoid biasing or constraining the answers unnecessarily.

4.2.3.4 Preparation

As many aspects of the experiment as possible should be prepared before the first participant shows up. For an experiment that includes a free-description task, this includes deciding precisely how the data will be analyzed (what categories will be used to parse the data, what are specific examples of each category, etc.?). By preparing everything in advance, potential biases in the interpretation of the results can be avoided.

4.2.3.5 Written Questions and Answers

The maximum separation between the identity of participants and their answers is afforded by having them write their answers down. The mere presence of someone else in the room while participants are answering questions can decrease the sense of anonymity. Written questions also ensure that every participant gets the exact same questions; no variations in intonation, emphasis, or body language are possible. This precaution will increase the validity, reliability, and repeatability of the experiment.

4.2.3.6 Follow-up Experiment

Once the experiment has been run and analyzed, (hopefully) some conclusions will have been reached. More specifically, the data should allow interpretations to be formulated. We strongly advised that the validity of these statements—the degree to which they reflect the real world—be double-checked in a more objective, reliable experiment. Since the interpretations or conclusions can be quite specific, it is easy to design a specific research question to test them. This will naturally lead to the use of a more constrained experimental task, such as one of those outlined in Chapters 5 through 9.

4.3 Specific Variants

4.3.1 Interviews

As noted earlier, interviews are a two-way dialogue. Due to the social nature of the interview, it is nearly impossible to avoid contaminating the participants' responses with the personal opinions and beliefs of the experimenter. The one place where interviews are still used in perception research is the *debriefing*, in which the goal is to get a vague, general idea of the participants' impressions of the experiments and collect information that would not have been measured by the task itself. For more on interviews, see Section 4.2.2.

4.3.2 Questionnaires

If one provides a participant with many written questions and expects a written answer for each, the free-description task essentially turns into a questionnaire. Quite often the questions in a questionnaire are very specific and the answers are usually expected to be short. There are a host of issues that are specific to the design of a questionnaire (including balanced distribution, objectivity, clarity, validity, and repeatability). A very large number of questionnaires have been developed and validated for many different applications. We recommend that existing questionnaires be used before a new one is developed. If a new questionnaire is needed, we recommend that a recent reference on how to design and validate questionnaires be consulted, such as Oppenheim (1992).

4.3.3 Long and Short Answers

In its purest form, the free-description task has one question (although it may be asked about multiple stimuli). The question in a free-description task can be general or specific. It can require a long answer or a single-word answer. Two specific experiments where free-description tasks were used to learn more about the perception of facial expressions are presented next. In the first, a very general question was asked and long answers were expected. In the second, a more specific question was asked and the answers were explicitly limited to two words.

4.3.3.1 Example 1: Long Answer

The face is a powerful source of information. In common folklore, the eyes are supposed to be windows into the soul. We can learn a lot about people by the way they move, act, and react generally, and by what they do with their faces and eyes specifically. A facial expression can not only convey meaning by itself—a glance at someone who has a sad facial expression is sufficient to tell us how that person is feeling—but it can also control conversational flow. The influence of eye gaze on conversations can be seen, for example, in the fact that improper or absent eye-gaze information is an oft-cited problem with most video-conferencing technology. In general, listeners are believed to use *back-channel responses* (Bavelas et al., 2000; Yngve, 1970). That is, the listener informs a speaker what needs to be said next through the judicious use of facial expressions. Speakers confronted with a nod of agreement, for example, will probably continue talking, while a look of confusion, disgust, or boredom will almost certainly prompt very different behavior.

This led us to conclude that by providing a sequence of facial expressions from two people—alternating the two—it should be possible to construct a sort of "facial dialogue." Can we really learn a lot about people by watching their body language? If so, what can we learn? Can we really learn about the relationship between two people just by watching a series of expressions? If so, what do we learn? As a first step in this new line of research, we decided to see if alternating temporally between two series of unconnected recordings of facial expressions would give rise to the impression of a conversation, and what other sort of information observers might notice about the interaction (Mega and Cunningham, 2012).

Research question. Can facial expressions alone convey the impression of a dialogue? If so, how deeply will viewers infer meaning?

Stimuli. Facial expressions from the Max Planck Institute for Biological Cybernetics's facial expression database were used as building blocks. The individuals in the recordings had never meet each other. All recordings for a given person were made on a single day, and different people were recorded on different days over the course of many weeks. The expressions were recorded at several different intensities and from multiple viewpoints, using a method acting protocol. For more information on the expressions, see Kaulard et al. (2012); for more information on the recording system, see Kleiner et al. (2004). Static snapshots for a few expressions can be seen in Figure 4.1.

A total of six short films were constructed, each consisting of two people who alternated performing expressions. Consistent with filming techniques, person A was shown looking off slightly to the left and person B slightly to the right (to heighten the impression that they were looking at each other).

To create the films, a series of short texts or scripts were created. Each script featured a simple dialogue containing between six and eight *turns* or statements (between three and four turns per person). Then the expression from the database that most closely matched the meaning in each turn was chosen. For example, if someone was supposed to say, "Um, I am not sure I understand," then a weak-intensity confused expression was selected. Finally, the selected expressions were extracted from the database and edited into a video. Note that the duration of each expression was fine-tuned to best match the information specified in the script; some expressions were shown from neutral

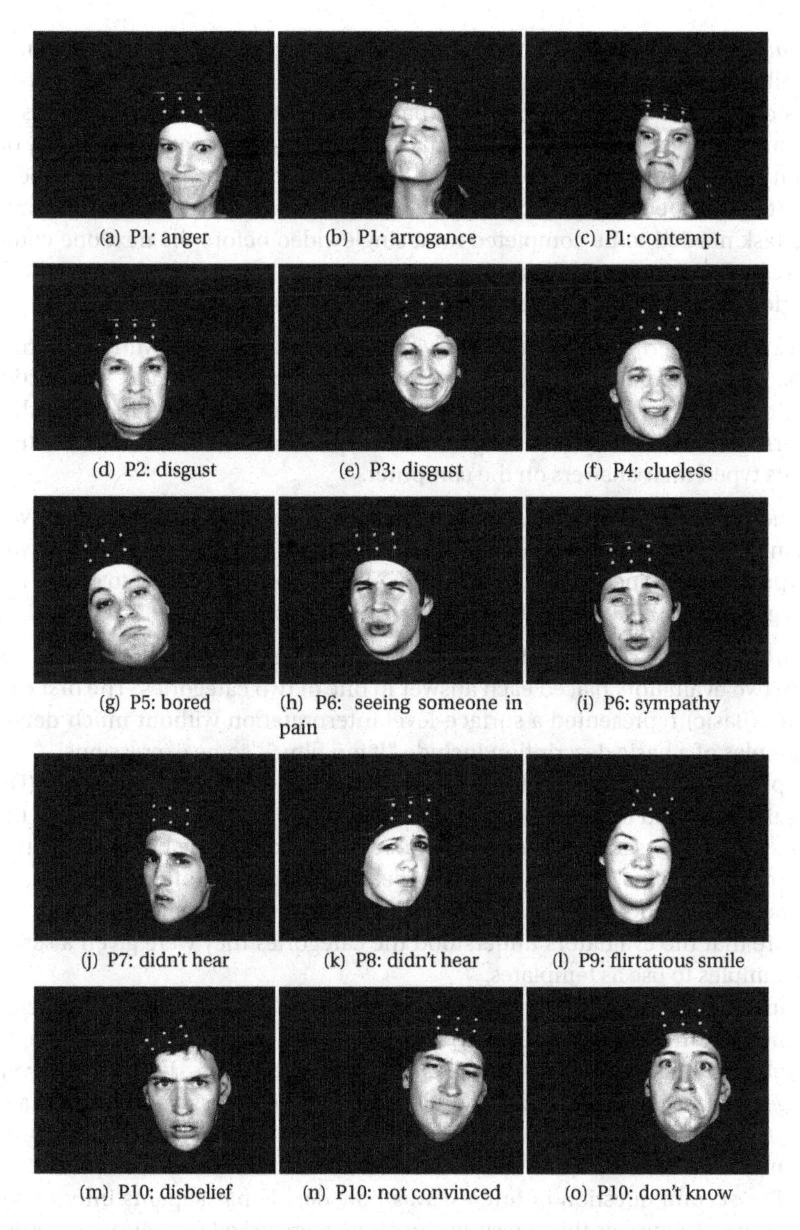

(a) P1: anger (b) P1: arrogance (c) P1: contempt
(d) P2: disgust (e) P3: disgust (f) P4: clueless
(g) P5: bored (h) P6: seeing someone in pain (i) P6: sympathy
(j) P7: didn't hear (k) P8: didn't hear (l) P9: flirtatious smile
(m) P10: disbelief (n) P10: not convinced (o) P10: don't know

Figure 4.1. Fifteen snapshots of thirteen expressions from ten people. Each expression is labeled with the person's number and the expression's name.

to peak, others from neutral to peak and back to neutral (i.e., the full length of the original recording).

Stimulus presentation. One at a time, each participant was placed alone in a small, enclosed, dark experimental chamber. A small amount of illumination was directed at the keyboard so that participants could enter their responses. Such conditions help to ensure that participants' responses are based solely on what is present on the screen and not on some ancillary aspect of the experimental chamber. The participants were shown each video one at a time, and the task needed to be completed for a single video before the next one could be seen. Participants could see the current video as often as they wished. Each participant received a different, random order of the videos.

Task. The experiment had two groups of participants, each with a different task. The participants in group one were asked to describe the film. The participants in group two were asked to describe the situation in the film. No further information about the task was given before the experiment was over. Participants typed their answers on the computer.

Participants. Twenty individuals participated in the experiment and received financial compensation at standard rates. All participants were naive to the purpose of the experiment. Each person was randomly assigned to one of the two groups, so that each group had ten participants.

Results. To see if participants saw a dialogue or interaction taking place in the film, two evaluators placed each answer in one of two categories. The first category (Basic) represented a surface-level interpretation without much detail. Examples of a basic description include "it is a film," "some expressions," "two people, one happy, then sad, the other disgusted." The second category (Dialogue) was used if there was any indication that participants interpreted the films as a conversation, dialogue, or interaction taking place between the two people in the film. This category included the attribution of one person's expression to the other person (e.g., he doesn't like the way she is acting). To ensure that the evaluators understood the categories they were given a series of examples to use as templates.

In general, 90% of the participants' responses were consistent with seeing an interaction or dialogue. This isn't too surprising, as putting isolated images together to make a movie is a standard filming technique. It does, however, suggest that facial expressions by themselves don't rely on speech or language to carry meaning, but can act as a primary channel of transporting meaning by themselves.

The second question of interest was how deeply participants interpreted the scene. To answer this question the raters were asked to indicate whether participants reported seeing something that was not in any given image (e.g., the cause of the behavior or the reason for the expression). The same two evaluators examined the responses for this question and again received examples for each type of category. On average, 70% of the participants saw some form of motivation or goal in at least one of the films (80% for group one participants,

60% for group two). This shows that the majority of participants projected details into the situations that were not present in the expressions themselves. Interestingly, many participants attributed a sentence or series of sentences to the films—as if they were trying to write the scripts for the scenes. This implies that participants might have seen the sequence as a coherent scene with a deep level of meaning.

Perhaps the most intriguing result is that some participants ascribed elaborate background stories for some of the films. One particularly striking example of this comes from a participant who wrote

> maria gets to know about the divorce of her friends parents. she is a bit sad and confused but her firend says it was somehow good because now there would be no more fights and they both would be happy. and then she unwillingly nods to it. [sic]

In short, it looks like facial expressions can convey a considerable amount of meaning in and of themselves.

Summary. Since this was the first experiment in a new line of research, it was not clear what participants could or would respond; there was no prediction as to how participants would interpret the sequences, let alone if participants would see them as a dialogue. Furthermore, the natural boundaries of the phenomena were not clear; it was unclear if people would infer meaning from the sequences and if so, how deeply. The use of a free-description task allowed us to probe the fundamental perception of the sequences and depth of information processing without biasing the participants. Subsequent experiments can now begin to use more quantitative methods to ask more detailed questions.

4.3.3.2 Example 2: Short Answer

It is very common to limit the response in a free-description task to one or two words. The goal of such a short-answer task is to provide some constraints on the possible answers, thereby reducing the variability of the responses. This in turn makes it easier for the experimenters to subsequently categorize the answers.

It has been well established that people easily recognize the six "universal" expressions (fear, anger, surprise, disgust, happiness, and sadness; Ekman, 1972) and several conversational expressions with varying degree of accuracy (Cunningham et al., 2003a, 2004a,b, 2005; Cunningham and Wallraven, 2009; Wallraven et al., 2004, 2007). There are, however, a vast number of expressions beyond the few exemplars used in these experiments.

Research question. How well do people recognize expressions in general? What sort of terms do they use to describe the different expressions? Do the descriptions imply some form of representational system for expressions?

Stimuli. The facial expressions from the Max Planck Institute for Biological Cybernetics's facial expression database were used. To reduce the length of the experiment, only ten of the twenty actors/actresses (five male and five female) and only the high-intensity version of each expression was used. All 55 expressions were shown (see Figure 4.1 for examples of several expressions from several actors and actresses). For more information on the expressions, see Kaulard et al. (2012).

Stimulus presentation. Each participant saw each expression from each actor/actress once. Each participant was shown the expressions in a different random order. Participants were placed in a small, enclosed experimental chamber one at a time and had to enter their response for each expression before they were allowed to see the next one. They could see the video for the current expression as often as they wished. Due to the large number of stimuli (510 stimuli per participant), the experiment was split into several sessions, which took place on different days. Each session lasted around two and a half hours.

Task. Participants were asked to describe the expression using a maximum of four words.

Participants. Ten individuals participated and received financial compensation at standard rates. All participants were naive to the purpose of the experiment.

Results. Three evaluators examined the results independently. Each was to determine whether the answer matched the expression or not. Overall, 60% of the answers were rated as being correctly named. This number is surprisingly close to the 70% average recognition rate (Cunningham et al., 2005) found using a non-forced-choice task on nine expressions (see Chapter 6). Participants had trouble identifying seven of the expressions: arrogant, annoyed/rolling eyes, contempt, I don't care, smiling sad/nostalgia, smiling winning, and doe-eyed.

Summary. Since this was the first experiment with a new database, we needed to establish a baseline for the expressions; how well do people recognize expressions in general? Since most of the expressions have never been used in experiments, it was unclear how they should be labeled so that everyone knew what was meant—most of the expressions were elicited using a short scenario and not just a word or two. Thus, the use of a short-answer task allowed us simultaneously to get a baseline without biasing the participants' answers very much and to determine what words are commonly used to refer to the expressions.

4.3.4 Partial Report

Have you ever gotten half way through answering a question and suddenly realized that you've forgotten exactly what the question was? Or, you think you saw something out of the corner of your eye, but by the time you look it is gone and you can't say what it was you saw? Sperling (1960) suggested that "we see more than we remember." His work is based on some experiments from the late-1800s by Cattell (1886). Cattell demonstrated that people can remember at most five items from a briefly presented stimulus, no matter how many items the stimulus has. This limit on memory has become know as the *span of apprehension.* In the late 1950s Sperling suggested that this limit was due not to perceptual processing, but memory: as soon as people see the stimulus, they begin forgetting what was there. More specifically, he suggested that we could remember almost all of an image—even if it was no longer present—but only for a very brief period of time. In general, this type of picture-perfect memory is referred to as *iconic memory* for visual stimuli and seems to last approximately 250 milliseconds.

To test his hypothesis Sperling invented a new task, which he referred to as the partial-report task. In both the partial report and the *full report* task (which is essentially a short-answer task; see Section 4.3.3) participants are presented with the stimulus array for a brief period of time (up to 500 milliseconds) and then asked to describe what they saw. In the partial-report task, participants are to report only a subset of the stimulus array. The advantage of the latter is that no matter how complex the stimulus is, the number of items (and therefore the length of the response) in the response array is held constant. The difficulty is to figure out how to tell participants which subset they need to report. If the subset is specified *before* the stimulus is shown, then participants may simply focus on that subset and not process the rest; this essentially reduces the size of the stimulus array. Sperling solved this problem by sounding a tone *after* the stimulus array was presented. The specific tone encoded which subset was to be reported. For example, if a 3 × 3 array of letters was presented, then a high tone could indicate that the top row is to be reported, a middle tone could stand for the middle row, and a low tone for the bottom row.

Since the partial-report technique focuses on a subset of the whole stimulus, it is essentially a sampling procedure. By running a number of trials with different subsets, one can determine how much information must have been processed from the whole stimulus. Sperling demonstrated that people have an astonishingly accurate representation of an image, even if the image is not physically present. As noted earlier, iconic memory lasts for about 250 milliseconds. Naturally, then, the longer one waits to present the tone telling participants what they should report, the more the stimulus fades from memory (Sperling, 1960).

We should mention that there is a fair amount of evidence that the full benefit of a partial-report technique is only seen when the participants are very well practiced. Following standard psychophysical practice, Sperling used the same few participants for all of his experiments and used a large number of repetitions. In an attempt to quantify the practice effect, Chow (1985) has found that at least 96 practice trials are needed to obtain the partial-report benefit.

Beyond allowing one to study reading or aspects of memory and attention, the partial-report task can be used to give a measure of how much information people are aware of in very complex scenes. For example, this technique has been used to determine how many other vehicles a driver actually is aware of in a driving simulator. That is, it has been used to get a clearer measurement of situational awareness (Gugerty, 1998).

4.4 Conclusions

The free-description task is a qualitative task: it focuses written or spoken answers and provides vague but broad answers to a question. The results of a free-description task are strongly influenced by social factors and are difficult—if not impossible—to uniquely interpret. They are not common in perception research, reserved usually for the first experiment in a new line, pilot experiments to derive more details for a subsequent quantitative experiments, or for debriefing participants after an experiment.

The free-description task is also a meta-task: it provides a measure of what people think or believe. Although the remaining tasks in Part II are all quantitative (they provide numerical answers), many of them are still meta-tasks.

Chapter 5

Rating Scales

In a rating-scale task, participants place a numerical value on some aspect of each stimulus. This value can be used to determine how that stimulus compares on that dimension to other (similar) stimuli. While some tasks provide only the order of stimuli along a dimension, most also indicate precisely how large the difference between two stimuli is (e.g., A is twice as good as B, B is four times as good as C, etc.). Note that the task measures the subjective, perceived difference along the scale between two stimuli and not the objective distance between them.

To conduct a rating-scale task, the aspect of the stimuli that is to be rated must be explained to the participants. Often examples are given to highlight the relevant stimulus aspects. Participants are then either given a set of numbers (i.e., a fixed scale) or are asked to create their own set of numbers or scale. If examples were used to highlight the relevant stimulus aspects, they may also be used to highlight important areas of the scale (such as the endpoints). Subsequently, the participant is shown a series of stimuli and is asked to rate them.

5.1 Overview of Rating Tasks

Here, we provide a brief overview of the highlights of a rating task.

5.1.1 Typical Questions

Rating tasks are designed to give an insight into how elements of a class of stimuli (e.g., expressions, paintings, cities) vary along a given dimension (e.g., sincerity, aesthetic value, size). The results range from a simple rank ordering (A is best, B is second best, etc.) to a full, multidimensional perceptual or semantic map (a representation of the distance between concepts along several perceptual or semantic dimensions, see Chapter 13). Some examples of typical questions follow.

- What are people's preferences among the following paintings?

- Do people tend to prefer cubist, surrealist, impressionist, or pop art paintings?
- Which of the follow expressions do people find to be more attractive or appealing?
- How do the following computer generated animations compare in terms of realism?
- How sincere do people think these expressions are?
- How similar are these expressions?
- Which elements of this human-machine interface did people like and which elements did they not like?

5.1.2 Data Analysis

For quantitative tasks in general, the simplest form of analysis is *descriptive statistics*, which measures central tendency (e.g., mean, median, mode) and variance (standard deviation, standard error of the mean). There are also *inferential statistics*, starting with the simple t-test and ANOVA (both of which will be discussed more in Chapter 12) and gradually becoming more involved and complex. Care must be taken when either descriptive or inferential statistics are used to analyze rating data, since many rating scales are either not linear, not consistently applied across participants, or not normally distributed. That is, the structure of the results may violate many assumptions of standard statistical tests (for more information about the statistical concepts and different scale types, see Chapter 12). In some cases, transformations can be performed on the data to correct some of these violations.

Rating tasks have several specific analysis options. When there is a one-to-one mapping between the stimuli and the ratings for each participant, so that each person associates one and only one stimulus with each number (which always happens for ordered ranking, often happens for magnitude estimation, and is rare for Likert scales), one can determine what stimulus most people placed with each number. For example, maybe seven out of ten people indicated that the national debt was his or her number one concern in the current political election. If either the magnitude of the ratings is important (which is usually the case for everything except ordered ranking) or any given number can and has been given to multiple stimuli (which is almost always the case for Likert scales), then for each stimulus the average value assigned across all participants is calculated.

5.1.3 Overview of the Variants

The techniques that are either derived from or related to rating scales are listed briefly here, and discussed in more detail later in the chapter.

5.1.3.1 Ordered Ranking

Ordered ranking is the simplest rating task. In essence, participants are simply asked to list the stimuli in order along the relevant dimension. After the dimension is clearly defined, the stimuli are presented to a participant—usually all at once—and the participant is asked to place the stimuli in order. To put it another way, participants are given as many numbers as there are stimuli and each stimulus is to be assigned one number. The stimulus that has the most of the relevant dimension (e.g., which expression is the most intense?) gets the first number (first place), and so on.

5.1.3.2 Magnitude Estimation

In a *magnitude-estimation task*, rather than being given a fixed number of labels, participants are asked to assign any number they want to the stimuli. These numbers should represent a more or less intuitive indication of precisely how much of the dimension each stimulus has. Different people will come up with wildly differing scales. Often the stimuli are presented sequentially and the participant assigns the first stimulus any number. The value placed on the second stimulus is based upon the perceived distance along the intended dimension to the first. For example, if the second stimulus seems to be twice as intense as the first, it receives a number that is twice the value of the first.

5.1.3.3 Likert Ratings

A *Likert rating* is similar to magnitude estimation, except that the experimenter determines the scale. Participants are given a range of numbers and are asked to assign each stimulus the number that represents its location within the allowed range. It is extremely common for multiple stimuli to share a number.

5.1.3.4 Semantic Differentials

The *semantic-differential task* is a special variant of the Likert task where the endpoints of the fixed range of numbers are assigned bipolar opposite terms (e.g., good and bad, fast and slow). Each of the stimuli are rated along many of these bipolar scales. The results of the ratings are analyzed with a technique called *factor analysis*.

5.2 Task Description

The greatest advantage of rating tasks is that they provide a numerical value for the participant's answer directly, while still allowing the experimenter to ask a wide variety of questions. Since they are more constrained than free-description tasks, they provide more precise and reliable insight into the research question. This comes, however, at the cost of being (potentially) less informative. The endless range of information possible with a free-description task is not possible with a single rating. On the other hand, the results of rating tasks are more reliable, and it is possible to be much surer that the resulting scores and the conclusions drawn from them represent the intention of the participants. Moreover, combining different rating tasks (each along different dimensions) can approach the insight of free-description tasks, albeit at the cost of a greatly increased number of participants.

5.2.1 The Underlying Dimension

At its core, a rating task requires experimenters to pick a single dimension to represent the research question. Rating tasks also often require that experimenters place numbers strategically at critical points along that dimension. Participants then place each stimulus somewhere along the dimension (usually by assigning it a number). Thus, rating tasks provide more control over what the participants do—including through being able to choose the relevant dimensions and the specific wording of the task question, and by being able to decide where the points along the scale go.

Technically, only one dimension can be used per scale. For example, we may ask participants to rate—on a scale of 1 to 9—how artistic they find a painting to be. Or we might ask how expressive they think different images of facial expressions are (see Section 5.3.1.2 for an example). In most cases, experimenters have a very specific idea about what they wish to ask, which is encapsulated in the research question. This means that they also will have a very clear idea about the underlying dimension that they wish to investigate. This is also almost always the dimension that underlies the rating scale. Quite often, the dimension to be investigated is also the one along which the stimuli vary (although this is not always the case; see Sections 5.3.3.5 and 5.3.3.6). The selection of the dimension for the scale, then, emerges from the formulation of the research question. If there is some difficulty in determining what the rating scale dimension should be (or exactly how the task question should be phrased and explained to the participants) then it is possible that the research question is too vague. Naturally, multiple scales can be used per stimulus (see the example in Section 5.3.3.5).

5.2.2 Example Trials

Since any stimulus (whether it is an image, concept, sound, or something else) can be described along multiple dimensions, we strongly advise that the experimenter try to ensure that participants understand which specific aspect of the stimulus they should be attending to. This can be accomplished with example trials.

Caution must be applied when using example trials. It can be very tempting to provide the participant with feedback on the example trials. In some cases, this is fully justified. In all cases, it is extremely dangerous. In addition to providing information about the underlying dimension, feedback may let the participant know what the "correct" answer is. More specifically, it provides participants with concrete evidence of what we want them to say in specific cases. Many participants will ignore the hint and continue to answer with their ideas and opinions—but as mentioned in Chapter 3 some people will try to actively provide us with the answer they think we want to hear (or an answer we do not want to hear).

There are some situations where the experimenter wants to see what participants believe is critical, and others where the experimenter simply does not know what the underlying dimension is. One might, for example, ask participants how similar two stimuli are without specifying what "similar" means. Rating tasks can be used here as well, and with the aid of special statistics one can actually extract what dimensions the participants used (see the example in Section 5.3.3.6).

5.2.3 A Hypothetical Example

To provide a generic example of a rating scale, let us return to the painting preference example from Chapter 4. In the rating-task version of the experiment, we would first decide which dimension interests us. In essence, the research question is that we want to learn people's preferences for specific paintings. While "like" is rather vague concept, it is certainly a valid dimension, and in fact most people will have an intuitive understanding of it. Unfortunately, it is by no means certain that every will one have the *same* intuitive understanding, which makes drawing conclusions from such a task tricky: what were people actually reporting when they said they liked something? Were they asking themselves if they would buy it? If they would display it at home? If they would pay someone just to look at it? It is quite likely that like decisions will follow these different sub-dimensions for different people. In fact, one person may well use more than one of these sub-dimensions in a given experiment.

There are two alternatives here. First, to obtain clearer insights, we might choose to be more specific in our task (by refining the question we ask people and in describing the relevant scale dimension to them). Since the choice of a dimension is tightly tied to the specific research question, the ability to ask a focused rating-task question is based on how focused the research question is. Second, we might choose to perform some rather complicated statistical analysis to try and pull out the different dimensions that people used. Of course, this kind of analysis requires that certain assumptions are met, and will require a rather specific experimental design (see the examples in Sections 5.3.3.6 and 5.3.4).

As with free-description tasks, rating scales can be used with concepts as well as concrete images or objects. We can ask people to rate how much they like anger, and we can also ask them to rate how much they like impressionist paintings. The problem with rating abstract concepts is that the experimenter has neither knowledge of, nor control over, how participants will interpret a particular term. For example, some people will associate impressionist paintings with a specific artist or even a specific period of a certain artist—and of course the specific artist chosen can vary quite a bit across participants as well. Other participants might think of the core goals of the impressionist movement (which also varied quite a bit). In short, there will be a large amount of variance in the results since different people will think of different things for the same term. We can be more certain that all participants are in fact rating the same thing if we use concrete examples of the general class; for example, by showing participants a set of impressionist paintings and asking them to rate this type of painting in some way, The downside of this version of the experiment is that most people will rate *only* the images they saw. If the images are not chosen well—that is, if they are not representative of the more general class—then the results will not generalize to other images in that class.

In addition to rating general concepts, groups of stimuli, or even single stimuli, participants can be asked to compare several stimuli directly. The traditional example here is to ask participants how "similar" two images or objects are (see the example in Section 5.3.3.6), although one can ask nearly anything.

As mentioned in Section 5.1.3, there are three major variants of a rating task: ordered ranking, magnitude estimation, and Likert scales. Within the class of Likert scales there is a sub-variant called semantic differentials that is popular enough to deserve special mention. Before we discuss these variants (Section 5.3), we will first look at some general guidelines that are valid for any rating task.

5.2.4 General Guidelines

In general, there are five concepts that are critical in the design of a rating scale: *scale dimension, anchoring, resolution, scale usage,* and *cultural bias.* The specific warnings and tips differ depending on the type of rating task one is using. These will be discussed in more detail in the appropriate sections. There are, however, some general guidelines.

5.2.4.1 The Underlying Dimension

As noted in Section 5.2.1, the scale dimension should be as representative as possible of the research question. It should also be as specific as is allowed by the task. Contrariwise, care should be taken that the chosen dimension is not more specific than is meant in the research question. Finally, the task question should be phrased as clearly as possible so that participants understand what aspects of the stimuli they should rate. Wherever possible, examples should be used. Unless absolutely necessary, feedback should not be given at any point during the experiment.

5.2.4.2 Anchoring

Just as it is important that participants understand what the dimension is, they must understand where along that dimension the specific values on the scale are. In some tasks this is inherently clear (e.g., ordered ranking), while in others participants create the values themselves as they need them (e.g., magnitude estimation). Even in these cases, it is still advisable that participants understand how the scale works. The typical method for marking the meaning of critical points clearly on a scale (which is called *anchoring*) is to provide examples at the extreme ends of the scale.

5.2.4.3 Resolution

There are two sides to the issue of resolution. On the one hand, trying to extract more information than exists in the results will only lead to inaccurate conclusions. If there are only three numbers on the scale, then discussing differences to a degree of three significant digits is not likely to be statistically reliable. If a higher resolution is needed or desired, then more numbers should be placed on the scale (or the participants should be encouraged to use more numbers). On the other hand, using too many numbers is neither advisable nor does it help. People generally tend to use no more than ten points on any scale, and if they do, the results will not be reliable (i.e., repeating the experiment with the same participant will not produce the same results; Cox, 1980).

5.2.4.4 Scale Usage

Be aware of the type of information present in the scales. Some scales provide only ordinal information (that is, it is only possible to say that $A > B$: we can only *order* the ratings according to magnitude). Other scales provide only interval information (that is, a difference of two between ratings of five and seven is the *same* as the difference of two between ratings two and four: all *intervals* are the same). The results for some scales are normally distributed, others are not. Some participants use only the extremes of a scale, most do not use the extremes at all.

5.2.4.5 Cultural Bias

Rating scales are essentially a self-report task: the participants are being asked to volunteer potentially personal information about themselves. Obviously, some people are more willing to talk about themselves and their opinions than others are. This will show up as between-participant noise in the measured responses. It has also been demonstrated repeatedly that ethnicity, culture, and gender strongly influence how willing people are to talk about specific classes of opinion or belief. For example, there is a long history of research on the self-report of pain and how such self-reports differ systematically between different cultures (see, for example, Joint Commission Resources, 2003; Lasch, 2000; Melzack and Katz, 2001; Narayan, 2010). A significant part of this debate has focused on whether the measured differences in pain thresholds and pain tolerance are physiological (i.e., members of culture A can tolerate more pain than members of culture B) or cultural (members of culture A are less willing to admit that they are in pain than members of culture B). The bottom line is that even if the scale is understood the same way by all participants and also the anonymity of the results has been ensured, there may still be critical, non-random sources of variance. Many conclusions can be drawn from this, including the reminder that the participants should be representative of the population that is to be investigated. A corollary of this is that if a between-participants experiment is performed, it is unwise to have groups segregated by culture as this will almost certainly insert systematic bias. In other words, rating tasks are a form of meta-cognitive task: they ask people what they think or believe.

5.3 Specific Variants

5.3.1 Ordered Ranking

Ordered ranking is sometimes also called *ordinal ranking*, or simply ranking. The basic idea is that since we want to see how the different stimuli compare along a given dimension, we can simply ask participants to place the stimuli along that dimension in some order. This is the perhaps the simplest procedure for attaching a numerical value directly to the participants' beliefs and opinions.

To perform an ordered-ranking experiment, we must first—as usual—have a research question. Based on the research question, a single rating-scale question or dimension is defined. This should be as defined as possible and as closely related to the real-world phenomenon of interest as possible. After the scale dimension (or question) is explained to the participants, all of the stimuli are presented. Naturally, each participant must be given a given a chance to view all of the stimuli at least once before being required to start ranking. While this is usually achieved by presenting the stimuli simultaneously, it is certainly possible to present them one at a time. In fact, a sequential viewing might be desirable to prevent direct comparisons, or preferable if the stimuli (such as movies) are dynamic. If a sequential presentation is chosen, remember to be careful about how that order may affect participants' responses (for more, see

Section 2.1.10). Finally, participants are asked to decide the order in which the stimuli should lie along the dimension.

Another way of thinking about this task is as follows: participants are given as many numbers as there are stimuli, each stimulus is to be assigned one number, and each number can be used only once. The rule for assigning numbers to the stimuli is that the single stimulus that has the most of the relevant dimension (e.g., which expression is the most intense) gets a number from one end of the scale (usually the number 1) to represent first place. The number with the second most of that dimension gets the next number in the sequence (if the first number is 1, then the second would be 2), and so on.

At this point, the obvious question is, what should the participants do if they believe that two stimuli have the same amount of that dimension? There are two mutually exclusive options. First, each number can be used only once, in which case the participant must somehow decide which of the two equal stimuli is more equal (consequently, with five stimuli, an order of 1, 2, 3, 4, 5 will always be obtained). Alternately, each number can be used more than once; in this case, there will be three alternatives as to how the numbers are used. First, a *dense packing* of numbers can be used: that is, the two equal stimuli share a number, and the next stimulus gets the next available number. With five stimuli, one possible order is 1, 2, 3, 3, 4. Second, a number could be skipped before the double entry (so the result would be 1, 2, 4, 4, 5). Third, a number could be skipped after the double entry (1, 2, 3, 3, 5).

5.3.1.1 Guidelines

Anchoring. Do not forget to inform participants which end of the scale is the high end and which is the low. For example, there is considerable cross-cultural variance (and even some within-cultural variance) as to whether the number 1 represents the high end of a scale (e.g., first place) or the low end (e.g., in a Cartesian coordinate system, a score of 1 is lower than a score of 2).

Scale usage. By definition, ordered-ranking tasks provide only ordinal information. They can never be construed as interval scales.

5.3.1.2 Example: Rating Stylization Techniques

In our work investigating the perception of facial expressions, we directly compared several evaluation methodologies (Wallraven et al., 2007). In the first experiment, participants were shown pairs of video sequences and asked which animation they felt captured the essence of the expression better. This *direct comparison* experiment is a form of *two alternative* forced-choice task (see Chapter 6). In the second experiment, participants were shown individual expressions and asked to identify them, using a non-forced-choice task (see Chapter 6) as well as a Likert rating task (see Section 5.3.3.5).

After they had completed the main experiment, participants in each study were asked to complete an additional, much smaller experiment. In this experiment, they were shown a static photograph of the happy expression from each stylization condition and asked to place the conditions in an order that reflected how well they met specific criteria. This is the ordered ranking example we will examine next.

Background. A vast number of artistic styles have been developed over the course of human history. Different styles take the same scene and show it in different ways. One way of thinking about this is to say that each of the different styles are presenting an abstracted version of the scene. The process of abstracting consists of emphasizing or de-emphasizing certain classes of information. Given that we use specific information to recognize facial expressions, it is reasonable to ask whether different artistic stylization methods might either enhance or obscure that information, thus affecting our ability to recognize expressions. The series of experiments in Wallraven et al. (2007), which are discussed in part here, provide a first examination of the issues surrounding the what information that is emphasized in different stylistic renderings and how that relates to facial expressions.

Research question. How effective do people think these three stylization techniques are in conveying the intended expression? Do these opinions change as a function of the amount of detail preserved by the stylization technique?

Stimuli. Participants were shown high-quality printouts at different resolution levels of several stylized renderings featuring computer-generated facial expressions. In the technique used to generate the facial expressions, high *spatial* resolution three-dimensional scans of peak expressions are combined with high *temporal* resolution motion-capture of the facial deformation (captured with 72 markers) taking place during these expressions Breidt et al. (2003). Scans of peak expressions are put into correspondence using a manually designed control mesh in order to create a basis set of morphable meshes. From the motion-capture data, non-rigid motion is extracted and used to specify linear detectors for the expression-specific deformations in the face. The detectors provide the weights that drive the morph channels. Finally, geometry for the eyes and teeth are added to the scans and anchored to the rigid head motion (for more information on the animation system, see Wallraven et al., 2005).

In addition to the *standard avatar* animation sequences (see the top row in Figure 5.1), three forms of stylized rendering were used. The first was a *brush stroke* stylization (the second row in Figure 5.1). This is a painterly style in which the output images are composed of a number of small brush strokes (based on the algorithm in Fischer et al., 2005a). While this technique preserves local colors in the image (albeit with a limited random color offset added to the each pixel), it masks small or medium-sized regions (depending on the brush stroke radius). The discrete sampling of input pixels and the typically rather large sampling point distance result in limited frame coherence or motion continuity for animated image sequences.

The second style was a *cartoon* stylization (the third row in Figure 5.1(c); described in Fischer et al., 2006). In general, the technique processes the two-dimensional images and filters them so that the final images consist of mostly uniformly colored areas enclosed by black silhouette lines. Increasing the number of filtering iterations results in highly simplified, blurred image (as well as less distinct edges). The technique stresses high-contrast edges in the image and preserves the dominant color in larger image regions. It does, however, remove small details as well as low-contrast details as an effect of the nonlinear filter.

The third and final technique is an *illustrative* stylization (the bottom row in Figure 5.1; described in Fischer et al., 2005b). It reproduces the brightness of the input image using black-and-white hatching and renders the high-contrast edges as black lines. The size of the hatching pattern determines how much information is preserved. Thus, this style emphasizes intensities in the image and high-contrast edges. All color information is removed from the image along with some small surface details.

In this experiment, we used only the happy expression, although all participants had previously completed an experiment with all seven expressions (confusion, disgust, fear, happy, sad, surprise, and thinking). This impairs the generalizability of the experiment some. It does, however, allow us to have a greater precision when measuring the effect of stylization on this expression. Future work should definitely sample a larger range of expressions.

Each image sequence was contrast-normalized in order to provide a consistent input to each of the three stylization techniques. As with the focus on a single expression, this helped to ensure that the differences among the four types of stylization were due solely to the stylized rendering technique used. We chose default parameters (derived from standard applications) for rendering each technique and took care to ensure that the three resolution levels produced as similar a loss of information (in terms of information detail) as possible across the four techniques. Three different resolution levels were created by varying the parameters appropriately. For the avatar condition, the different resolution levels were produced with a Gaussian blur (for more details, see Wallraven et al., 2007).

Stimulus presentation. The high-quality printouts of the 12 stimulus images (four rendering styles with three resolution levels each for one expression) were given all at once to the participant.

Task. Participants were asked to rank the 12 images according to three different criteria. The first criterion asked *how artistic* participants thought the different stylization techniques were. The second criterion asked which of the techniques was the *most effective* in rendering facial expressions—the same question asked for the stimuli using a direct-comparison task, which is

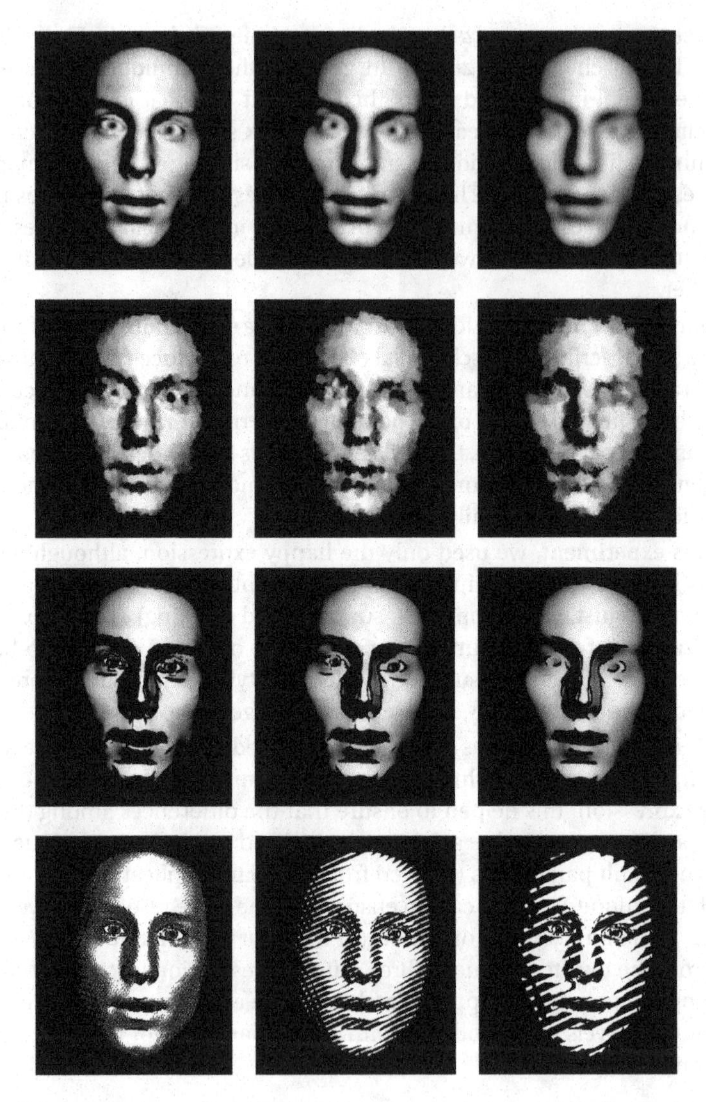

Figure 5.1. Stylization techniques used in the experiment, shown for the fearful expression at all resolution levels. Top row: standard avatar; Second row: brush-stroke; Third row: cartoon; Fourth row: illustrative. Note that although the images are reproduced here in black and white, the originals for all but the bottom row were in color. See Wallraven et al. (2007).

discussed in Chapter 6. Finally, we asked participants to rank the techniques according to which one they *liked best*.

Notice that the first and third scales are clearly asking for participants' opinions and as such are obviously meta-tasks. Although the second task may seem to ask a question that is objective—how effective is the stylization technique in conveying expression information—it is in fact still a meta-task. While the task is used to investigate the recognition of expressions, it does not in fact ask people to recognize anything, but instead to talk about which one they *think* would allow them to recognize expressions better.

Participants. Two groups of ten individuals who were naive to the purposes of the experiment participated and received financial compensation at standard rates. Each individual began the experiment immediately after completing a separate experiment with these stimuli. On the one hand, this ensured that the participants were familiar with the entire range of stimuli, even though only one expression was used in the experiment. On the other hand, it meant that the starting level of knowledge was different for the two groups (since the two groups had performed different main experiments). Thus, the data had to initially remain separated by group, since the prior experience in the first experiment might have biased or influenced the results in this experiment.

As with all of the experiments presented in this chapter, the participants were between 18 and 40 years old and came from a variety of social and educational backgrounds. To ensure that the culture-specific aspects of the expression were correctly perceived, all participants were either German (just as all the expressions came from German actors) or had lived in Germany for a long period of time.

Results. For each of the 12 possible ranks (one for each condition, with four rendering techniques and three resolution levels), the winning condition was determined (i.e., the mode of the distribution was chosen for each position). Initially the results were analyzed separately for the two groups of participants (who had performed different initial experiments to begin with and therefore had systematically different initial experience with the stimuli). The two sets of results showed similar trends, and since performing different initial experiments did not statistically influence the results of the second, two conclusions can be drawn. First, the difference in task in the two main experiments did not seem to influence the ranking of the different techniques. Second, although this is more speculative, the results *might* generalize to people who had performed a completely different main experiment or even performed no main experiment at all. This latter conclusion should be double-checked with a direct experiment to be absolutely sure.

Since it was clear that the two sets of data did not differ from each other, we pooled the answers for the final analysis. These results are summarized in Table 5.1.

The clear winner in terms of *artistic preference* is the standard animation. Second place obviously goes to some form of stylization. Of the three stylizations, the illustrative stylization was judged as most artistic, followed by the cartoon stylization and then the brush stylization.

For the *effectiveness* dimension, the standard animation was again ranked highest, followed closely by the illustrative and cartoon techniques, with the brush-stroke method judged as least effective. This pattern mirrors the one found in the direct-comparison experiment (see Chapter 6) rather closely, although the degree to which the techniques are separated in terms of their preference was much less pronounced here. This pattern is, however, different than the one found with a direct-recognition task, demonstrating that meta-tasks and direct tasks do not always produce the same results. What people *think* is more recognizable is not necessarily the same as what *is* more recognizable.

For *subjective preference*, the ranking order changes—here, the illustrative style wins, followed by the standard and cartoon renderings. As with the previous measures, the brush-stroke technique comes in last. It is interesting that the result from the subjective preference differs from the artistic preference results; stylization is strongly preferred to the original animation even though it is thought to be less effective and less artistic.

Finally, as should be expected, there is a clear ordering of the resolution levels from high to middle to low for all three dimensions. A comparison of the numbers in Table 5.1 using the general analysis procedure with those from the direct-recognition version of this experiment demonstrates clearly that the ranking technique allows only for a rather coarse analysis and interpretation. A much clearer and more precise answer to the research question can be obtained with the direct-recognition technique.

Rank	Artistic	Effectiveness	Subjective
1	Std-1	Std-1	Ill-1
2	Std-2	Ill-1	Std-1
3	Ill-1	Std-2, Car-1	Ill-1
4	Std-1	Std-2	Ill-2
5	Ill-2	Car-2	Ill-3
6	Car-2	Ill-2	Std-2, Car-2
7	Std-2, Car-2, Ill-3	Ill-2	Std-3
8	Car-1	Car-3	Std-2, Car-3, Ill-2
9	Car-3	Std-3, Bru-1	Std-3
10	Bru-1	Bru-1	Bru-1
11	Bru-2	Bru-2	Bru-2
12	Bru-3	Bru-3	Bru-2

Table 5.1. Ordered ranking: results for artistic, effectiveness, and subjective preference judgments. Abbreviations indicate stylization technique and resolution level, respectively.

Summary. This experiment demonstrates some of the analysis options for a ordered ranking task, as well as how one can deal with issues of combining several groups. It also demonstrates some of the resolution limits of an ordered ranking task.

5.3.2 Magnitude Estimation

This task was first used by Stevens (Stevens, 1946, 1951), and is an attempt to obtain interval data for basic perceptual phenomena. In short, a magnitude-estimation task asks participants to assign any number they want to the stimuli, as long as the numbers follow an interval scale. The first few steps in using a magnitude-estimation task are similar to those of an ordered-ranking task: devise a research question, create a rating dimension based on that question, and explain the dimension clearly to the participants. In general, the stimuli are then presented sequentially, and after each one is presented participants rate it. Participants are usually allowed to assign the first stimulus any number they like. No bounds are given, but the participants are encouraged to use numbers that are intuitive. The value of the second item is based upon the perceived distance between it and the first item along the intended dimension (e.g., if the second stimulus seems to be twice as intense as the first, it receives a number that is twice the value of the first). It is possible for two stimuli to share the same value.

5.3.2.1 Guidelines

Anchoring. Since the participants have no idea whether the first stimulus is nearer the high end of the scale or the low end, there can be some uncertainty as to what the first number should be. If the number is too low, then either fractions or negative numbers must be used. While this is certainly allowed—no bounds were placed on the numbers that could be used—it can rapidly become non-intuitive for the participant. For example, if a participant rates the first stimulus as a 1, only to realize later that this was the maximum stimuli, then all subsequent values will have to rated as less then 1. This is, in effect, an anchoring problem and as such can be solved by providing two examples, one at each end of the scale. It is critical that the experimenter provide no number with the example, as to do so would affect the range of numbers that the participant uses. An alternative solution to the anchoring problem is to use a repeated-measures design (each stimulus is rated more than once; a blocked design would usually be used, in which all stimuli are seen once before any is seen a second time).

Resolution. A two-point scale has just as much resolution when the numbers are 1 and 2 as when they are 1 and 100. Do not try to get more resolution out of the scale than is reliably there.

Scale non-uniformity. A second issue with magnitude estimation is that different people will use different ranges of numbers. One person might choose to use values between 0 and 1, while another might use values between 100 and 1,000, in steps of 10. In these instances one cannot simply calculate the mean of the participants' responses: one must first transform them into a common range. Since this is easy to do, it is not a serious difficulty.

Non-linearity. Some people see the change from 1 to 2 as larger than that from 100 to 101, even though the two intervals are by definition identical. For example, doubling the perceived intensity of a sound that was rated as 1 will produce a sound that should be rated as 2. Providing the same perceptual increase in sound intensity to a sound that was rated 100, may or may not be rated 101. Thus, the response may be non-linear. Just as with the presence of different ranges, this can be taken care of with a simple transformation.

Order effect. Not everyone is good at internal consistency. Some people will easily remember the choices they previously made, and will make the same decision when subsequently presented with similar stimuli, others will not. Moreover, as people see more and more stimuli, the criterion they use to make a judgment might well change. That is, two identical stimuli might receive vastly different numbers depending on where they are presented in the sequence. This is essentially an order effect, and the solutions are the same: repetition or randomization (or both).

Controversy. As one might expect, there are some controversies surrounding what magnitude estimation really measures, how it measures it, etc. For the most part, these will not affect the use of the scale. For more information, the following articles provide a starting place: Kranz (1972); Luce (1990); Marley (1972); Narens (1996); Shepard (1978, 1981).

5.3.3 Likert Tasks

In the 1930s, Rensis Likert (pronounce with a short "i" as in "LICK-ert") worked on measuring attitudes (Likert, 1932). Undoubtedly his most famous contribution is his measurement technique, which has been named after him. In essence, a Likert scale is a special case of magnitude estimation. The primary difference is that in the case of a Likert scale, the experimenter determines the scale. That is, a fixed set of numbers (e.g., 1 through 5) is provided to the participants. The individual numbers represent steps between two extremes along a dimension. For example, people might be were asked to rate—on a scale of 1 to 7—how much they like a series of paintings, with a score of 1 representing, "do not like it at all" and 7 representing "really like it." Participants are asked to assign each stimulus the appropriate number to represent where the stimulus lies along the dimension.

5.3.3.1 Nomenclature

A short word on jargon is necessary here. There is an ongoing conflict between the official and the common usage of the term *Likert scale.* Likert's original aim was to provide a quantitative measurement of attitudes. To do so, he gave the participants five numbers (1 through 5, inclusive) and assigned each number an adjective (strongly agree, agree, neutral, disagree, and strongly disagree, respectively). Participants were to rate how well they agreed with different statements. Moreover, Likert did not use the values from individual statements, but instead averaged the responses over a set of similar statements. Some purists, then, insist that only the average of a set of statements can be referred to as a Likert scale. Moreover, they insist that only the terminology used by Likert is allowable; specifically, that any given 1 through 5 rating should referred to as a *Likert-item* and only the averaged score should be called a Likert-scale.

The common usage of terminology in Likert tasks is much more relaxed. First, the range of possible numbers (e.g., 1 through 5) associated with each rating question is referred to as a scale (this is arguably more in line with the mathematical usage of the term scale). To the degree that individual ratings are averaged, the average scores are referred to as the averaged (or summed, or aggregate) score. Second, any range of numbers can be used in a rating. Most commonly, one end of the scale is 0 or 1 and the entire scale is usually also positive. In referring to such expanded Likert tasks, the number of available steps is listed. For example, the scales 0 through 4 and 1 through 5 both have 5 steps, and would both be referred to as 5-point Likert scales.

In this book, we will follow the common usage of the term Likert scales.

5.3.3.2 Labeling the Scale Points

When using a Likert scale, there are a number of alternatives for labeling the scale points. At one extreme, only the endpoints will be labeled (as the warnings about anchoring have hopefully made clear, the endpoints *must* be labeled). At the other extreme, each number on the scale would be labeled. Wells and Smith (1960) showed that participants are able to produce a finer grained (and more reliable) discrimination when adjectives are present at all scale points than when only the end points are labeled. Moreover, extensive research has shown that using certain adjectives (for example, "extremely," "quite," and "slightly"; or "always," "often," and "seldom" for a 7-point scale) can ensure that the Likert scale is an interval scale (that is, that the perceptual or conceptual distance between 1 and 2 is the same as the distance between 3 and 4, etc.; Cliff, 1959; Howe, 1962, 1966a,b).

There is also evidence that the exact name of the endpoints is less critical than Likert purists would have us believe. Asking, "I would buy this painting" and using Likert's agreement adjectives (strongly agree, agree, etc.) produces results similar to asking, "I would buy this painting" and using the terms extremely likely, quite likely, and so on up to extremely unlikely. The fact that using slightly different forms of either the end points of a Likert scale or the question can produce nearly identical results is sometimes called *equivalent forms*, and is part of the basis of semantic differentials (see Section 5.3.4). In short, when constructing a Likert scale caution must be used in order to ensure that the scale is an interval scale (through using the appropriate adjectives); but being overly strict (by limiting the nouns, or the number of points used) does not seem to be warranted.

Note that Likert's original "items" were bipolar opposite terms (strongly agree and strongly disagree). In most Likert scales, this bipolar structure is retained, although it may be a different set of opposites (such as like versus dislike, fast versus slow, or likely versus unlikely). On a bipolar task it is advisable to use an odd number of terms and to have the middle number represent the neutral point. A unidimensional scale can be constructed with one end of the scale being neutral and the other end being the desired extreme.

5.3.3.3 Constructing a Likert Task

The simplest method to construct a Likert scale is, naturally enough, very similar to the procedure involved in magnitude estimation. First, a single statement or question is generated to match the research question as closely as possible. Then, the number of steps on the scale should be decided. In general, there should be five or more steps and 11 or fewer. Given that an odd number of steps should be used, this limits the options to using five, seven, nine, or eleven steps. Subsequently, the individual steps are assigned adjectives. A fair amount of carefully designed research has shown that if the terms "extremely" (or "strongly"), "very," and "neutral" are used, then the scale can be considered to be an interval scale. That is, for most native English speaking people, the term "very" is precisely halfway between "neutral" and "extremely." If a language other than English is used, we recommend that the existing psychological literature be consulted to see if appropriate adjectives have been researched. Finally, after the statement or question has been explained to the participants, they are shown the stimuli one at a time and asked to rate them. For example, imagine that we wish to measure people's preferences for a set of paintings. We might give participants the statement "I would hang this painting on my wall" and ask them to use a scale of 1 to 5 to indicate their agreement (with the labels: extremely likely, very likely, neither likely nor unlikely, very unlikely, and extremely unlikely for the numbers 1, 2, 3, 4, and 5, respectively).

It is, of course, possible that the research question is not adequately reflected in one scale statement. To ensure that we are really measuring what we are interested in, we may choose to create several similar scale statements and then use the averaged or cumulative score across them. To do this, we would either generate the statements ourselves or have a group of people generate them. As a rule, many more statements should be generated than are actually needed in the final experiment. The initial statements would be provided to a group of participants in a pilot study who will then rate (perhaps using a 5-point Likert scale) the appropriateness or relevance of the statements to the research question. The results of these ratings are run through a correlation analysis to find the scales that co-vary best with each other and with the research question. Those scales that are judged to be particularly relevant or appropriate can then be used in the final experiment. It is not necessary to perform this pilot validation, but it does help to increase the validity and precision of the final results. To continue with our painting example, in addition to asking people if they would hang the painting on their wall and if they would buy it, we might add a few more statements that measure similar things (e.g., "I would pay to see this painting in a museum," "I enjoy looking at this painting," "I would like to see more paintings like this"). Each participant would rate each stimulus on each of the five scales each of which has five steps with the terms "extremely likely," etc. Note that if a large number of scales are used (e.g., more then ten), then order effects may well start to appear; that is, the way people answer the last few questions may be biased by the first few. Either standard order-effect solutions, or some of the special techniques for the designing of questionnaires, can be used to address this problem.

5.3.3.4 Guidelines

Anchoring. This is one of the central difficulties with Likert scales. When participants first show up for the experiment, they do not know what range of stimuli they will see. Since the rating scale has definite ends, participants are often hesitant to use the extreme values since a stimulus might show up later in the experiment that is more extreme than what they have seen so far. To avoid this, we highly recommended that examples of what the end points be provided. Note that this is not always easy to do explicitly (e.g., not everyone will have the same reference painting in mind for an "extremely likely to hang on my wall" rating) and may have to be done with verbal descriptions. Alternatively, a repeated measures design in which each stimulus is seen once before any stimulus is seen a second time could be used. Here one would either use a fixed number of repetitions or—more elegantly—keep repeating until two (or more) subsequent set of ratings are similar enough to be considered stable. If a repeated measures design is chosen, then only the final repetition's results should be used in the final data analysis (if the stimuli were repeated until the results were stable) or a test should be performed to see if the repetitions differ significantly.

Criterion changes and order effects. At the onset of an experiment, people will have some criterion or set of criteria that they will use to make their decisions. It is very likely that these will change over the course of the experiment as a result of actually seeing the stimuli. This is intimately related to the issue of anchoring. To ensure that the measurements accurately represent the participants' opinions and that the results are reliable (i.e., that a replication of the experiment would provide similar results), it can be advisable to use a repeated measures design. Likewise, the standard order effect techniques or some advanced questionnaire techniques can be helpful here.

Cultural bias. The cultural bias on scale usage is perhaps most noticeable with Likert scales. Some cultures inherently avoid using extreme values to rate anything; others are reluctant to discuss certain issues, and so may also avoid extreme values. On the other hand, some cultures prefer to use extreme values. If all of the participants come from the same culture, then beware about drawing conclusions about other cultures. If a between-participants design is used, try to make sure that all groups are equally representative of the target population (and that, for example, two groups do not happen to differ with regard to culture).

Scale type. The main controversy surrounding Likert scales is really not the nomenclature, but whether the scale is ordinal, interval, or something in-between. Considerable research has shown that under many circumstances, it can be regarded as interval (Cliff, 1959; Howe, 1962, 1966a,b). Nonetheless, care should be taken when analyzing the data (especially since the distribution may not be Gaussian).

5.3.3.5 Example 1: Rating the Sincerity and Intensity of Stylized Facial Expressions

Automatically generating computer animated facial expressions that are as well recognized as real expressions is still an unsolved challenge. There are many reasons why it is proving to be so difficult. This is even more true for dynamic facial expressions (those generated as a video sequence instead of as a photograph). Yet, even if we managed to create an expression that everyone identified correctly, we would still not necessarily have a useful animation system. This is because there are many other dimensions beyond recognizability that are important to the use of expressions in a natural conversation. Two of the most important are sincerity and intensity. For example, it is easy to fake an expression—but very often it is clear from looking at someone if he or she is sincere or is pretending. While sincerity is not currently a big concern in the automatic synthesis of facial expressions, it will undoubtedly become a central issue in the future. Who would buy anything from a virtual salesperson if the sales agent seems dishonest and insincere? Who would employ digital actors, if no one ever believed any lines they delivered? The same is true, albeit perhaps to a lesser degree, for intensity. If we wish to have a virtual sales agent indicate gently that the customer's wish cannot be fulfilled, generating a "strongly disagree" expression would not be appropriate.

Over the course of many experiments, we explored the perceived sincerity and intensity of facial expressions (for real and synthetic expressions) as well as the factors that play a role. The most common methodology for doing this is a Likert scale. In Wallraven et al. (2007)—see also Section 5.3.1.2—we performed several experiments comparing different evaluations methodologies. In the next section, will discuss the rating scale portion of the second experiment.

Research question. To what degree are the facial expressions generated with the four rendering techniques able to convey the proper amount of intensity and sincerity?

Stimuli. The animation sequences discussed in the ordered ranking example (see Section 5.3.1.2) were used here. We used all seven expressions (confusion, disgust, fear, happy, sad, surprise, and thinking). This number of expressions is large enough and diverse enough to provide a moderate degree of generalization to other expressions, but small enough to allow the experiment to be completed before fatigue effects set in.

Stimulus presentation. The participants were placed in a small, dark, enclosed room. A single trial of the experiment consisted of the video sequence being shown in the center of the screen repeatedly. This allowed participants the opportunity to view all aspects of the video until they felt they were as familiar as necessary with it to answer the questions. Detailed analysis of the results of this experiment (as well as previous research in our lab) has shown, however, that participants usually view the video only once and that there is little difference in results between experiments with repeated viewings and experiments in which a single viewing was used (see, e.g., Cunningham et al., 2005), so a repeated viewing is not strictly necessary. A blank screen that lasted 200 ms was inserted between repetitions of the video clip to ensure that the beginning and the end of each sequence were clearly delineated. When participants were ready to respond (which they indicated by pressing the space bar, allowing us to measure their reaction time) the video sequence was removed from the screen, and the participants performed three tasks. The experiment used three repetitions of each sequence. Crossing seven expressions × four stylization techniques × three resolution levels × three repetitions yields a total of 252 trials. The order of the trials was fully randomized. Each sequence was seen once before any sequence was seen a second time.

Task. Participants performed three tasks for each stimulus. The first task was always a non-forced-choice recognition task (see Chapter 6). The second and third tasks were rating tasks using 7-point Likert scales. The three tasks were always performed in the same order, and the full experiment was explained (including examples of the three tasks) to each participant before the experiment began. No feedback was given at any point on how the participants performed. Consequently, the participants knew precisely what would be asked of them before they viewed each trial. Pilot work indicated that participants generally knew their answer to all three questions before they terminated the playback of the videos. Thus, little order effect was expected for the three tasks, and randomizing the order of the tasks (or even controlling for order) was not deemed necessary.

In the first rating task, the participants had to rate the intensity of the expressions on a scale from 1 (not intense) to 7 (very intense). Note that this is a unidimensional scale (the term on one end is neutral and the term on the other end is the extreme of the dimension of interest). Likewise, it should be mentioned that the intensity of the different expressions was not manipulated in the stimuli. The physical intensity of a given expression was the same in every rendering condition. Thus, any difference in intensity ratings cannot be due to actual differences in intensity, but instead must be due to the emphasis or removal of information (which is the basis of stylized rendering). This is a demonstration that although the dimension that is being measured is usually also the dimension that is being manipulated, that does not need to be the case. This does have the side effect that it is impossible to explicitly anchor the ends of the scale with images of "very intense" expressions and "not intense" expressions. The scale was anchored with verbal descriptions of what was meant by "very intense" and "not intense." Notice also that only the end points were labeled and that the end adjective was "very," not "extremely." This makes it somewhat less than certain that the scale is in fact interval. Some attempt was made to encourage an interval scale with the spacing desired by explicitly instructing participants to use a value of 4 for normal intensity and to try and use the whole range of the scale during the experiment.

For the second rating task, participants were asked to rate the sincerity of the expressions, with a rating of 1 indicating that the actor (upon whom the animations were based) was clearly pretending and a value of 7 indicating that the actor really meant the underlying emotion. As with intensity, we did not manipulate sincerity, but were instead interested in whether the information selectively highlighted by the different stylization techniques altered the apparent sincerity. Thus, no images for the extreme values of the scale were possible. Instead, the scale was verbally anchored with examples and participants were encouraged to try and use the whole range of the scale during the experiment.

Participants. Ten individuals who were naive to the purposes of the experiment participated and received financial compensation at standard rates. As with all experiments in this chapter, the participants were between 18 and 40 years old and came from a variety of social and educational backgrounds. To ensure that the culture-specific aspects of the expression were correctly perceived, all participants were either German (since all the expressions came from German actors) or had lived in Germany for a long period of time.

Results. Only the results for those trials where the participant correctly recognized the expression (which was the first task) were analyzed. For intensity ratings, an ANOVA was run with expression, stylization, and resolution as within-participants factors. All three main effects were significant as were the two-way interactions between expression and method, and stylization and resolution. For more on the ANOVA, see Chapter 12.

Overall, emotional expressions such as disgust, and fear were rated as more intense than expressions such as sadness or thinking. This shows that the intensity scales were sensitive enough to pick up the expected variation.

In terms of stylization, the brush-stroke stylization was rated as much less intense than the other three methods. One reason for this is that the brush-stroke pattern masks both rigid and non-rigid head motion (which are both highly correlated with ratings of perceived intensity Wallraven et al., 2005). Analysis of the interaction between expression and method revealed in particular that happy, sad, surprise, and thinking were rated as much less intense for the brush-stroke technique than the remaining three expressions. Finally, there was a large decrease in perceived intensity for the standard and the brush-stroke techniques at the lowest resolution level, whereas this decrease was less pronounced for the cartoon technique, and virtually absent for the illustrative technique. In short, the illustrative technique provides the highest and the best impression of intensity (even at low resolutions).

For the sincerity ratings, another ANOVA was run. Again, all main effects were significant, as was the interaction between expression and method. The results are consistent with other work in suggesting that lower resolutions provide less sincere expressions across all techniques. This interaction of expression and method can even be seen in Figure 5.1, in the third row.

Summary. This example demonstrates some of the issues involved with a simple Likert rating tasks, including a few concerns on analysis (when the rating task was combined with a forced-choice task).

5.3.3.6 Example 2: Rating Similarity

Research question. How similar are facial expressions? What is the metric structure of the perceptual space for facial expressions? Are there natural clusters of expressions?

Stimuli. The stimuli for this experiment came from the small MPI facial-expression database (Cunningham et al., 2005). These expressions have been well investigated using a variety of methods and questions (Cunningham et al., 2003a, 2004a,b, 2005; Cunningham and Wallraven, 2009; Wallraven et al., 2004, 2007). The database was briefly introduced in Chapter 2, and a few examples can be seen in Figure 2.9. In slightly more detail, the database consists of nine expressions recorded from six individuals (four male, two female) using a method-acting protocol. For this protocol, a set of real-world scenarios designed to elicit the desired expression were developed. After this scenario was explained, the person controlling the cameras and the actor/actress searched jointly for a similar experience in the participant's life. Once found, participants were to try to place themselves in the emotions of the experience and then react accordingly. Each actor/actress performed each expression three times in a row. The best-recognized and most believable version of each expression for each individual was selected for all future experiments (Cunningham et al., 2005). Research has shown that presenting the full sequence (from neutral expression to peak expression and back to neutral) provides results similar to just showing the neutral to peak. To save time, only the neutral to peak versions were shown. For sample static snapshots of the expressions from one actress, see Figure 2.9. While using nine expressions and six individuals might be near the lower limit of expressions necessary to be able to say something about expressions in general, it is small enough that it allows a number of manipulations without causing the experiments to become too long. For more on the recordings and the database, see Cunningham et al. (2005).

Stimulus presentation. The image size was reduced from its original size of 768 × 576 to 256 × 192 pixels (10 × 7.5 degrees of visual angle). Previous research (see Section 2.1.13) has shown identical recognition accuracy for these two sizes, as well as for whether people saw the expression just once or several times. Pairs of expression were shown, one after the other. A blank screen that lasted for 200 ms was inserted between the video clips. This is sufficiently larger than the temporal-integration window of the visual system that it guarantees that participants will see the two videos as two separate sequences. Only expressions from the same actor were shown, to ensure that participants are rating the similarity of the expression and not the similarity of the actor or the way the actor performed that expression. Moreover, each participant saw all pairs from one actor before seeing any pairs from another actor. This helped to ensure that the similarity ratings were due to differences in expression and not actors. The order of pairs was random for each actor, with each participant receiving a different order. The order of actors was also random, with each participant receiving a different order. For each of the six actors there were 45 pairs of sequences, for a total of 270 trials.

Task. The participants were to indicate if the two expressions were identical (a rating of 1) or completely different (a rating of 7). Verbal examples were used to anchor the scale.

Participants. Ten individuals who were naive to the purposes of the experiment participated and received financial compensation at standard rates. To ensure that the culture-specific aspects of the expression were correctly perceived, all participants were either German (since all the expressions came from German actors) or had lived in Germany for a long period of time.

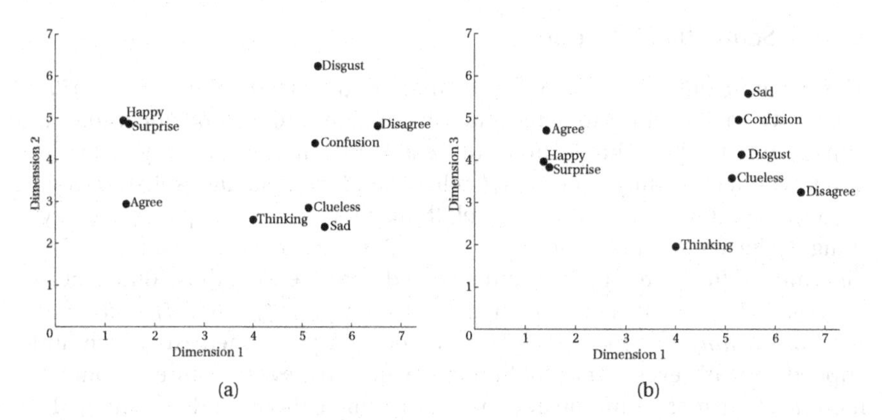

Figure 5.2. Results of the multidimensional scaling analysis. Two-dimensional plots of the three semantic dimensions, (a) Dimension 1 versus Dimension 2, (b) Dimension 1 versus Dimension 3.

Results. The mean ratings for each pair of expressions were calculated across all participants and submitted to a *multidimensional scaling* (MDS) analysis. In short, an MDS analysis takes an $N \times N$ matrix of similarity data and returns the coordinates of the N data points in the d-dimensional space. There are different procedures for determining how many dimensions are needed. For details on MDS, see Chapter 13. Initial analysis of the 9×9 matrix for this experiment indicates that three dimensions are needed to appropriately describe the data. Since visualizing a three-dimensional space is tricky, we have chosen here the relatively standard approach of showing all the two-dimensional plots (see Figure 5.2). The plots show some structure for semantics space. For example, the first dimension clearly separates the positive expressions (happy, pleasantly surprised, and agree) from the rest. Likewise, the most negative expressions (disagree) show up on the other end of the scale. While we have no clear idea what this dimension really captures, it is reasonable to assume it is an evaluative (good versus bad) dimension. The second dimension seems to also split the expressions into two groups, although the properties of this dimension are less clear. These plots clearly demonstrate the advantages and disadvantages of MDS. It is clear that three dimensions describe the similarity space. Likewise there is a clear structure within that space. Unfortunately, we have little to no idea what the dimensions of that space are.

Summary. This example demonstrates how a simple task—such as an unconstrained similarity rating—can reveal a tremendous amount of information with the proper assumptions and statistics.

5.3.4 Semantic Differentials

The semantic differential is a special variant of the Likert scale. In the 1950s, Charles Osgood decided to determine why humans are so good at comparing apples and oranges. That is, how are we able to compare two very disparate and unconnected things to produce reliable and meaningful results? Osgood suspected that we must, at some level, think about (or even represent) everything in the same set of conceptual dimensions. He developed a method to determine if this were the case, and if so to determine what those dimensions are. His 1957 book detailing the method is appropriately called *The Measurement of Meaning* (Osgood, 1957). The method has been refined extensively and applied to nearly every aspect of human life in nearly every culture or country. It is one of the most common and most powerful methods for determining the semantic and conceptual space underlying any group of stimuli.

5.3.4.1 Standard Procedure

In the standard procedure, a large number of participants (the more the better) are presented with a series of stimuli or concepts (again, more is better, but too many leads to fatigue) and are asked to rate each concept on a large number of bipolar scales (at least 12, often more than 100). The ends of the scales are—as is typical for Likert tasks—anchored with two terms. These terms must be semantically opposite and should be equally powerful. Common examples are big–small, fast–slow, and transparent–opaque. Notice that big and small are clear, very common terms that are relatively equal (but opposite). As such, they make good endpoints. In contrast, gigantic and small are not equal (gigantic can be seen as being much larger than a normal "big" size). Tiny makes a more appropriate opposite to gigantic.

Once the data have been collected, they are averaged across all participants (for stability and accuracy). Thus at the end, a $N \times M$ matrix is obtained, where N is the number of stimuli and M is the number of scales. A factor analysis is run on the data. Briefly, the factor analysis is a derivative of principle component analysis and is a way of examining the covariation in the data. The goal is to determine the minimum number of dimensions that are needed to explain the data. Naturally, 100% of the variance in the data can be explained or described by using a number of dimensions equal to the number of scales. If, however, two scales co-vary strongly, then they are probably measuring the same underlying concept and can be safely collapsed into a single dimension. By removing or combining scales that are highly correlated, one can explain the data with fewer dimensions with only little loss in accuracy.

5.3.4.2 Semantic Space: The EPA Dimensions

It has been found consistently that the same three dimensions explain most of the variance in semantic-differential responses (usually around 70%), regardless of the concepts or stimuli involved. In order of importance, the dimensions are usually called *evaluation* (examining things like good–bad), *potency* (e.g., strong–weak), and *activity* (e.g., fast–slow). Quite often, a fourth dimension—*predictability*—plays a significant role. In one of the largest cross-cultural applications of semantic differentials, Osgood (1964) gathered 100 basic concepts from everyday life and translated them into 14 different languages (English, Finnish, Japanese, Kannada, Dutch, Flemish, French, Arabic, Swedish, Cantonese, Farsi, Serbo-Croatian, Hindi, and Pashto). The 100 concepts were broken down into ten subsets. Each subset was rated by 20 different people for each language (all native speakers, all male between 12 and 16 years old) along 50 bipolar scales. With 20 participants, 10 subsets, and 15 countries (Farsi data were collected for two separate countries) this amounts to over 3,000 participants! The results were subjected to PCA analysis. The first—and most dominant—factor in every language was the evaluation dimension. The second factor was either the potency or activity dimension, and the third factor was the remaining dimension. This structure has been verified repeatedly since then for nearly every possible type of concept, despite large variations in both experimental methodology and statistical analysis.

5.3.4.3 Scale Selection

In selecting the scales to be used, care must be taken to ensure that they sample semantic space representatively. This is very difficult and costly, as it requires large numbers of trials, participants, and stimuli. Many different scales must be tried to see how well they vary with the concepts one is interested in capturing. Obviously, in order to be able to measure the semantic space with as few scales as possible, it is best to use only *pure* scales—scales that correlated uniquely and very highly with their appropriate dimension. Using weakly correlated scales dilutes and confounds the results (Koltuv, 1962; Mitsos, 1961).

Determining which scales are truly pure can only be done empirically. To this end, early studies used hundreds of scales to determine which ones were pure for a range of concepts. Since this is very time consuming and expensive, subsequent experiments tended to use only the pure scales from these early studies. One of the most common sources of pure scales is Osgood's thesaurus experiment(Osgood, 1957). Osgood created 76 scales using pairs of adjective from *Roget's Thesaurus*. The scales were chosen to represent a very large range of concepts. One hundred college students rated 20 concepts. The first three dimensions reflected usual EPA structure and account for more than 67% of the variance. While the large number of ratings makes this a good source of scales, the low number of concepts makes it unclear if the scales are appropriate for all research domains. That is, it is best to use scales that are meaningful to the current application domain (Triandis, 1959). For example, asking if a facial expression is transparent–opaque will yield reliable but very noisy results. Thus, new studies occasionally are needed to determine which scales are pure for the new application domain.

Since research has shown that three pure scales per dimension is sufficient to produce reliable results (adding more scales rarely changes the results), subsequent studies usually used the best 9 to 16 pure scales from these earlier studies—three or four scales for each of the three or four primary semantic dimensions.

5.3.4.4 Presentation Order

Since there are many stimuli as well as many scales, the question naturally arises as to how they should be presented. There are, logically, three alternatives:

Stimuli-dominant. All scales for one stimulus must be completed before the next stimulus is presented.

Scale-dominant. All stimuli must be rated along one scale before the next scale is presented.

Paired. All possible pairs of stimulus and scales are generated and then shown in random order. Thus, although only one scale will be presented with a stimulus on any given trial, each stimulus will eventually be paired with all scales.

Research has shown that there is very little difference in the results for these three alternatives in terms of accuracy. There is, however, some difference in terms of efficiency (Osgood, 1957; Wells and Smith, 1960). If all stimuli are rated on a single scale before the next scale is seen (scale-dominant presentation), then there is a stronger possibility of an order effect (ratings on one stimuli will tend to have a stronger impact on other stimuli). Depending on the number of scales and the number of stimuli, as well as the similarity or divergence of the stimuli, presenting all the scales together for a single stimulus (stimuli-dominant presentation) might be slightly faster than showing each stimulus-scale pair in random order.

5.3.4.5 Guidelines

Anchoring. Since each semantic differential scale is in essence a Likert scale, the usual anchoring warning applies.

Fatigue. It has been suggested that participants' patience or endurance may be strained if they are asked to do more then 400 rating; this effect depends on the experience of the participant and how much they are paid (Hiese, 1970).

Repeated measures and order effects. If a repeated measures design is used—that is, if each stimulus is to be rated more than once—the order in which the stimuli are presented might cause a serious problem. A repeated measures design might be useful when ratings are done, for example, before and after some treatment. For example, Coyne and Holzman (1966) asked participants to rate aspects of their own voices both before and after hearing themselves on a tape recorder. When the exact same set of questions was used, no significant differences were found between the before and after ratings. When different but equivalent scales were used for the two tests, then significant differences appeared. This and other work (see, for example, Miron, 1961) suggests that participants tend to remember their previous answers and base all subsequent answers on that.

Scale type. As mentioned previously, there is some controversy about whether Likert scales are ordinal or interval. Research has shown that using appropriate adjectives for each point on the scale (such as "extremely," "quite," and "slightly," for a 7-point scale) can ensure that an interval scale is created (Cliff, 1959; Howe, 1962, 1966a,b).

5.3.4.6 Example: Semantic Structure of Facial Expressions

An extensive experiment was conducted in 2007 to determine the semantic structure of emotional words (Fontaine et al., 2007). Twenty-four terms describing emotions were presented, and participants were asked to imagine that someone had just used the emotional term to describe an event. The terms were presented to people from three different cultures (198 Dutch-speaking students in Belgium, 144 French-speaking students in Switzerland, and 188 English-speaking students in the UK). Note that since each participant saw only four terms in the experiment, there were effectively only about 20 participants per term. The participants were asked to rate the likelihood that other words (i.e., those on the semantic scales) could also describe the event. For example, assuming that someone had just described an event as "happy," on a scale of 1 to 9 how likely is it that the event could also be described as "excited." The study used 400 scales chosen from a very wide range of potential semantic dimensions. Although the endpoints of the scales were bipolar (likely–unlikely), the scales themselves were always unidimensional (e.g., "feel constrained"). Thus, the scales were not, strictly speaking, semantic differentials. This difference, however, is relatively unimportant. Research has shown that the results with this form of scale (likelihood of a single term) can be converted into proper semantic differential scales with little trouble, especially if both poles have been measured. This is related to the concept of equivalent forms discussed in Section 5.3.3.2. Fortunately, for most of the scales, Fontaine et al. (2007) had a matched scale that was its semantic opposite. Thus, in addition to a "how likely is it that they were also excited" scale, there was a "how likely was it that they were also calm" scale. The study found the four EPA dimensions in the expected order.[1] The four factors explained approximately 75% of the variance, although there was some differences between the cultures.

[1]For technical reasons, Fontaine et al. (2007) used a PCA rather than factor analysis, as did Osgood in many of his experiments.

We were interested in seeing if these results also hold for facial expressions—both emotional and conversational—when the facial expressions were actually seen, as opposed to being merely imagined. As a first step we replicated Fontaine et al. (2007)'s results with our cultural group (German speakers in Germany) using a more traditional semantic-differential design (Cunningham et al., 2012). To do this, we selected the three scales that correlated best with each of the four primary dimensions (all correlations were very high; i.e., $r > 0.8$). The semantic opposite of the selected scale was then defined. This opposite was usually also a scale in the original experiment and usually correlated just as well with the underlying dimension as the original term. This highlights the fact that unipolar and bipolar scales are relatively equivalent procedures. Initial attempts to replicate the experiment showed very unreasonable results. Eventually, it became clear that participants were having strong difficulty with the instructions. After altering the instructions to increase clarity and to make them more appropriate for the experiments with facial expressions, the results matched those from the original experiment very well. We found the four EPA factors in the expected order, and they explained 87% of the variance.

Once convinced that the methodology was sound, we proceeded to examine the central question.

Research question. What is the semantic structure of facial expressions?

Stimuli. The stimuli for this experiment came from the small MPI facial-expression database (Cunningham et al., 2005). The database was briefly introduced in Chapter 2, and a few examples can be seen in Figure 2.9. All nine expressions for all six actors were used.

Stimulus presentation. The 54 videos were shown in a random order, with each participant receiving a different random order. Each video was shown repeatedly, in the center of the screen while the participant rated it on the 14 scales. Once all 14 answers had been entered for a particular stimulus, the participant could move on to the next stimulus.

Task. The participants were asked to rate each expression on 14 scales. All scales were presented in German and English. Participants were instructed to ignore the person performing the expression as much as possible and to focus on the communicative intent. Due to technical difficulties, data from one scale were not included in the data analysis.

The first, fifth, ninth, and thirteenth scales were chosen to correlate with factor 1. These scales were:

- felt positive–felt negative
- felt liberated or freed–felt inhibited or blocked

- wanted to be near or close to people or things–wanted to keep or push things away
- wanted to be tender, sweet, and kind–wanted to do damage, hit, or say something that hurts.

The second, sixth, tenth, and fourteenth scales were chosen to correlate with factor 2. These scales were

- felt strong–felt weak
- felt dominant–felt submissive
- wanted to tackle the situation–lacked the motivation to do anything
- wanted to take initiative himself or herself–wanted someone to be there to provide help or support.

The third, seventh, and eleventh scales were selected to correlate with factor 3. These scales were

- felt restless–felt restive
- felt heartbeat getting faster–felt heartbeat slowing down
- felt breathing getting faster–felt breathing slowing down

The fourth, eighth, and twelfth scales were selected to correlate with factor 4. These scales were

- caused by an unpredictable event–caused by a predictable event
- emotion occurred suddenly–experienced the emotional state for a long time
- caused by chance–person had control over cause

Participants. The video sequences were shown to 15 participants who participated as a requirement for a class exercise. While this is a common source of participants in psychology, it should be mentioned that it does more or less ensure that the participants will not be necessarily representative of the entire population. For one thing, most of the participants tend to be studying psychology. Additionally, most participants tend to be between 18 and 25 years old. For more on the problems with basing research on this type of participant pool, see Henrich et al. (2010b). Since the experiment was part of a class experiment, the participants were *not* fully naive to the purposes of the experiment. This is a second reason to be uncertain of the generality of the results, although it must be said that the participants had no idea what the data should look like, and the purposes of the experiment were simple enough that normal participants would be able to guess the purpose within a few trials anyway.

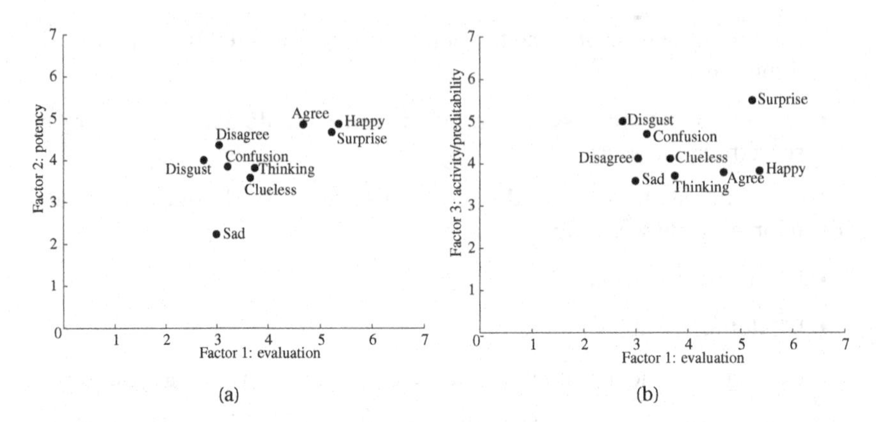

Figure 5.3. Location in semantic space of the facial expressions. The average score of each expression along each of the three semantic dimensions is plotted: (a) Factor 1 versus factor 2, full scale, (b) Factor 1 versus factor 3, full scale.

Results. The results were averaged across all participants, yielding a single 54 × 14 matrix (54 stimuli × 14 scales), which was subsequently submitted to a factor analysis. Only the first three factors had *eigenvalues* (see Chapter 13) over 1, suggesting that three dimensions are sufficient to explain the results. The three factors jointly explain 88.9% of the variance.

The first factor, accounting for about 32.2% of the variance, is a fusion of dimensions 3 and 4 (activity and predictability), with scales 3, 4, 7, 8, 11, and 12 loading onto it. There appears to be insufficient variation in the stimuli along the activity and predictability dimensions for them to require separate factors. The second factor, accounting for about 31.4% of the variance, is dimension 2 (potency), consisting of scales 2, 6, 10, and 14. The final factor is dimension 1 (evaluation) accounting for 25.3% of the variance, consisting of scales 1, 5, 9, and 13.

For each expression, the scales for each of the three factors were averaged together over all actors to give a three dimensional vector describing the location of the average version of that expression in a three dimensional semantic space. These average scores are plotted pairwise in Figures 5.3 and 5.4. Figure 5.3 plots the results along the full scale (1 to 7). A quick glance at Figure 5.3 shows that the scores all tend towards the middle. Participants did not use the extremes all that often. This is one of the most common findings with any form of Likert scale. The semantic structure of the expressions is more clear in the closeup plots seen in Figure 5.4.

The evaluative factor (factor 1) splits the expressions into two groups. The smaller group consists of positive expressions (happy, pleasantly surprised, and agree). Interestingly, these expressions are all also clustered close together

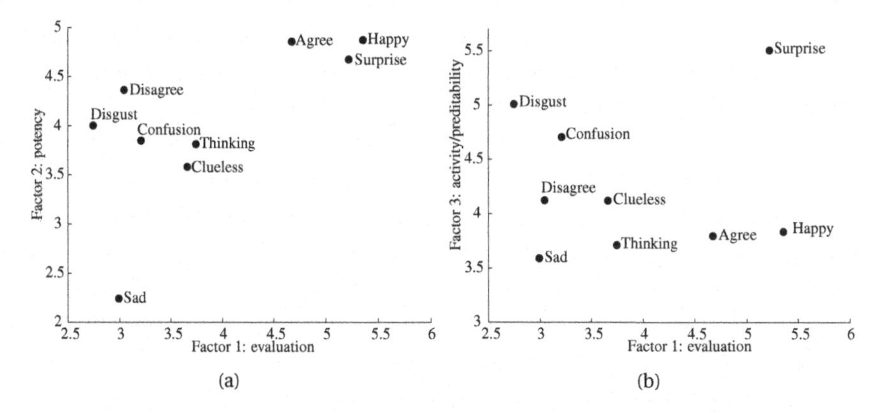

Figure 5.4. Location in semantic space of the facial expressions. The average score of each expression along each of the three semantic dimensions is plotted: (a) Factor 1 versus factor 2, closeup, (b) Factor 1 versus Factor 3, closeup.

on the high end of the potency dimension. The remainder of the expressions lie more towards neutral or slightly negative end of the evaluation factor. The potency dimensions seem to split the expressions into three groups. Happy, pleasantly surprised, and agree stay together as not only positive expressions, but also strong expressions. Disagree is on the lower border of this group. Sadness shows up as an outlier at the weak end of the potency dimension. The remaining expressions are more or less neutral. Finally, in terms of activity and predictability, surprise lies very high as an active, unpredictable expression. Disgust and confusion are only slightly less active or predictable. The cluster containing agree and happy remains close together on this dimension as well, joined by sadness and thinking as the slowest or more predictable expressions.

Summary. This example shows how a simple set of Likert scales can be used in semantic differential task. By analyzing the results properly–either through a PCA or factor analysis—as much or more information can be obtained as through multidimensional scaling a similarity rating task.

5.4 Conclusions

In this chapter we presented the first of the quantitative tasks. Hopefully it is clear that they are much more useful at arriving at clear, precise answers than qualitative tasks. It is probably also clear that they are more limited in the scope of questions that can be addressed.

Chapter 6

Forced-Choice

Forced-choice tasks measure which of a limited number of potential answers participants choose for different stimuli. At its core, it is a discrimination task that shows how well participants can perceive a specific difference between several stimuli. If constructed carefully, the task can measure recognition or even full psychometric functions. In a forced-choice task, participants are required to choose from a limited number of explicitly listed alternatives. The alternatives describe some aspect of the stimuli (such as its color, shape, location, painting style, etc.). The alternatives are almost always mutually exclusive (i.e., if one alternative is true, then the others cannot be) and are often mutually exhaustive (i.e., they cover all possible options). They are specifically tailored to the experiment and thus require prior knowledge about the expected results.

6.1 Overview of Forced-Choice Tasks

Here, we provide a brief overview of the highlights of a forced-choice task.

6.1.1 Typical Questions

Forced-choice tasks are designed to give a qualitative measurement of how distinct or discriminable several stimuli are from one another. The results range from a simple threshold (i.e., how much of a change is needed to distinguish two stimuli) to a full psychometric function (i.e., a curve showing how changes in the physical stimuli correlate with changes in the percept). Some examples of typical questions follow.

- How well can people recognize, identify, or discriminate among the following paintings?
- Which style of painting do people prefer?

- Which of the following expressions do people find to be most attractive or appealing?
- How easy are the following expressions to recognize?
- How much do these specific stylization methods affect the recognition of facial expressions?
- How much does the addition of real eye motion play in the recognition of facial expressions?
- How good are people at noticing changes in the geometry of static faces? Of dynamically changing faces?

6.1.2 Data Analysis

Since the forced-choice task is one of the most common in psychological research, it should not be surprising that a very wide variety of statistical analysis techniques have arisen for it. Beyond standard descriptive and inferential statistics, various forms of curve fitting are common, especially for psychophysical experiments. More details on the analysis of forced-choice data can be found in Chapter 14.

6.1.3 Overview of the Variants

The techniques that are either derived from or related to forced-choice tasks are listed briefly here, and discussed in more detail later in the chapter.

6.1.3.1 N-Alternative Forced-Choice

N alternatives are given and the participants must choose one of them for each stimulus. The participant is free to choose the same alternative for different stimuli. Although the stimuli are usually selected so that each alternative will be chosen equally often, this is not a requirement. The alternatives might be absolute descriptions (e.g., happy, blue, impressionist) or relative to some standard (faster than the standard, darker than the standard, more attractive than the standard). In a special variant of this task, multiple stimuli may be shown simultaneously and the participant would be required to pick one of them (which of these three expressions is the happiest?).

6.1.3.2 N+1-Alternative Non-Forced-Choice

This is the same as above, except that one of the alternatives allows the participant to refuse to make a choice (for example, one of the alternatives alternatives might be "none of the above").

6.1.3.3 N-Interval Forced-Choice

This is a special variant of the N-alternative forced-choice task in which N stimuli are shown sequentially and the participants are required to choose one interval based on some criterion (e.g., which of the following two stimuli is the happiest? which of the following three stimuli is the fastest?).

6.1.3.4 N+1-Interval Non-Forced-Choice

This is the same as the N-interval forced-choice, except that one of the alternatives allows the participant to refuse to make a choice (for example, one alternatives might be "none of the intervals").

6.2 Task Description

Do you like this painting or not? This question requires that you take a specific item (in this case, "this painting") and place it into one of two categories. The first category is, roughly, "things that I like." The second is "things that I do not like." This is a very simple two-alternative forced-choice task or a 2-AFC.

Several things are immediately obvious in this example. First, there are two alternatives (like and do not like) and the participant *must* choose one (and only one) of them for the stimulus. There can be any number of alternatives. We might ask people to recognize the style of the painting and provide two, three, or more alternatives. The number of alternatives listed shows up in the name of the task. If we ask people to indicate which of the following five art styles best describes the painting, then the task is a 5-AFC.

Second, since there is a single painting, the results provide information solely about that painting. If we asked many people this question for this painting, we could begin to get an idea about how popular the painting is in general, but we would not be able to say anything about what kind of paintings the participants do like, or make any predictions about any other paintings. If we added more paintings and asked the participants to decide for each painting whether or not they like it, the task itself would not change (it is still the same 2-AFC) but the type of information we would obtain would change. This highlights the fact that the same task can be used to ask fundamentally different questions, depending on what stimuli are shown. If, for example, we used many impressionist paintings and asked many people to perform the task, we would have concrete information about how popular those paintings are and should be able to draw conclusions about the popularity of impressionist paintings in general. That is, based on the results, we would have a fairly good idea what results we would get if we were to show a new impressionist painting. If, on the other hand, we were to show an equal number of paintings from different styles (e.g., 30 impressionist, 30 cubist, and 30 pop-art paintings), we would be able to talk about the relative popularity of the paintings (e.g., are impressionist paintings preferred over pop art? If so, by how much?).

Third, the second alternative (do not like) is the pure negation of the first (like). These alternatives were not chosen randomly; they have two important characteristics. First, the two alternatives are by definition mutually exclusive: it is not possible for both to be true at the same time. It is not possible to simultaneously like and not like the painting. The two alternatives are also mutually exhaustive: no other alternative is possible. You either like a painting or you do not. What if someone has no strong feeling one way or the other towards a painting? Might one argue that indifference represents a sort of neutral category? Note that the second of the two alternatives is a logical negation of the first. If a person does not *like* the painting (and thus cannot choose the first alternative), then by definition that person *does not like* the painting (and should choose the second alternative). In many cases, however, there is in fact a third possible alternative. If, for example, the task were to decide if you love or hate a painting, it could be reasonably argued that many paintings fall somewhere in-between. In a pure forced-choice task, a neutral or "none of the above" alternative is never listed. Participants must choose one of the listed alternatives. Allowing participants to refuse to categorize the stimuli turns the task into a non-forced-choice task, which has a few of its own special characteristics (see Section 6.3.4).

The fourth thing that is immediate obvious is that the question, "Do you like this painting or not?" provides two alternatives for a single stimulus. One could just as easily use the stimuli as the alternatives and provide a single category. For example, one could show two paintings and ask "which do you like more?" If the paintings are shown at the same time—such as side by side—then this would be a 2-AFC and the two alternatives would be "the painting on the left" and "the painting on the right." If the paintings are shown sequentially, then this is a two-interval forced-choice task (2-IFC) with the two alternatives being "the first painting" and "the second painting." Note that this task provides a direct measure of people's ability to make a specific discrimination between two stimuli.

6.2.1 Discrimination

If the research question deals with whether people can discriminate between two stimuli, the clearest way would be to show both stimuli (either sequentially or simultaneously) and ask them to choose one of them (based on some criterion). If the results are tied then it can be concluded that participants cannot discriminate the two stimuli along that dimension.

6.2.2 General Guidelines

In the design of a forced-choice task, there are a few critical points that should be emphasized. These are discussed next.

6.2.2.1 Non-Mutually Exhaustive Alternatives

The alternatives do not have to be mutually exhaustive. There is, however, a very real chance that the results will not be uniquely identifiable if the alternatives are not mutually exhaustive. For example, one might ask if a particular painting is impressionist or surrealist. If the painting is in fact pop art, participants will not be allowed to choose the *best* answer since it is not listed as an alternative! Instead they will have to decide if that specific pop-art painting is more impressionist or more surrealist. Although the task will produce data, the results might be random (participant might just press buttons randomly) or they might be systematic. If, for example, the majority of the participants choose the alternative "impressionist" for most pop-art paintings, one might be tempted to say that people think that pop-art paintings are the same as impressionist paintings—but this would not be a correct interpretation of the data. It would only be possible to say that people think that pop-art paintings are more similar to impressionist than surrealist paintings. To say that they could not distinguish pop art from impressionist paintings, requires the addition of a pop-art alternative, making the task a 3-AFC. If the participants *still* placed the pop-art paintings in the impressionist category, then it can be concluded that the participants felt that the pop-art paintings were impressionist.

To help ensure that results are interpretable, one of three things should be done. First, we could try to use mutually exhaustive alternatives. This is not always possible. We cannot, for example, list all possible colors. Second, we could choose alternatives that clearly lie along and cover a single dimension. Finally, we could ensure that all the stimuli fit easily (and obviously) into one or the other alternative. For example, imagine that we present a facial expression and ask if it is happy or sad (a 2-AFC). As long as the expressions are rather happy, or rather sad, the participants should be able to perform the task. If, however, we present a pleasantly surprised expression, the interpretation of the results becomes less clear. If 30% of the participants rate a pleasantly surprised expressions as sad, what precisely does that mean? Does it mean the expression is 30% sad? Or that 30% of the participants cannot determine what the expression is? Neither interpretation is necessarily correct, but either or both might be.

6.2.2.2 Asymmetry

Participants will usually assume that the alternatives show up in the stimulus set. In fact, they will usually assume that the alternatives occur equally often. If one of the alternatives never appears in the stimulus set, participants will still choose it on some trials simply because they believe that it *must* be in there somewhere. As the frequency with which the different alternatives occur in the stimulus set begins to differ (e.g., alternative 1 occurs twice as often as alternative 2), some participants will begin to try to guess what the relative frequency is in the stimulus set. Thus, the results will reflect not only the participant's ability to perform the task, but also what they *think* the frequency of the different alternatives is in *this* experiment. In other words, the experiment no longer provides a pure measurement of the phenomenon that we are interested in.

A commonly used task, for example, is the same–different 2-AFC. Technically, in the real world, there is one and only one item that is truly the same, assuming that "same" is interpreted as "identical." When "same" is interpreted as "very similar," then the set of items that are similar is very small. In contrast, however, there is a nearly infinite number of items that are different—the real-world distribution of the two options will essentially be 1 : infinity. As a consequence, if participants is unsure whether the current stimulus is the same as, or different from, the standard stimulus, then they are best off guessing "different" since in all likelihood the item *is* different. Given the limitations of an experiment (such as length), few experimenters would choose a stimulus set where only one option is the same and the rest are different, and the participants know this. As a result, most participants will assume that half of the answers should be same and half should be different (i.e., each alternative will occur equally often). If the participants have chosen "different" several times in a row they may choose "same," even if they believe the answer should be "different," just to ensure that the distribution of their answers matches their expectations. Again, the results no longer reflect a pure measurement of the phenomenon of interest, but are contaminated by unrelated and avoidable cognitive strategies. Asymmetries can almost always be avoided, although it may require rethinking the task (for example, by using a *matching-to-sample* task—see Chapter 7—instead of a same–different task).

6.2.2.3 Order

An issue that is closely related to asymmetry is the order of the correct answers. People have very clear expectations about random distributions. People are willing to accept a short sequence of identical stimuli, but not a long one. For example, few people would have any concerns if a coin were flipped three times and all three times the same side of the coin landed face-up. Most people, however, would feel that something was wrong with the coin if the same result occurred ten times in a row. This is just as true for people's willingness to choose the same option in a forced-choice task. If the answer should be alternative 1 three times in row, in most cases this will not be too problematic. If the answer should be alternative 1 ten times in a row, however, most people would start to choose alternative 2 to ensure that their pattern of answers matches their expectations.

6.2.2.4 Category Level

Another form of asymmetry is category level. We could present several pictures of animals and ask participants to indicate if the animal is dog or a Siamese cat.

The two alternatives are in different category levels. Dog is in an *entry-level category* while Siamese cat is in a *subordinate-level category.* Since an entry-level category is at a higher level in the taxonomic hierarchy of categories, it has many more items in it, with a larger variance, than the subordinate level. In other words, any picture of a dog would go into the entry-level "dog" category, but only one specific type of a cat can go into the subordinate-level "Siamese" category. Since there are many natural examples of the first and few of the second, what sort of a distribution should participants expect? Many participants would ignore real-world frequencies and expect 50% of each, since this is an experiment and there are two alternatives. Moreover, placing something into a subordinate-level category requires one to be much more specific. What should participants do if a picture of a cat is presented, but it is not a Siamese?

6.2.2.5 Chance Level and Task Resolution

In any forced-choice task, there is a performance level referred to as *chance.* This level reflects the percent of the time that any given alternative would be chosen if participants had no knowledge of what the alternatives or the stimuli would be and therefore choose an alternative blindly. Chance performance is, obviously, directly tied to the number of options. Specifically, *chance performance* is defined as $1/N \cdot 100$, where N is the number of alternatives. For a 2-AFC task, chance performance is 50%. For a 10-AFC, chance performance is 10%. Any performance by the participants that can be said to be statistically different from chance level shows that participants were able to make the discrimination. As a consequence, the more alternatives there are, the better the resolution of the task. This means that it is easier to find smaller differences or discrimination with more alternatives.

6.2.2.6 Identification versus Discrimination

Quite often, forced-choice tasks are incorrectly considered to be identification tasks. For example, imagine that we show a series of expressions to participants and ask them choose a name for each one from ten possible alternatives (a 10-AFC). If participants are able to choose the correct name, we would be tempted to say that they identified the expressions. Such an interpretation is, however, not uniquely supported by the data. The reason for this is simple and is based on the process of elimination. Only the alternatives listed can be chosen. There are, then, three ways to choose the correct answer. The first is through random chance. If people are choosing randomly, it will become obvious once the results are averaged across stimuli, participants, or repetitions. The second method would be to actually know the answer and then find the alternative that is correct. The final method is choose an option because the stimulus is not likely to fit into the other category. For example, let's say we are using a 3-AFC for the recognition of facial expressions: happy, sad, and angry. On the first trial, we provide participants with a rather ambiguous expression.

They are certain that it is not a happy face, and fairly certain that it is not an angry face. They do not think it is a sad face, but since they are more certain that it is not one of the others, they choose the sad face. Even if this is the correct answer, we cannot say that they identified the expression as sad. If they were left to create their own name for it—as in a free-description task (see Chapter 4)—they would certainly have called it something else (such as "thinking"). Since the task *can* be solved by the process of elimination, we can never be sure that it *was not* solved this way. This problem is sometimes referred to as *inflated alpha* (correct answers have a higher frequency than they really should). The addition of a none-of-the-above option—which transforms the task into an N+1-alternative-forced-choice task—can help to alleviate this problem, but does not completely eliminate it.

6.2.2.7 Clarity of the Alternatives

Not only must the participants understand the alternatives, they must understand what we mean by the alternatives. Many words, especially in English, carry a number of different meanings. If the participants understand the alternatives differently than what we do, then misinterpretations of the results are likely. Even everyday terms can be easily misconstrued. Although the use of specially defined terms (such as jargon) can help to increase the precision of the alternatives, jargon is by its very nature not universally understood. Thus, the use of jargon or area-specific terms is really only encouraged when either all of the participants come from the field being studied (and thus are guaranteed to know the term), or when the term is clearly and fully explained at the onset of the experiment.

6.2.2.8 Uncertainty

A forced-choice task does not actually measure any physical property. It measures a participant's *judgment* or *belief*, and thus the participant's response will include an measurement of uncertainty. If, for example, we ask about the presence or absence of a stimulus, and the stimulus is near the absolute threshold (and thus difficult to see), participants will be uncertain about their judgment. As a result, some of the time they will choose one alternative, and some of the time they will choose a different alternative. This fact is critical for measuring full psychophysical functions, and will be discussed more in Section 6.3.1.1.

6.3 Specific Variants

6.3.1 The Two-Alternative Forced-Choice Task

In its most basic form, the N-alternative forced-choice task has two alternatives. It requires participants to classify the stimuli into one of two categories.

This task—the *2-alternative forced-choice* or *2-AFC*—lies at the core of traditional psychophysics, where it is usually used to determine thresholds (Falmagne, 1985; Gescheider, 1997). There are two forms of threshold: absolute and difference. *Absolute thresholds* are the absolute minimal amount of a dimension that is needed for people to just barely detect the dimension. How much contrast is needed before an edge can just barely be seen? How much noise can we place in an audio track before people just begin to notice it? In an absolute threshold task, participants are asked to discriminate between the presence and the absence of a stimulus on each trial (note that the discrimination between presence and absence is essentially a detection task). A *difference threshold* is also called the *just noticeable difference* (JND). It measures how much we need to change the stimulus along a given dimension before people just begin to notice the change. JNDs are measured by asking participants to discriminate between two stimuli along some dimension (e.g., which of two weights is heavier). Such tasks can help answer questions such as, "By how much does the resolution of current night vision goggles need to be improved before people would just begin to notice the improvement?" Note that since an absolute threshold can be thought of as discriminating between the presence and absence of a stimulus, it can be seen as a special form of a difference threshold.

There are two primary requirements of the two alternatives that are selected. First, the two alternatives cannot be simultaneously true. Second, they need to cover the manipulated space well. In fact, quite often, the two alternatives jointly exhaust all possibilities.

Generally, in an experiment with a 2-AFC task, a single stimulus is presented with two alternatives. For example, we might present an image through some night vision goggles and ask participants if they see an airplane in the image or not. Other times there are two stimuli and a single category. We might present two images from two different night vision goggles and ask in which one the airplane is easier to detect.

The 2-AFC can measure a wide variety of things depending on the categories chosen, the question asked, the stimuli used, the manner in which the stimuli are presented, the participant populations used, and so on. In the next few sections, we will discuss a few of the more common configurations of a 2-AFC.

6.3.1.1 Absolute Threshold

As a simple example from classic psychophysics, let us assume we wish to determine the absolute threshold for luminance. We ask our participants whether they see a specific target. As our two alternatives we choose "present" and "absent." Since something cannot be both present and not present at the same time, these alternatives are mutually exclusive. Furthermore, since presence and absence are the only two possible states an object can have, they are mutually exhaustive. For the experiment, we place the participants in a dark room one at a time, in front of a calibrated monitor,[1] and present them with a series of trials. On each trial, the participant must indicate if a square (which is presented in the middle of a black screen for 200 ms, and subtends two degrees of visual angle) was present or not. The stimuli are varied along a single dimension: the intensity of the light coming from the square. For this hypothetical experiment we have decided to use ten levels, which were selected—using pilot data—so that half of them will be below where we expect the threshold to be and half above. This ensures that the two alternatives will occur roughly equally often.

At least one of the luminance levels is chosen to be clearly below threshold and one to be clearly above. There are two reasons for this. First, it will demonstrate to the participant clearly that both alternatives occur in the stimulus set. Second, and perhaps more important, it will help prevent frustration. Trying to see things that are just barely visible is very difficult. Participants can get upset and frustrated if they are never sure if their answers are correct or not. In their frustration they may decide that the task is impossible and then give up (start answering randomly). Since we cannot give them feedback about the correctness of their responses on any trial (this would bias the results), we instead choose to provide occasional trials that are easy. These easy trials also serve as a reminder of what a "present" (or "absent") answer is supposed to look like.

In this experiment, we will use a repeated-measures, within-participant *method of constant stimuli* design. The first part of this description—repeated measures—means that each stimulus will be seen many times. Since this is a psychophysical experiment, the number of repetitions will be high. Since it is only a test experiment and the results are more or less known, we have not set the number too high. We will use 80 repetitions. The second part of the description—within-participants—means that each participant sees all the stimuli. This allows each participant to serve as his or her own control and allows us to directly compare performance between stimuli. The final part of the description—method of constant stimuli—refers to how the stimuli will be presented. Each stimulus will be presented for the full number of repetitions. The order in which the stimuli will be presented is randomized, with each participant getting a different random order. We decide for this experiment to use *blockwise* randomization: each stimulus will be seen once before any stimulus will be seen a second time. The order of the stimuli within each block is randomized for each block.

[1] It is absolutely critical that the monitor be calibrated for a psychophysical experiment, since it must be certain that what is physically present on the screen has precisely the luminance and color that the computer was supposed to present. Sadly, people often forget to calibrate.

Uncertainty and psychophysical functions. As mentioned previously, the task does not actually measure the physical presence or absence of the square, but the participant's *judgment* or *belief* about its presence, and thus participant's response includes an measurement of uncertainty. If a participant is 100% certain that the square was there (as should be the case for the brightest condition), then they will answer "present" on all 80 trials. Likewise, when they are 100% certain that nothing was there (as should be the case for the darkest condition), they will answer "absent" on all 80 trials. If the participants are completely unsure, then they will guess randomly, answering "present" half of the time and "absent" the other half. Critically, for those conditions between fully certain and fully uncertain, the percentage of the time the participant chooses an answer over the repetitions for a given condition turns out to be an accurate measure of their uncertainty! In general, then, the results will form a sigmoid curve—also called an ogive—as can be seen in the hypothetical results plotted in Figure 6.1. The percentage of the time that the participants chose "present" is plotted as a function of the ten lightness conditions. Since we chose the darkest square to be obviously below threshold, the fact participants never chose "present" for that condition is reassuring. Likewise, the 100% "present" rating for the lightest square had been expected. The horizontal line at 50% represents the absolute threshold. This luminance level is, on average, just barely enough to be detected. For more detail on how the curve fitting should be performed, see Chapter 14 and Wichmann and Hill (2001a,b).

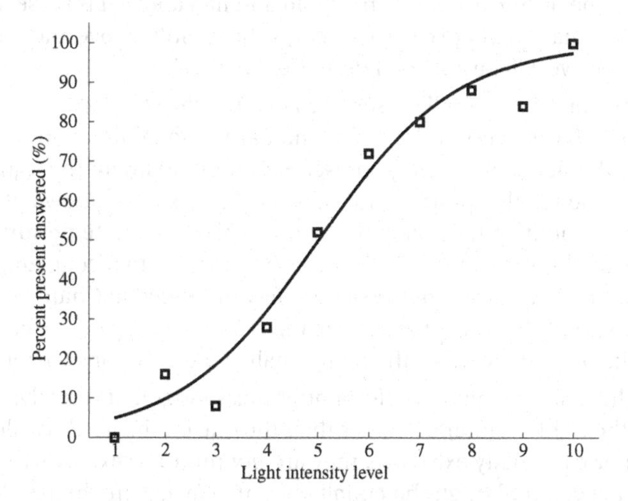

Figure 6.1. Hypothetical results for an absolute threshold luminance experiment. The percentage of the time that the participants chose present is plotted as a function of the ten lightness conditions. Note that the data were generated to illustrate an absolute-threshold experiment and may not reflect real brightness discrimination thresholds.

6.3.1.2 Difference Threshold

In a difference-threshold experiment, participants are asked to compare two stimuli and choose one of them. One of the two stimuli is the same on every trial and represents the baseline. This stimulus is referred to as the *standard* stimulus. The other stimulus—referred to as the *test* stimulus—is systematically varied from trial to trial.

In the last section we measured the absolute threshold. More specifically, we measured the minimum amount of light that must come from a small square presented to the fovea in order for people to just barely detect the presence of the square. Now we want to know how much we would need to increase the lightness of a visible square before participants would notice the change. For example, imagine that we have a video projector that is capable of projecting an image with 1200 lumens. How much would we have to increase the luminance of the projector before customers would notice? One way to test this is to present a bright white square at the 1200 lumens level on a black background (100% contrast). This represents our standard. Next, we choose ten test stimuli. Since we are only interested in how much we need to increase the lightness before people notice, none of the test stimuli will be darker than the standard. To anchor our psychometric function properly, at least one of the test stimuli must be clearly brighter than the standard. The other will be very close to the standard. In fact, we could choose to have the darkest test stimulus be the same as the standard. Note that, as in the last example, half of the test stimuli should be below the expected threshold and half above. Likewise, the more stimuli that we have that are close to the threshold (both above and below), the more precisely we will be able to determine the threshold.

On any given trial, we will present two stimuli: the standard and one of the test stimuli. We will present the two stimuli at the same time, one on the left and one on the right. If we always presented the standard on the right, participants might notice this pattern after a while, and as a result always respond with "right" without actually looking at the display. Accurate performance in such a case might be due to a correct discrimination or to recognizing the role of position. This is a variant of the order effect discussed in Chapter 2, and we must control for it. We must randomize the side on which the standard is displayed so that it is on the left or the right in half of the trials for each participant.

As for the task, we might ask for a brightness judgment directly: "Was the square on the right brighter or darker than the one on the left?" While the two alternatives are mutually exclusive, they are *not* mutually exhaustive. The two stimuli on a given trial might be equally bright. We also might ask, "Which is brighter, the square on the left or the one on the right?" The two alternatives—left and right—are mutually exclusive. They are still not mutually exhaustive (the two might be equally bright). The alternatives are symmetrical and balanced. That is, with the exception of equal brightness, one of the two alterna-

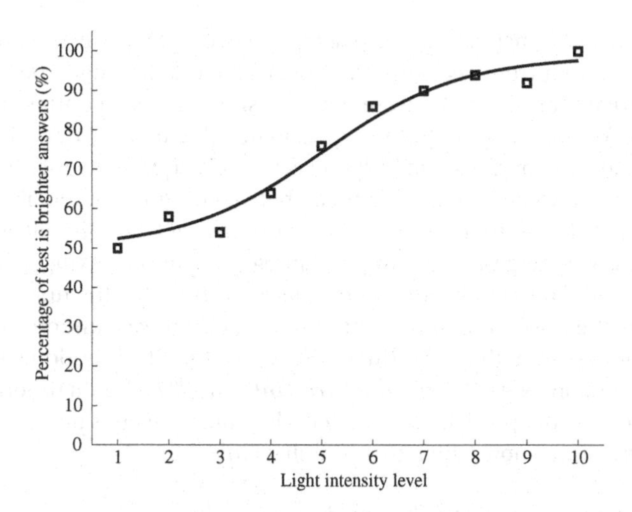

Figure 6.2. Hypothetical results for a difference threshold luminance experiment. The frequency at which the participants chose the test as being "brighter" is plotted as a function of the ten brightness conditions. Note that the data were generated to illustrate a difference threshold experiment and may not reflect real brightness discrimination thresholds.

tives *must* be true on every trial and the expected frequency of the two alternatives should match the frequencies used (half of the correct answer will be "right").

Hypothetical results can be seen in Figure 6.2, where the percent of the time that the test stimulus was chosen as brighter is plotted. Since the darkest test stimulus is identical to the standard, participants should be fully uncertain as to which of the two stimuli is brighter and thus will have to randomly choose an answer. Since there are two options, chance performance is 50%. Thus the graph in Figure 6.2 starts at 50%. The lightest test stimulus was chosen to be obviously brighter. This is reflected in the results with a 100% average score. The 75% point represents the halfway point between being uncertain which is brighter to knowing which is brighter, and thus is defined as the difference threshold (the degree to which the brightness of the stimuli must be increased before people can just notice the difference).

6.3.2 Two-Interval Forced-Choice Task

As mentioned in Section 6.2, rather then presenting one stimulus and two alternatives, one might present two stimuli and ask participants to chose the one that most closely matches some criterion. If the two stimuli are presented *sequentially*, this task is referred to as a *two-interval forced-choice* or 2-IFC task. Let us imagine, for example, that we wish to remeasure the brightness JNDs for our square. We suspect that by placing the two squares next to each other on the screen we introduced some extraneous factors that made people seem more sensitive than the really were (such as simultaneous contrast or lateral inhibition). We decide to redo the experiment using a 2-IFC design. That is, we present the standard square and one of the test squares sequentially, with a short break between them. This break between the two intervals is, naturally enough, referred to as an *inter-stimulus interval* (*ISI*). The ISI should be at least 250 ms to help prevent temporal integration and masking effects. The task, then, would be to indicate which square was brighter, the first or the second. Just as the position (left or right) of the standard was randomized in the simultaneous version, the order (first or second) should be randomized in the sequential version. Note that the *inter-trial interval* (*ITI*) should be longer than the ISI to ensure that participants can tell when one trial ends and the next begins, and therefore know which two stimuli go together.

6.3.2.1 Example: Difference Thresholds using a 2-IFC

Difference thresholds can be measured for complex stimuli just as with the more traditional simple stimuli. To demonstrate this, we present next an experiment measuring the motion JNDs for several conversational expressions.

Humans are very sensitive to changes in motion. Motion plays a critical role in the perception of facial expressions (Cunningham and Wallraven, 2009). In fact, when facial expressions have the proper motion (and are not simple photographs), the loss or impoverishment of static information (such as shape or texture) has little or no effect on the recognition of the expressions or the perception of intensity (Cunningham and Wallraven, 2009; Wallraven et al., 2006, 2007). What is unclear, however, is how accurately the motion must be represented. Are 24 frames a second enough? Or do we need an even higher temporal resolution?

Research question. How sensitive are we to changes in facial motion?

Stimuli. Since we want to generalize the results to all facial expressions, we needed to use as many expressions as possible. At the same time, the large number of repetitions means that we need to limit the number of expressions so that the participants do not become fatigued. Finally, since we will be manipulating the spatiotemporal characteristics of the expressions directly, we need computer animations. Therefore, we chose five facial expressions from our facial animation system (disgust, fear, happy, sad, and surprised; Breidt et al., 2003). These expressions have been previously validated as producing the same recognition and intensity perception performance as the video sequences they are based on (Griesser et al., 2007; Wallraven et al., 2006, 2007).

Each expression has its own normal motion. Some are fast (such as agreement, which consists of a rapid nodding of the head) and others are slow (such as sad, which has a slow, downwards tilting of the head and closing of the eyes).

Rather than try to equalize the motion physically, we allow each expression to represent a different standard *S*. To help maintain some degree of consistency across the different standards—and using insights from psychophysics and motion perception—we constructed the test stimulus set as a percentage of original motion rather than at some absolute measure of speed. Thus, for each of the expressions, the standard stimulus has 100% of the original speed. The eight remaining stimuli have 60%, 80%, 90%, 95%, 105%, 110%, 120%, and 140% of the original speed. Thus, there are nine test stimuli: one equal to the standard, four below the standard, and four above it. With five expressions, nine speeds, and twenty repetitions (which is actually on the low end for measuring thresholds), we have a total of 900 trials. The experiment lasted about one-and-a-half hours per participant.

Stimulus presentation. On a given trial, the original and one of the test stimuli were presented sequentially (a 2-IFC task) with a 250 ms blank screen between the two stimuli. The order in which the standard and a test stimuli were presented was randomized on each trial, with each participant getting a different random order. Each trial was presented 20 times. This number is somewhat below what one would normally use in a true psychophysical experiment, but should be enough to obtain reasonably accurate data. Additionally, the trials for the different expressions were interleaved. That is, trial 1 might use the standard stimulus and a test stimulus from the happy expression, trial 2 thinking stimuli, trial 3 sad stimuli, and so on.

Task. Since the animations may differ in length, and since people can only attend to one animation at a time, we decided to present the two animations sequentially rather simultaneously. Thus, the task is a 2-IFC. Since the independent variable is motion, the task is to determine which of the two animations was faster. Participants were explicitly instructed to ignore the emotion and emotional intensity and focus purely on the physical motion.

Participants. Since we are interested in psychophysical thresholds, we assumed that thresholds are similar for most people. Thus we used a large number of repetitions and a within-participant design for a few participants (four is typical number). We wished to be sure that the thresholds measured truly represent the absolute best performance of the participants. For this reason, we used participants who are well practiced as psychophysical observers.

Results. The results can be seen for two participants in Figure 6.3, in which the percentage of the time that the test stimulus was chosen as being faster is plotted for each of the five different expressions as a function of the test stimulus's speed. The first thing that one can see is that the five curves are very close together (which can make it difficult to see the different plot lines). Thus, the decision to parametrize the stimuli by the percentage of original speed seems to have successfully equated the different expressions.

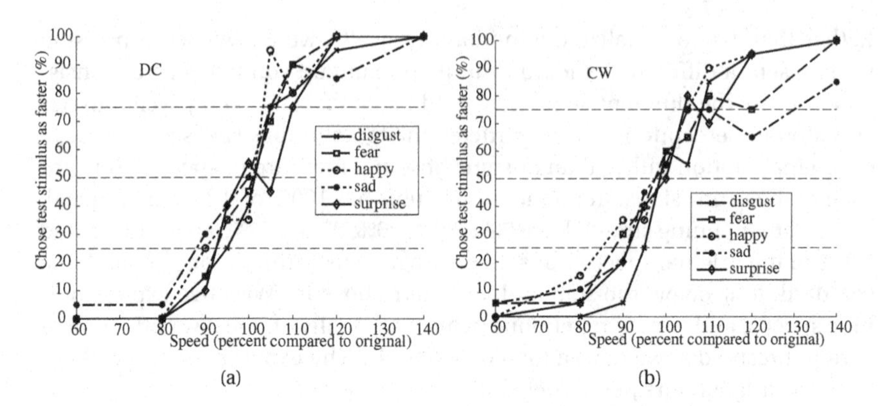

Figure 6.3. Facial motion JNDs. The percentage of time participants answered that the test stimulus was faster is plotted as a function of the test stimulus's speed for two participants.

The second thing that can be seen is that for nearly every expression the point where the test and standard are seen as being equally fast (i.e., the 50% point) is at 100% of the original speed. That is, people are very accurate at this task. This is seen more clearly in Figure 6.4, where the results for both participants are shown collapsed across expressions. There is some systematic difference across expressions in terms of speed thresholds, with the entire happy curve generally lying slightly on the slow side and disgust generally on the fast side. Overall, when the speed decreases to somewhere between 85% and 95% of the original speed, people will start to notice it. Likewise, speed increases to around 105% will be noticed.

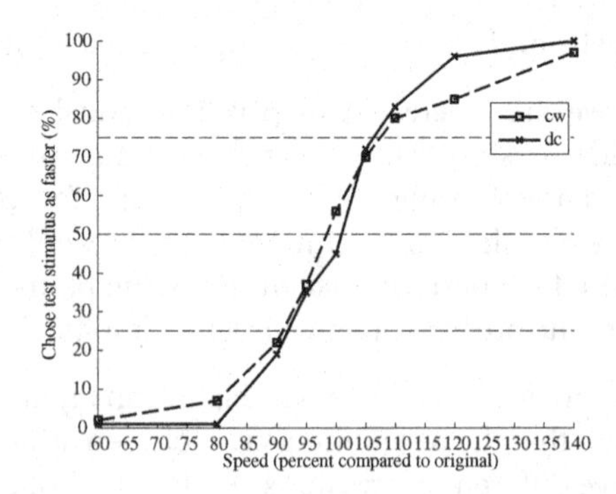

Figure 6.4. Facial motion JNDs. The results for the two participants are shown averaged across expressions.

Summary. This experiment demonstrates a classic 2-AFC task to measure difference thresholds. It also shows how classic psychophyics procedures can be used with complex stimuli.

6.3.3 N-alternative Forced-Choice Task

When the two alternatives in a 2-AFC are not mutually exhaustive, the alternatives should lie along along a single dimension (and this should be obvious to the participants). Failure to take this into consideration will lead to problems for the participant and to potentially uninterpretable results. For example, one might ask if a colored square was red or blue, or if a facial expression was happy or sad. Since neither color space nor facial expressions can be represented easily by a single dimension, it is unclear where the correct answer lies for many stimuli (e.g., should I choose red or should I choose blue when the color is really green? Is green closer to being blue than it is to being red? Is angry closer to being happy or sad?). While one might restrict the stimulus set to ensure that all stimuli fall easily into one of the two categories (this is recommended anyway), the data analysis can still be somewhat ambiguous; for example, what does a 30% happy rating mean on a happy–sad discrimination? If one cannot choose two alternatives that lie along a single continuum, one might alternately choose to sample the response space and provide more alternatives.

When more than two alternatives are used, the task is the very familiar *multiple-choice* task. As with a 2-AFC task, one can have a single stimulus with many alternatives or many stimuli with a single alternative. In general, each of the alternatives listed show up in the stimulus set somewhere, usually with equal frequencies. This is not required, but participants, especially experienced ones, may be operating under the assumption of equal frequencies and thus might choose options just to have chosen them. Note that if the options are more or less equally spaced along a single dimension, the task can be strongly related to a rating task (see Chapter 5). For example, we could present a facial expression and ask participants to choose the best of the following alternatives to describe its speed: (1) extremely fast, (2) quite fast, (3) somewhat fast, (4) neither fast nor slow, (5) somewhat slow, (6) very slow, or (7) extremely slow. As soon as we allow participants to make a choice from a number of alternatives, a number of specialized variants of the task become possible. These will be addressed in Chapter 7.

6.3.4 N+1-Alternative Non-Forced-Choice Task

Forced-choice tasks require participants to discriminate between several options. As mentioned, in some cases the answer they wish to give is not listed. If the desired answer is along the continuum between the listed options, then the statistics of a repeated-measures design can extract these answers, as was seen in the absolute and difference threshold examples. If the answers are not along a continuum and the answer is not listed, then there are problems. What if I *think* I see a green square but am asked if it is red or blue? Any answer I give will not represent my perception, and it is not clear how statistical analysis of the answers will ever recover my perception. In other words, the results will be biased and will probably not reflect the underlying perception. To correct for the artificially increased recognition rates, and allow the answers to more correctly reflect the participant's perception, we might add a "none of the above" option. Results on this task are known to correlate well with other methods and to produce reliable results (Frank and Stennett, 2001).

6.3.4.1 Example: Ten-Alternative Non-Forced-Choice

The physical differences between an expression that is recognizable and one that is not can be very subtle. Moreover, there are a number of different ways that humans express any given meaning, and not all of the resulting expressions are easily recognized (Cunningham et al., 2003a,b). In Cunningham et al. (2005), we examined several questions that are related to how well people can recognize nine conversational expressions.

Research questions. How good are people at recognizing facial expressions? Which recordings of the expressions are the most easy to recognize? Does the inclusion of the decay of an expression (when the face relaxes from a peak expression back to neutral) alter the perception of that expression?

Stimuli. The full set of recordings from the small MPI facial-expression database were used (for more on the how the database was recorded, see Section 5.3.3.6 or Cunningham et al., 2005). In short, a brief scenario was described to someone in detail and that person was asked to place himself or herself in that situation and then react accordingly. A single recording of an expression consisted of three repetitions of that expression performed in rapid succession, with a neutral expression preceding and following each repetition. While at least one recording per person was recorded from each of the nine expressions, some difficult expressions had more than one recording. In total, 213 individual repetitions were recorded. To provide an objective basis for selecting the sequences, all 213 recordings were shown to the participants.

Half of the participants saw a video sequence that went from the neutral expression through the peak of the target expression and back to neutral ("full" video). The other half saw video sequence that went from the neutral expression and stopped at the peak expression ("clipped" video).

Stimulus presentation. The size of the images was reduced to 256 × 192 pixels and the participants sat at a distance of approximately 0.5 meters from the computer screen (i.e., the images subtended approximately 10 × 7.5 degrees of

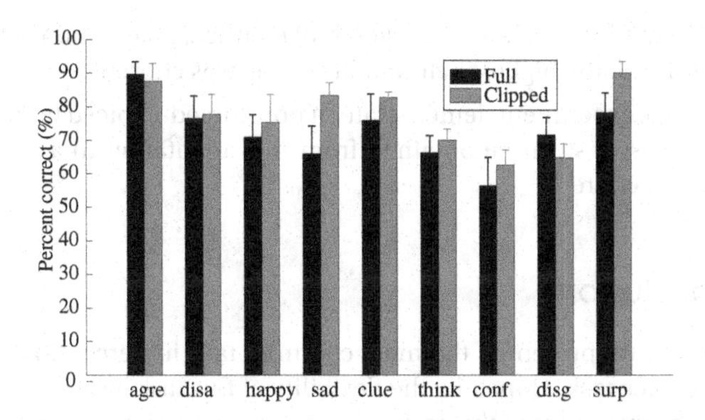

Figure 6.5. Overall recognition accuracy. The percentage of the time that each type of expression was identified correctly is shown for each of the two groups. The expressions from left to right are agree, disagree, happy, sad, clueless, thinking, confused, disgust, surprised. The error bars represent the standard error of the mean.

visual angle). The order in which the 213 trials were presented was randomized for each participant. A single trial consisted of a video sequence being shown repeatedly (without sound) in the center of the screen. A 200 ms blank screen was inserted between repetitions of the video sequence. When participants were ready to respond (which they indicated by pressing the space bar), the video sequence was removed from the computer screen.

Task. Each participant performed two tasks. The first task was to recognize the expression using a ten-alternative, non-forced-choice task. The name of each of the nine expressions, along with the option "none of the above," was shown in German and English on the side of the screen at all times. The second task was to rate the expression for its sincerity using a 7-point Likert scale. The data from the second task will not be discussed here.

Participants. Eighteen individuals who were naive to the purposes of the experiment participated and received financial compensation at standard rates. The participants were randomly split into two groups of nine individuals each. One group of participants saw the full videos, the other saw the clipped videos.

Results. In general, the expressions were well—but not perfectly—recognized; average recognition performance was 74.6%; see Figure 6.5. The expressions were seen as relatively believable, since the average believability rating was 4.91. Clearly, the visual information present in the video sequences is sufficient to identify these conversational expressions. There were few performance differences between the full and clipped groups (although the clipped sequences occasionally had higher recognition rates). Based on the results, the repetition of each expression for each recorded individual that was most often recognized correctly was chosen for use in all future experiments. In cases where two or more repetitions of an expression had identical recognition accuracies, the repetition with the highest believability rating was chosen.

Summary. This experiment demonstrates a non-forced-choiced task, showing how general answers can be obtained from this adaptation to a classic psychophiscal procedure.

6.4 Conclusions

In this chapter, we presented the most common task in perception research: the forced-choice task. Hopefully the flexibility of this task has become apparent. The extensive use that this task has seen in perception research has lead to many advanced variants being developed. A selection of these advanced forced-choice methods can be found in Chapter 7.

Chapter 7

Specialized Multiple Choice

In essence, multiple-choice tasks are still discrimination tasks; that is, how well a specific difference can be perceived among several stimuli. By carefully arranging the stimuli within a trial, as well as the stimuli across trials, one can ask some amazingly sophisticated questions, all of which essentially can be reduced to issues of detection and discrimination. Although the details vary, in all these tasks participants are required to choose from a limited number of explicitly listed alternatives.

7.1 Overview of Specialized Multiple-Choice Tasks

Here, we provide a brief overview of the highlights of a few of the specialized multiple-choice tasks.

7.1.1 Typical Questions

Over the years, a number of tasks based on the well-known forced-choice procedure have arisen. Usually these derivative tasks were designed to address special questions, but they have often become standardized techniques in their own right. Typical questions that can be asked with these techniques include:

- How long do specific higher cognitive processes take?
- How well can people search for a specific target in noise?
- Is this kind of visual feature processed pre-attentively?
- How well and how quickly can people (or animals) learn a (new) rule?

- How much and what kind of information do we process in an image (consciously or unconsciously)?
- How does the perceptual system select what gets further processing?

7.1.2 Data Analysis

Each of the different techniques has its own special analyses. Since in each case there is an objectively correct answer, all of the variants allow one to describe how often the participants were correct and how long it took them to enter their responses (i.e., the reaction time) and thus to run standard inferential statistics.

7.1.3 Overview of the Variants

7.1.3.1 Go/No Go Tasks

If the stimulus currently present meets certain, prespecified criteria, then the participants are supposed to issue a simple response (such as pressing a button). This would be a "go" trial. If the stimulus does not meet the criteria, the participant does nothing. That would be a "no go" trial. Although this is a very small modification of the standard 2-AFC (one of the two possible responses has merely been replaced with a non-response), the change allows it to serve as a *gateway* task (where the secondary task is only performed on the "go" trials).

7.1.3.2 Matching-to-Sample Tasks

The matching-to-sample technique is based on a very specific stimulus presentation. Specifically, a standard stimulus (called the *sample*) is shown together with more than one other stimuli (called the *comparison stimuli*). The participant is asked to choose the comparison stimulus that most closely matches the sample. If comparison stimuli are presented after the sample has been removed, then the task is referred to as a *delayed matching-to-sample.*

7.1.3.3 Visual Search Tasks

A specific *target* (such as the letter *T*) is presented simultaneously with a series of other items called *distractors* (such as the letters *L* and *I*). The distractors are similar to the target along the dimension of interest. The number of distractors is manipulated across trials systematically. The participant is to indicate whether the target is present or not (usually the target is present on half of the trials). The primary result is a measurement of how much the addition of each distractor increases the amount of time it takes to find the target.

7.1.3.4 Rapid Serial Visual Presentation Tasks

The Rapid Serial Visual Presentation (RSVP) technique presents a number of stimuli one at a time in a very rapid sequence. As with a visual search, one of the stimuli is the target and the rest are distractors. The participant's task is to indicate when the target stimulus is present. In many respects, this technique is similar to the more general go/no go task.

7.1.3.5 Free-Grouping Tasks

In this task, participants are presented with large set of stimuli and are asked to arrange them into groups. In the simplest variant, the participants decide how many groups there should be and what the defining characteristics of the different groups are. More constrained variants where the number of groups is defined by the experiment are also possible. If both the number of groups and the characteristics of the different groups are defined by the experimenter, the task is an N-alternative forced-choice task.

7.2 Task Description

Multiple-choice tasks examine how well participants can perceive a specific difference between the stimuli and how they choose to categorize the stimuli. All of the tasks presented here can be derived from the basic principles of experimental design presented in Chapter 2 and for forced-choice tasks presented in Chapter 6. Most of the tasks presented here require not only a specific task, but also specific stimulus presentation rules. The particular experimental constellations not only have an important place in the history of psychology, but have been employed so often that they have become standards. They are also flexible enough that one can ask a wide range of questions with them.

7.2.1 General Guidelines

Most of the general issues with forced-choice tasks apply here as well. In many of the variants, however, the alternatives are set, so some issues do not apply. If the alternatives are not preset, they should be mutually exclusive and where possible mutually exhaustive. They should also be constructed so as to avoid asymmetry, category-level effects, and frequency-of-occurrence issues. This last point is particularly tricky for visual search, RSVP, and free grouping. The order of the responses is still a concern. As with all tasks, all aspects of the experiment should be clear to the participants. In these specialized tasks, this includes the task instructions.

7.3 Specific Variants

7.3.1 Go/No Go Tasks

The go/no go task seems to have been originally developed in 1868 by Donders (1969/1868) as a way of objectively measuring how long higher cognitive processes (such as planning and making decisions) take. The go/no go task is a very simple variant of the standard two-alternative forced-choice task (2-AFC), where participants respond *only* when the stimulus meets predefined conditions. For example, imagine that our stimuli are all strings of letters. The task might be to find out how well people can discriminate words from non-words under certain conditions. In a 2-AFC version of the task, one key would represent "is a word" and the other would represent "is not a word." In a go/no go task, there is only one button. Pressing it represents a decision that the stimulus is a word. The lack of a button press, then, is an indication that the stimulus is not a word.

Note that while one of the two alternatives is defined explicitly (in the string of letters examples, the response criterion was "the string is a word"), the other alternative is the logical negation of the first (in this example, "the string is *not* a word"). Since the two options are always *A* and "not *A*"—where *A* can be nearly anything—the two options are always mutually exclusive and mutually exhaustive.

Measuring the length of higher cognitive processes requires that we have two conditions. One condition should have the cognitive process and the other should be almost perfectly identical, lacking only the cognitive process. A comparison of reaction time in the two conditions should tell us how long the cognitive process takes (for more on this type of experimental logic and some of the assumptions it requires, see Chapter 2).

Donders used three conditions, each with a different task. The first was a simple reaction-time task, where participants were asked to press a key when they saw anything (in an otherwise completely dark room). The second condition used a go/no go task involving a linguistic discrimination: participants were to press the key *only* if what they saw (i.e., the stimulus) was a word. Donders reasoned that the first task measured simple perception-response loops. The difference between the two tasks is that the go/no go task required participants to process the stimulus and make a decision. As a result the difference in reaction time between the two conditions (assuming that identical stimuli were used) reflects the amount of time it takes to process the stimulus and make a decision. The final condition used a standard 2-AFC task: participants were to press key A if the stimulus was a word and key B if it was not a word. Since the difference between a go/no go task and a 2-AFC task is merely the decision of which button to press, then the reaction time difference between them should represent how long it takes to choose the response key. Donders's logic and the validity of his assumptions have always been controversial, from 1880 up to the present (Gomez et al., 2007; Wundt, 1880). A number of researchers have tested Donders's assumptions (which are spelled out in detail in Luce, 1986; Ulrich et al., 1999).

Obviously, reaction times are indeed shorter in the go/no go task than in a 2-AFC. Interestingly, the reaction times in a go/no go task also have a larger distribution. Even more interesting is that the two tasks have very similar accuracies and response biases. Although some researchers have claimed that there are fundamental differences between what is measured with 2-AFC and go/no go tasks (Jones et al., 2002), most research points to the opposite conclusion. The difference between a go/no go task and a 2-AFC task does not lie in the how stimulus is processed, but in how the response is chosen and executed (Chiarello et al., 1988; Gomez et al., 2007; Measso and Zaidel, 1990; Ulrich et al., 1999). In a very thorough modeling and testing of the two procedures, Gomez et al. (2007) came to the conclusion that the only difference between 2-AFC and go/no go tasks is in fact the response criteria (and not in the other aspects of processing). Thus, the go/no go task can be used as a replacement for a 2-AFC in most cases.

The go/no go is also sometimes used as a gateway task. That is, in the primary task (the go/no go task), participants are asked to discriminate the stimuli within some dimension. If the stimulus matches the criterion, this is a "go" trial and a button is pressed. Following the button press, a secondary task is performed. The secondary task could be anything, but is often a free-description. For example, we might show a large variety of images to participants and ask them to press a key if the image contains an animal. This would be a go/no go task. I could then ask them to describe the animal they think they saw (which would be a free-description task). Such a procedure can be found in Intraub (1981).

7.3.2 Matching-to-Sample Tasks

In the simplest and most common form of the matching-to-sample task, three stimuli are presented simultaneously. One of them is the sample stimulus and the remaining two are the comparison stimuli. The sample is usually presented spatially separated from the comparison stimuli. One of the two comparison stimuli is physically identical to the sample, and participants are asked indicate which one that is. That is, they must choose the comparison stimulus that is identical to (or *matches*) the sample stimulus.

Although traditionally one of the comparison stimuli was identical to the sample, this does not need to be the case. In fact, the task first becomes really interesting and powerful if the comparisons vary from the sample along specific stimulus dimensions instead of being identical to the sample. For example, we might present a banana as a sample, and an apple and a chair as comparison. The correct answer would be the apple, since both are fruit. By systematically varying how the comparison stimuli differ from the sample—as well as how the comparison stimuli differ from one another—one can study many different perceptual and cognitive processes. Premack (1976), for example, used some very ingenious matching-to-sample techniques to examine the ability of apes to learn human-like languages.

The first matching-to-sample experiment is thought to be Thouless (1931), who examined the perception of shape in humans (and the effect of orientation). Since then, matching-to-sample tasks have been used to study a wide range of phenomena, including identity perception (e.g., when are two things perceived to be identical? Cumming and Berryman, 1961), memory processes (Santi and Roberts, 1985), rule learning (Premack, 1976), and category formation (Fields and Nevin, 1993). Although matching-to-sample tasks can used with humans, they are most commonly found in animal research, especially in research using pigeons (see, e.g., Ferster, 1960).

A very common form of this task is the delayed matching-to-sample task. In this variant, the sample is presented alone for a brief period of time and is then removed. After a delay of N seconds (referred to as the *retention interval*), the comparisons stimuli are presented. By systematically varying the length of the retention interval, one can study many different things, especially various aspects of memory.

7.3.3 Visual Search Tasks

Just as with the measurement of absolute thresholds (see Section 6.3.1), in visual search tasks participants are asked to decide if a specific stimulus (called the *target*) is present or not. The uniqueness of the visual search task lies in the stimulus presentation: stimuli other than the target are also presented on every trial. The stimuli that are not the target represent the standard visual clutter of the everyday world and are called *distractors*. On some of the trials (usually on half of the trials, in order to maintain the assumption of a balance between the stimulus set and response options that accompany forced-choice tasks; see Chapter 6) the target is present. On the other half it is not. Even in the most difficult condition, accuracy is close to 100%, which means that differences between conditions are hard, if not impossible, to find (this is referred to as a *ceiling effect*). The primary dependent variable in a visual search task, then, is reaction time: how long does it take for participants to enter their responses?

Imagine, for example, that we have generated a new medical-visualization technique that should make it easily visible to the doctor whether or not patients have tumors in their lungs. The data for this hypothetical technique comes from magnetic resonance imaging (MRI) data. In some of the datasets (scans of individual patients) there will be a tumor (the target). In other datasets, there will not. In all datasets, there will be a variety of distractors (organs, blood vessels, etc.). The participants' job is to examine the visualization of MRI scans and decide if a tumor is present or not. If our new visualization technique is well designed, then this will be easy to do. Note that since the technique is designed to be used by trained medical professionals, it would be very useful and fully justified to use only trained medical professionals as participants.

Although the visual search task can be used to study many things, it is generally used to study attention. It is also very often used in the visualization literature. Section 7.3.3.5 briefly discusses the relevant theories and assumptions.

7.3.3.1 Set Size

In a visual search task, two things are manipulated. The first is the complexity of the scene. In a visual search experiment, just as in the real world, there will always be clutter and thus always some distractors. The number of distractors present on any given trial is varied systematically as a part of the experiment. For any given number of distractors, the target will be present on half of the trials. For example, imagine that we have trials with two, four, six, or eight items. On half of the trials with two items, one will be a target and the other a distractor. On the remaining half of the two-item trials, both items will be distractors.

7.3.3.2 Target-Distractor Differences

The second thing that is manipulated in a visual search task is, naturally enough, the target itself—or, more specifically, the difference between the target and the distractors. In the simplest case, all the distractors are identical and the target differs from them along a single dimension. For example, we could present a series of squares where the target square is blue and the distractor squares are all be red. The task would be to indicate if a blue square is present or not. Such a search is often referred to as a *feature search*, since the participant is looking for a single visual primitive or feature (such as a single color, or a single orientation). Naturally, we could allow the distractors to differ from one another. For example, keeping a blue square as the target, we could use squares of different colors (but not blue!) as distractors. Obviously, this asks a slightly different research question (if it is not clear why, see Chapter 2).

Finally, the target might differ from the distractors along multiple dimensions. For example, with a blue square as a target, we could have blue triangles, red triangles, and red squares as distractors. In this case, the target is defined by a *combination* of differences: it is the only item that is *both* blue *and* a square. This form of a search is called a *conjunction search*. Differences in results for feature searches and conjunction searches have lead to several theories about how perceptual processing occurs (see Section 7.3.3.5).

7.3.3.3 Display Time

In most versions of the visual search task, the displays are presented for as long as the participant wishes. This is perhaps most closely related to real-world conditions. The doctor in our tumor example above would—in principle—have as much time as he or she wants to look at the visualization. It is possible, of course, to limit the display time. There are several reasons for limiting display time. The first is because the relevant real-world situation also has time limits. For example, when you are driving down a highway rapidly, you will pass a number of signs. If you are looking for a specific sign (such as an exit sign with the name of your destination), then you will have a limited time to find it among the other signs and other distractions present on the highway.

The second reason to limit display time is to control for eye motion. If, for example, we wish to see if our display works well in non-foveal locations (i.e., can people see the information out of the corner of their eyes?), then we will want to insert a *fixation cross* at the start of the trial (the spot people should focus on) and then present our trial for less than 200 ms (to help ensure that eye motions are not possible). It is also a reasonable idea to include eye-tracking in such an experiment to be absolutely sure that participants were focusing on the fixation cross and that no eye motions occurred. For more on eye tracking and eye motion, see Chapter 9.

The third reason to limit display time is to focus on particular cognitive or neural mechanisms (i.e., the task must be performed based on information present in memory after the stimuli are removed). If a mask is presented directly after the stimuli are removed, then iconic memory will also be blocked from contributing to the task. For more on iconic memory, see the partial-report task in Section 4.3.4.

The limits on display time assumes that all of the items for a given trial are present at the same time. In nearly all visual search tasks, this is the case. One can, of course, present them sequentially. Some differences in the results may or may not be expected, depending on which theory of attention one believes to be true (Huang and Pashler, 2005). A special form of sequential visual search is the Rapid Serial Visual Presentation technique (see Section 7.3.4).

7.3.3.4 Analysis

The most common analysis in a visual search task to measure the cost of adding distractor items. This is called the *search rate* or *display-set-size* analysis. This analysis assumes explicitly that the distractors and targets are of similar complexity and that it takes roughly the same amount of time to process items of similar complexity.

To determine the search rate, we measure the amount of time it takes to respond (correctly) as a function of how many items were present. There are two likely results. First, the search slope might be flat, indicating that reaction time is independent of the number of items present. Second, the slope might increase with an increasing number of items. The first pattern of results is precisely what one would expect if the participants processed everything at once (parallel processing). In such cases, the target would seem to be visible immediately. It would "pop" out of the display.

If the participants are not processing everything at once, then they must be processing the items serially. In any serial search, adding items will increase reaction time, yielding a non-flat slope (Carter, 1982; Carter and Carter, 1981; Nagy and Sanchez, 1990). The actual grade of the slope is dependent on two things. First, it is dependent on the task difficulty—that is, the amount of time it takes to reach a decision for a single item. Second, is it dependent on the type of serial search. There are two important types of serial search: *exhaustive* and *self-terminating.* In an exhaustive search, participants look at all items, store the decision for each, and then once all items have been examined look to see if one of the items was the target. This is useful if you wish to know how many targets were present. The total search time will be proportional to the number of items. Dividing the reaction time by the number of items provides one with a good estimate of the cost of adding a single distractor. Since, however, it is sufficient for the actual task if one item is a target, looking at all the items is inefficient. It might be more efficient to simply stop searching once a single target has been found. Such a search is referred to as a *self-terminating search.* Since each item either is or isn't the target, and the order in which the items are searched is (usually) more or less random, then on average half of the items will need to be searched before the target is found. Thus, the distractor cost in a self-terminating serial search is based on half of the total number of items.

7.3.3.5 Theories of Attention

The world is full of information, but we only seem to be aware of a small amount of it at any given moment. To use everyday language: we only pay attention to a few things at once. To help explain this selectivity, Helmholtz introduced the metaphor of a spotlight to help explain attention (von Helmholtz, 1962/1867). He suggested that there is some mechanism, that chooses a subset (usually a small spatial region) of the information available and allows that subset to be processed while the rest is ignored. The processed area is "visible" because it is "in the light" of the spotlight and the rest is "in the dark." This form of attention is now more commonly referred to as *selective* or *focal attention* (Broadbent, 1958; Kehneman, 1973; Neisser, 1967).

Such a spotlight would, of course, represent a form of serial search. We shine the light on one area of a scene, process what is there, and then move the spotlight. A normal spotlight moves from point A to point B by covering all spaces in between. Selective attention on the other hand seems to be spatially discrete; as the spotlight is gradually diminished in one area, it is gradually increased in another (Sperling and Weichshelgartner, 1995). Moreover, selective attention appears to be independent of eye motion.

If selective attention is moved from location to location, some mechanism is needed to explain how it is decided what the next location must be. If the shift is non-random (and it is indeed non-random), then the information in the display must be pre-processed in some way in order to decide where the further processing allowed by attentional selection should occur. This phase is often referred to as *pre-attentive* processing.

In the late 1970s, several researchers suggested that the presence of low or negligible search slopes in a properly designed visual search experiment can be taken as evidence that humans can process information in parallel (Hoffman, 1979; Wolfe et al., 1989). This supported the assumption that perceptual processing has two stages. In the initial phase, a number of features (e.g., color, orientation) are processed at all spatial locations in parallel. Based on the results of this initial processing, decisions are made (unconsciously) about which items should be processed in more detail in the second stage, which is the serial processing stage (Broadbent and Broadbent, 1987; Duncan, 1980; Treisman and Gelade, 1980). Perhaps the most famous two-stage model of attention is the feature integration theory (FIT) from Anne Triesman and colleagues (Treisman and Gelade, 1980; Treisman and Sato, 1990). In short, Triesman suggests that the pre-attentive processing yields a number of *feature maps,* one for each stimulus dimension. These are subsequently merged to become a form of saliency map. The *saliency map* is then used to direct selective attention. In this model, it is explicitly argued that any search for a target defined by more one stimulus dimension (i.e., a conjunction search) requires the slower, serial processing of selective attention.

A wide variety of evidence has emerged showing that complex differences—such as those present in faces and surfaces—can be found in parallel (He and Nakayama, 1995; Nakayama and Silverman, 1986; O'Craven et al., 1999). That is, some conjunction searches are done in parallel. Moreover, some feature differences (such as colors that are very similar) are processed very slowly, seemingly done in serial. The evidence suggests rather strongly that there are not two separate stages of processing—a parallel stage followed by a serial stage—but just one. People seem to always process in parallel (Eckstein, 1998; Eckstein et al., 2000; Townsend, 1976, 1990; Wolfe, 1998). Under these theories, the non-negligible search slopes are not an indication of serial search, but an indication of physiological noise and Gaussian combination rules.

The most common one-stage model is a *simple noisy parallel model* based on signal detection theory from classical psychophysics (Eckstein, 1998; Eckstein et al., 2000). Briefly, the theory points out that the response pattern for neurons is essentially a Gaussian distribution for two reasons. First, each neuron has a preferred property. That is, it responds optimally for some specific characteristics such as an orientation of 45 degrees. The response of the stimulus decreases as the difference between the stimulus and the preferred property increases. Thus a neuron might respond most to a line orientation of 45 degrees, slightly less for a 40 degree orientation and much less for a 5 degree orientation. The second reason for a Gaussian response distribution is the physiological noise mentioned in Chapter 2. If a stimulus with a neuron's preferred property is presented to the neuron a number of times, the response distribution will still be Gaussian due to internal noise (Bradley et al., 1987; Newsome et al., 1989). Thus, in a signal-detection version of visual search theory, each feature of every item in the display is processed in parallel but represented in the visual system by noisy variables. The mean of the Gaussian response distribution for any neuron is the preferred property and the variability of the distribution is related to noise in the visual system and the degree of specialization of the neurons. Some features have very little noise, while others have a lot. If the distributions of neural responses to a target do not overlap the distributions for the distractors, then the difference can be detected quickly and by very few neurons. The target will seem to "pop out." If the distributions are narrow, then the feature will not often overlap with other distributions, meaning it will pop out often. If the distributions overlap, then the summed response of more neurons is needed to be certain that the difference is meaningful, and this kind of search will take longer.

For our purposes, it does not matter if the difference in search slopes is due to a parallel mechanism versus serial mechanism, or to the difference in Gaussian distributions in a parallel mechanism that has limited resources. The bottom line is that some differences can be detected more or less immediately, and others cannot. Which differences are easy to detect is immediately apparent from the task results. Thus, a simple atheoretical analysis of the results will show which set of stimuli or parameters or conditions share a search advantage. For our medical-visualization example, then, we could simply run a visual search task varying the type of visualization as well as the number of distractors and see which visualization has the lower slope (and overall faster reaction times).

7.3.3.6 Example: Visual Search

Obviously, anticipated events can be reacted to with more accuracy, speed, and ease than events which are a complete surprise. Forewarned is forearmed. Within psychology, one special type of forewarning is called *priming*. In general, once you process information about something, it is easier to continuing processing that type of information. For example, if you see a picture of a dog and are asked to identify it, it will be easier to identify the next image if it is a dog than if it is a cactus. The processing benefit is more linked to the semantic aspects of the stimuli and less to the visual properties of the stimuli. That is, if you are presented with the *word* "dog," you will respond faster if the next stimulus is an *image* of a dog than if it is something else. The word "dog" in this example is referred to as a *prime*.

There is a very large literature on priming, pointing out in extensive detail what works as a prime, as well as when, where, and how it works. Among the many findings is one by Thornton and Kourtzi (2002) that a dynamic prime is more effective that a static prime (as long as both primes are presented for the same length of time). In another study, Pilz et al. (2006) showed that this holds true for identifying someone from a photograph.

Research question. The primary question in this experiment was to determine if learning a new person's identity by watching that person in a video makes it easier to find that person in a photographed crowd than if the person had been shown in a photograph initially.

Stimuli. The Max Planck Institute for Biological Cybernetic VideoLab (Kleiner et al., 2004) was used to record a set of expressions from eight people. The actors or actresses were asked to read short stories that highlighted a specific emotion. Subsequently, they were recorded saying a word that was appropriate to that emotion (for more, see Amaya et al., 1996). For this experiment, only the recordings of anger and surprise were used. The evocation words were "what!" for an anger expression and "wow" for surprise. Since previous research has shown that rigid head motion plays a significant role in the perception of facial expressions (Cunningham et al., 2005; Nusseck et al., 2008; Wallraven et al., 2007) and since such motion will make certain manipulations very difficult, the individuals were trained to keep their heads as still as possible during their recordings.

Each video sequence was edited to start with a neutral expression and end with a peak expression, and to be precisely 26 frames long. Where static pictures were used instead of video sequences, the last frame of dynamic sequence was used.

Stimulus presentation. Participants needed to familiarize themselves first with the target person so that they could search for the actor or actress later based on their memory. Each participant was randomly assigned two actors/actresses from the database (one male and one female). The surprise expressions were shown (dynamically for one actor/actress and statically for the other) and the participants were asked to try to memorize the actors/actresses. The two faces were presented in alternation for 1,040 ms each (1,040 ms was the length of the videos: each of which was made up of 26 frames at 25 Hz), with a two second blank screen between faces. Each face was seen 100 times. While watching the faces, the participants were asked to fill out a questionnaire regarding the personality of the person. This was an attempt to ensure that they actually paid attention to and processed the images.

Once the learning phase of the experiment was over, participants entered the search phase. During the search portion of the experiment, an array of static photographs of the faces was presented (all in an angry expression). The faces were placed at equal distances in a circular pattern (a radius of eight degrees) centered on the middle of the monitor. The search array consisted of two, four, or six photographs. Since at most one target could be present on any given trial the different number of items in the array means that the number of distractors was varied systematically.

Three types of target display were possible: identity 1 present, identity 2 present, or no target present. The number of times a given target was presented was equal to the number of times no target was present. Of the 150 trials, each target was present 50 times, and on 50 trials no target was present. As a result, a target was present two-thirds of the time. Although this represents an asymmetry between the present–absent answer alternatives, it is justified in the circumstances.

Task. The participants were asked to indicate whether one of the target faces was present or not.

Participants. Eighteen individuals who were naive to the purposes of the experiment participated and received financial compensation at standard rates. The participants were between 18 and 40 years old, and came from a variety of social and educational backgrounds.

Results. On average, the participants responded faster to faces learned from a dynamic source than from static ones (1,317.8 ms versus 1,595.5 ms). This difference is statistically significant ($F(1,17) = 4.7$, $P < 0.05$ in a repeated-measures ANOVA). No difference was found in the search slope. In other words, people were faster at finding a person in a photograph of a crowd if they had learned about that person from a video than from a photograph, but not more efficient at it. The efficiency of the search does not differ across learning materials (adding more people to the crowd increases the difficulty just as much for faces learned from a photograph as for faces learned from a video).

Summary. This example experiment demonstrates how a visual search task can be conducted with relatively broad questions.

7.3.4 Rapid Serial Visual Presentation Tasks

In a visual search experiment, several items are presented simultaneously and people are asked to examine them to find a specific item. In the case of a Rapid Serial Visual Presentation (RSVP) task, the procedure is essentially the same task with the exception that the items are presented sequentially (thus, the "serial" portion of the experiment's name). Each item is presented for an extremely short period of time (hence the "rapid" in the name). Since it is believed that people can pay attention to only a few things at any given time (Cattell, 1886), the goal of RSVP is to avoid this limitation by trading spatially separated presentation of items for a temporally separated presentation.

Initially, RSVP was used to study elements of visual attention (Eriksen and Collins, 1969; Eriksen and Spencer, 1969; Lawrence, 1971). Indeed, results from initial work with RSVP suggested that humans can process certain features in parallel, but that more detailed processing must be done in serial. This was the basis for the well-known two-stage attention model, in which a few features are processed everywhere in parallel, and, based on that, some of the spatial locations are chosen for more in-depth processing. Modern theories suggest that there is only one stage—people always process in parallel. For more information, see Section 7.3.3.5.

In general, a number of items (anywhere from 12 to over 4,000 have been used) are presented one after another at a rapid pace. Each item is displayed for exactly the same length of time, with generally between 1 and 50 items per second being shown (i.e., each item is presented for somewhere between 1,000 ms and 20 ms). As soon as one item is removed, the next item is displayed. The participant's task is to detect whether any given item is the target. The target may be specified explicitly (e.g., "look for this picture") or indirectly (e.g., "look for a picture with food in it"), or even negatively ("look for a picture that does not have food in it"). The first few and last few positions in a sequence of items are never targets (to prevent recency and primacy effects).

In all cases, the participants must report whether they feel that a target is present (usually in a go/no go procedure). Since it can take time to press the button, it is rare that the image is still present when people finally manage to respond. Thus, it is not clear how the researcher should decide if a button press was correct or not. Generally, a response within one second after the *onset* of target presentation is considered correct. In some cases, the participant must subsequently perform a secondary task. For example, if we present a series of letters, and ask participants to indicate whether or not the letter *T* is present, we can also ask them to report the color of the target letter.

Detection performance is amazingly accurate (over 80%, often as high as 95%) for even very short presentation times. Perhaps just as astonishing is that performance on the secondary task is not as accurate. That is, the participant can accurately say whether the letter *T* is present, but will often not know its color (note that a part of this may be the partial report effect discussed above in Section 4.3.4). More interestingly, participants will sometimes report the color of the item that was shown just before or after the target. This is known as *intrusion*, and has important theoretical implications for object perception (see, for example, Botella et al., 1992).

More often than not, RSVP has been used to investigate reading performance. The stimuli typically have been letters, words, or digits. In several cases, however, pictures have been used. Intraub (1981), for example, presented a series of pictures at rates of 114, 172, and 258 ms per image. The target image was identified by name ("look for a butterfly"), by superordinate category ("look for an animal"), or by negation of a category (e.g., "look for a picture that has no food in it"). One group of participants was asked to report when they felt a target was presented (by button press) and to briefly identify it. The participants were astonishingly accurate.

Recently, there has been interest in taking advantage of RSVP-like tasks in interface design (Spence, 2002). Many current information devices have limited screen real estate. Information about the ability to read words one letter at a time or process images in the blink of an eye could be very useful when applied to things like increasing the amount of information present in small displays, and other applications that trade off time and space.

The two biggest hidden problems in RSVP are *repetition blindness* and *attentional blink*. Repetition blindness refers to the fact that when a series of items is presented in the same location, people can often fail to see some prominent items. In other words, if your attention wanders for a brief instant, you may miss something critical. Attentional blink refers to the fact that processing something ties up resources. Thus, if the visual system is busy processing a target, then any additional target presented within the next half a second has a strongly reduced chance of being detected (Broadbent and Broadbent, 1987).

7.3.5 Free-Grouping Tasks

All of the multiple-choice tasks are essentially discrimination tasks: can people see a certain difference between the stimuli? Such a formulation of multiple-choice tasks focuses on the perceptual side of the perception-action loop. One might just as easily focus on the cognitive or even the action side of the loop. For example, all multiple-choice tasks force participants to decide which alternative to choose for each stimuli. Another way of looking at this is to say that each alternative represents a bucket. Participants place each stimulus in one of the buckets. By doing so, they are not saying that everything in a given bucket is identical, but they are saying that they are similar. That is, they are performing a categorization task. The concept behind free grouping is to take this way of looking at multiple-choice tasks to its logical extreme: let the participant decide what the categories are and even how many there are. We lose some control over the perceptual side in that we can no longer force participants to make a specific perceptual distinction, but we gain the possibility of learning a lot about how participants perceive or think about the stimuli. In that sense, the free-grouping task is more akin to the free-description task than a forced-choice task.

7.3.5.1 Example: Categorizing Facial Expressions

During the course of a conversation we see, recognize, and respond to facial expressions very rapidly and accurately. Rarely do we explicitly label the expressions we see. If you tell someone that today is your birthday and he or she looks pleased and surprised, you would not usually think "Ah, that was a pleasantly surprised expression, of intensity level 5, but only a sincerity level 2 . . . he is really surprised!" In fact, there are certainly many expressions that we understand easily but which do not have short or easy descriptions (such as the expression that people have when they just emerged from a very important exam, thinking that they did well but suddenly realize that they completely misunderstood the questions and have in fact probably failed). Despite this lack of labels for some expressions, most experiments with facial expressions ask the participants to label them (in part because full, closed-loop experiments with facial expressions are still nearly impossible; see Chapter 8). Whether or not the results of such studies generalize to the real world is still unclear.

Although many people assume that facial expressions are categorized into discrete states (for a review see Adolphs, 2002), there is some evidence that facial expressions can be members of multiple emotional categories. What these categories are depends to some extent on the conceptual relation to other expressions with which they are compared (Russell and Bullock, 1986; Russell and Fehr, 1987). Perhaps, facial expressions are represented in some semantic space in which several dimensions encode important properties of expressions (Valentine, 1991; Adolphs, 2002).

The categorization space of facial expressions could be examined with either a similarity-rating approach or a semantic-differential approach (see Chapter 5). Given that a large number of expressions are possible, this would result in a tens of thousands of trials (if not more). Moreover, such an approach would require people to explicitly label the expressions. In an effort to find a more direct measurement of categorization, we conducted a free-grouping experiment.

Research question. What natural categories do people have for expressions? Does providing dynamic information alter these categories?

Stimuli. The stimulus set consisted of 55 of the expressions from ten of the actors/actresses in the larger Max Planck database (see Chapter 4 for more information on the stimuli). All of the expressions were converted to movie files.

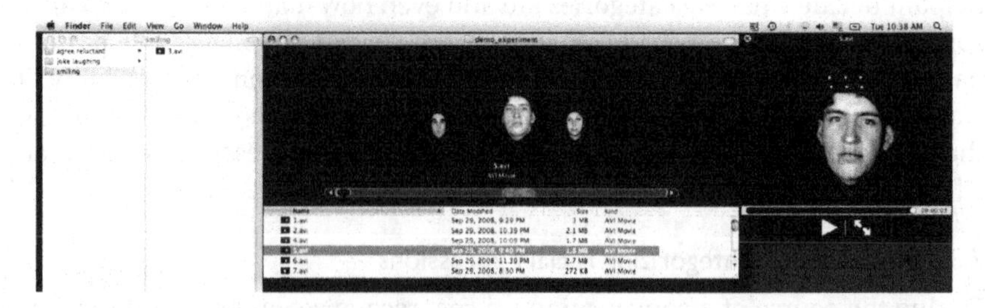

Figure 7.1. A snapshot of the stimulus presentation for the free-grouping experiment.

Stimulus presentation. The expressions were presented on an Apple MacBook Pro and the monitor was divided into three windows (see Figure 7.3.5.1). In the left window the created groups (folders) with their facial expressions were displayed. The central window showed the complete stimulus set and was the main working window for participants. The movies were presented using the "cover flow" view, which is a special view setting for the file browser in which each file can be previewed easily. The cover flow view enabled a preview of the neutral face of the selected actor in the upper half of the centered window. As soon as a movie file was selected, automatic playback was started in the separate right window. Participants were allowed to replay the movie by pressing the Play button in the right window as often as they wished.

The dynamic stimuli used the whole expression (from neutral through peak and back to neutral). The static presentation used special files that started with the starting frame, then immediately went over to the peak frame and ended with the ending frame. This allowed us to match the starting and ending frames and the presentation time for the static and dynamic conditions. It also ensured that the participants would not see the peak frame in the preview of the movies.

Despite all efforts to limit the length of the experiment, participants took around seven hours to finish. The experiment was broken down into several sessions, each at most two hours long. Thus, depending on how long a participant took, the experiment was split into three or four sessions. The sessions took place on consecutive days and participants were explicitly instructed to use similar strategies for grouping the facial expressions in all sessions.

Task. Participants were asked to watch the facial-expression movies and to group similar facial expressions by either creating new groups or by moving the movie file into an existing group. The whole experiment was finished as soon as all movie files were grouped. Afterwards, participants were allowed to revisit their groups and make final changes.

Participants. Forty naive participants (twenty female, twenty male) participated and received financial compensation at standard rates. The participants were randomly assigned to one of two groups: static stimuli and dynamic stimuli. All participants were native German speakers and were between 20 and 52 years old.

Results. A matrix of grouping frequency for all facial expressions across participants was created. The dynamic and static conditions were analyzed separately. On average, participants in the dynamic condition created 50.6 groups. It seems as if one category was created for each expression, regardless of the actors/actresses. Participants in the static condition created 37.6 groups on average. By examining which expressions showed up in the different groups, it became clear that the presence of temporal information increases the ability to group similar expressions together significantly.

Summary The research question was broad, essentially asking what the natural boundaries of the phenomenon are. While a free-description task would have been possible (see Chapter 4), the primary goal was to see what stimuli get grouped together to form categories.

7.4 Conclusions

All of the tasks mentioned in this chapter are variants of the classic 2-AFC, demonstrating the incredible flexibility of the task. There is little doubt as to why most perception experiments—indeed, most psychological experiments—use some form for forced-choice task.

Just as seen in the example provided in Section 7.3.5.1, some research questions involve a specific behavior. In these cases, it is possible to directly use that behavior in the experiment. There are a number of experiments where the real-world action is used. An overview of such real-world tasks can be found in Chapter 8.

Chapter 8

Real-World Tasks

In some cases, it will be possible to use a task that is for all intents and purposes identical to the real-world behavior we are studying. For lack of a better term, we will call these *nonverbal* or *real-world* tasks. Such tasks are most often used in virtual reality (VR) setups, where the loop between stimulus and action can be closed so that the action changes what is currently being seen. Such tasks can, of course, be found in the real world as well. For example, simply asking someone to point to a target is a real-world or nonverbal task.

8.1 Overview of Real-World Tasks

The details of nonverbal tasks are as varied as can be imagined. Here, we provide a brief overview of the highlights of nonverbal tasks.

8.1.1 Typical Questions

Non-verbal tasks are best at answering questions that have a very specific behavior associated with them. Some examples of typical questions follow.

- How well can people stay on a virtual road?
- How much does fog affect driving speeds?
- Does altering the temporal characteristics affect how people react to an expression in a conversation?
- How well do people know where they are and what information contributes to spatial orientation?

8.1.2 Data Analysis

This is perhaps the trickiest part of real-world tasks. One would think that since the technical challenges involved in generating, presenting, and controlling the stimuli in response to real-time interaction is so amazingly challenging, that actually generating useful measurements and then analyzing them should be easy. Sadly, many researchers forget to think about what their dependent measures are before starting the experiment and thus end up with little useful direct data. Moreover, since most real-world tasks are new, standardized analyses do not yet exist for most of them. Careful thought before the start of the experiment, however, can probably ensure that assumptions are met so that traditional analyses can be used.

8.1.3 Overview of the Variants

There are a near infinite number of things people can do in the real world, and a near infinite number of ways to measure those behaviors. Not surprisingly, then, there are a near infinite number of real-world tasks. So far, no classification of real-world tasks has really ever been attempted.

8.2 Task Description

The basis of a nonverbal task is to have participants actually perform the action that we are interested in. As we have seen in the Chapters 4–7, however, the vast majority of perceptual research questions ask people to either describe something or choose a label or name for something. That is, most perceptual tasks inherently require and are based on language.

8.2.1 Direct Tasks and the Linguistic Bias

Without the ability to at least speak fluently in a specific language, most of the tasks in the Chapters 4–7 would be impossible. Ignoring the obvious need to understand the spoken (or written) experimental instructions, most tasks require a common language for participants to communicate the answer to the experimenter. Obviously the free-description task and its variants simply cannot be performed without language. Even rating tasks or forced-choice tasks are extremely difficult without language. Naturally we can use images for the alternatives, but we will still need some form of common language to tell the participants what task they are to perform (and maybe even how to interpret the images). To fully understand how difficult it is to use a task without any common language, try to imagine running an experiment with a pigeon or an infant who is several weeks old. While many of the tasks in the specialized multiple-choice chapter do not require a common language (indeed, matching-to-sample is heavily used in animal research), using them without involving language usually involves a lot of training. For example, training a monkey to use a joystick so that the monkey can select answers can take several months! Likewise, training a pigeon to peck certain keys to give specific answers is a lengthy procedure.

The use of language as a central element of a task is not in and of itself a negative thing. It does, however, have ramifications. For one thing, it tends to turn any task into a meta-task. In Chapter 2, for example, we talked how accurately people could point to a target. We wanted to know, for example, if pointing behavior was affected by target contrast or target size and there are a number of possible tasks we could choose to investigate this question. We could run a free-description task and ask the participants what factors they think would affect their pointing. We could ask them to rate how difficult they think it would be to point at certain targets. We could present them with two targets and them to choose the one that they think would be the easiest to point to. All of these variants, however, are not only obviously verbal tasks, but also require participants to give us their *opinions*. Rather than actually pointing, people say what they *think* they would do. We would definitely get an answer in each of these tasks. The answer would almost certainly reflect what they really thought. The answer would probably also be very consistent for each person over time and probably even consistent across a group of people over time. That is, the results would be valid, reliable, and repeatable.

It is very unclear, however, whether the results would have anything to do with how people *actually* point in the real world. Indications that it is not always possible to generalize from the results of a descriptive tasks to a more direct task were seen in the comparisons of tasks for facial-expression recognition in Wallraven et al. (2007) and Chapter 6. In Wallraven et al. (2007)'s work, recognition of facial expressions was studied with a number of different measures, including an ordered-ranking task, a direct-comparison task, and a non-forced-choice task. For the ordered-ranking task, it was clear that most people thought that the *standard* animation was most effective. The forced-choice tasks, however, demonstrated that when people are forced to actually recognize the expressions, the illustrative style was most effective.

In a more direct demonstration of the difference between a direct and an indirect measure, Guy Wallis and colleagues examined how people change lanes when driving a car (Wallis et al., 2007, 2002). To change lanes properly, three phases are involved. First, the driver must first rotate the wheel in one direction so that the car is no longer aligned with his or her current lane. Second, once the driver is in the new lane, he or she must undo the initial heading change. That is, the driver must realign the car with the road by counter rotating the steering wheel well past its initial starting point. Third, once the car is once again heading down the street (aligned with the new lane) the driver moves the steering wheel back to its initial starting point. The simplest, most obvious way to measure how people change lanes is to ask them. This is obviously a verbal, meta-cognition task: people describe what they *think* they would do if they were to change lanes. In general it has been found that people do not perform this task well: they almost always skip the second phase and simply stop the car-changing heading once they leave the new lane. Since this invariably results in the car leaving the road and crashing, it obviously does not represent real-world behavior.

In Wallis et al. (2002)'s study, participants were placed in a driving simulator and asked to change lanes. They did so easily, demonstrating that the simulator can produce real-world behavior. Interestingly, when Wallis and colleagues arranged the simulation so that the lane changing started just as the participant entered a virtual tunnel (which meant that they could not see where they were going), the participants did precisely what they *said* they would do: they skipped the second phase and drove off the side of the virtual road into the tunnel wall. It seems that in order for lane changing to occur properly, people need to see the immediate consequences of their actions. The need to "close the loop" between action and perception will be discussed in more detail in the Section 8.2.2.

Although the forced-choice recognition task for facial expressions is more direct than ordered ranking, it is still linguistically mediated. People must name the expression. In contrast, the free-grouping task described in Chapter 7 is not as obviously as tied to language. This is true for many of the tasks in Chapter 7. In most cases, though, these tasks are still not real-world tasks. It is not clear, for example, how well free grouping generalizes to actually conversing with someone. Likewise, while matching-to-sample is very useful it is not clear how directly the results relate to the relevant real-world behavior. In fact, one of the central criticisms of Premack's work demonstrating that non-human apes could use language (Premack, 1976) hinges on his assumption that performance on a matching-to-sample task is representative of the abilities used in processing complex grammar rules.

It should be clear from reading Chapter 2 that the tasks presented in Chapters 4–7 are generally preferred over their real-world counterparts. There are very good reasons for this. Those tasks are well-defined, well-explored, and well-established. The advantages and disadvantages are well known. Their relationship to the real world is also often known. Moreover, they are easy to perform, always produce measurable and analyzable results, and provide the precise control and repeatability necessary for a reliable experiment.

Nevertheless, sometimes a more direct alternative is needed. In the pointing experiment, for example, wouldn't it be simpler to just let the participants point? Yes, it would. Not only would it be simpler, it would also be a much more direct way to answer the research question and much closer to the real-world behavior in which we are interested. The closer the experimental situation is to the real-world behavior of interest, the more likely it is that the experimental results will generalize to real-world situations. Unfortunately, the real world is extremely complex. Ensuring that all possible elements and factors are controlled for as laid out in Chapter 2 so that there are no unexpected or unwanted sources of variance or noise is simply not possible in the real world. Doing so in an experimental setting requires not only careful control of the stimuli, but also control of the response environment and how the two interact. The tasks in the previous chapters keep the response environment as simple as possible by relying on written responses. Using more direct, real-world behaviors such as pointing makes the response environment much more complex. Thus, all the lessons learned in Chapter 2 for stimulus control also apply to response control.

8.2.2 Perception-Action Loops

In order to survive, animals must be able to interact with the world around them rapidly and accurately. The environment, perception, cognition, behavior, and consequences are intimately tied together. The qualities and constraints of one affect the possibilities and processes in the others. What information is present in a display strongly affects what can be seen and how it might be seen. The specific perceptual processes used to examine a scene affect what information we extract and how it is represented (as was demonstrated with the RSVP task in Chapter 7). The choices that are possible and the actions that can be made are determined by the information we have. Thus, the whole pipeline, from input (the stimuli) to output (the response or behavior) are intimately intertwined. This was alluded to in Chapter 2, where it was shown why it is extremely difficult—and probably impossible—to separate the perceptual elements from cognitive or behavioral elements in an experiment. The tasks described so far have aimed at keeping the output side as simple as possible. The first big change that occurs with real-world tasks is that the possible actions become considerably more complex. The second big change is that they take seriously the claim that the perception-action pipeline is actually a closed loop and they implement it in the experiment: the actions that participants take directly affect the information that is present in the stimulus array!

To demonstrate the complexity introduced by closing the perception-action loop, consider the seemingly simple act of making a pot of tea. It involves a surprisingly large number of steps and behaviors, each of which requires a fair amount of new information as well as feedback regarding ongoing actions. In an interesting eye tracking study, for example, Land et al. (1999) demonstrated that making a pot of tea can be broken down into step such as "put the kettle on," "make the tea," and "prepare the cups," each of which can likewise be subdivided into actions such as "fill the kettle." The action "filling the kettle," can also be subdivided into "find the kettle," "lift the kettle," "remove the lid," "transport to sink," "locate and turn on tap," "move kettle to water stream," "turn off tap when full," "replace lid," "transport to worktop." Note that even though these last steps are not supposed to be subdividable functionally, they are are still quite complex. The action "transport to sink" requires route planning, the detection and avoidance of obstacles, and so on.[1] One of the central findings of Land et al. (1999)'s study was that even tasks which are thought to be more or less automatic and *open loop* (i.e., that do not require new information to perform them) in fact require constant tracking and visual feedback to be performed correctly.

Therefore when we wish to study real-world situations, it is advisable to use *closed-loop task* tasks. This poses a number of technical challenges, and the virtual reality (VR) literature is full of examples of what one should or should not do (see, e.g., Jack et al., 2000; Mania et al., 2005; Wann et al., 1997). Creating a VR setup takes a tremendous amount of time and money, especially when a fully immersive setup is being built. VR allows us to present complex stimuli while retaining full control. Regardless of the technical setup, the experiment still needs to be as precise, repeatable, and controlled as possible. Unfortunately, as soon as we allow the next stimulus to be chosen by the participant's current behavior, we lose a considerable amount of control over the experiment. Noise will invariable creep into our measurement. To help minimize the noise, extra care must be taken to ensure that the realm of possible situations that the participant can end up in through the feedback between action and response are anticipated and appropriate.

Creating a dependent measure in a closed loop task follows all of the same rules as the tasks described in Chapters 4–7. The list of actions that the participant can take—in principle—during their interaction needs to be specified fully. All possible actions need to be anticipated and the responses that the system should give for each one should be appropriate for the question we wish to ask. Perhaps the trickiest aspect of this, though, is to decide what we will actually take as a measure for answering our research question. For example, let us imagine that we have built an immersive VR setup based on a driving simulation to study how people act and react on the highway. Let us imagine that, in order to maximize the similarity to the real world, we actually place a real car on a motion platform so that the research environment is as close to the real world as possible. We have the stimuli (e.g., a virtual world including a number of highways), output devices (e.g., projection screens, loudspeakers, forced-feedback steering wheel and a motion platform) as well as input devices (e.g., force-feedback steering wheel, and gas and brake pedals). We seat the participants in the setup, let them get used to driving in a virtual world, and then let them loose on the virtual highway. What now? That all depends, as usual, on our research question. For the sake of demonstration, say that we're investigating how fog affects driving behavior. We can easily simulate different kinds of fog in the virtual world. We set up the system so that some-

[1] In an extended version, Johannes Zanker described a version of making a pot of tea that has well over 50 steps, some of which were still quite complex (such as "detect and avoid obstacles," "search for switch," "recognize and discriminate objects," and "plan route").

times there is fog and sometimes there isn't. We can even vary across trials how thick the fog is, and so on. Thus, we have a within-participants design, in which each participant experiences different thickness of fog (the independent variable) across a number of trials. What exactly happens on a given trial, though? We could choose to have them "just drive around." Some people will drive fast, others slow, some will not bother to stay on the highway, and so on. In other words, we have a situation just as unconstrained as a free-description task. On the other hand, we could choose to ask them to drive for five minutes and try to stay in their lane. Such a task is quite specific and constrained. We might also ask them to navigate from city A to city B and then analyze which choices they make when they come to a fork in the road (which presumes that we designed the virtual world to have forks in the road!). Note that this last task is, in essence, a series of forced-choices. At each intersection, the participant must choose to drive down one of several explicitly shown alternatives.

From this discussion it can be seen that often the direct tasks that can be used are often just camouflaged counterparts of the tasks described in Chapters 4–7. They have all of the advantages and disadvantages of these tasks are precisely the same. For example, in any forced-choice task (whether it is virtual navigation or facial-expression recognition), the dependent measure is clear: which of the alternatives was chosen? In a free-description or free-interaction experiment, however, the choice of a dependent measure is less clear. When our participant is driving around the highways, with no specific task, what is our measure of performance? We might choose to measure speed, lane-changing or lane-keeping behavior, how much of the world they explored, how well they followed normal traffic laws, how well they react to unexpected situations (like a cat running onto the highway) and so on. Each of these options has its advantages and disadvantages. The type of statistics that can be run on some of them, for example, can be very complicated. we strongly recommend that a method for analyzing the data be selected *before* the experiment begins!

In some cases, despite have a closed-loop setup, the dependent measure still tends to be open loop. For example, two of the biggest issues in virtual reality research are simulator sickness and presence. Simulator sickness is very similar to motion sickness. Some people become nauseated when they are riding on a boat or try to read in a moving car. Likewise, some people tend to become nauseated in driving simulators. Countless experiments have been run to determine the factors that affect the frequency and intensity of simulator sickness. Despite the use of closed-loop setups, with a real-world task, the most common dependent measure is a questionnaire that is filled out before and after the experiment. The same is true for presence (which is the feeling that one is actually present in the virtual world): The phenomenon is experienced interactively but measured with a questionnaire. Some argue that since both sickness and the sense of presence are very subjective, then only subjective measures are possible and thus the questionnaire is the best (or maybe even only) measurement method. Others argue that if these subjective phenomena do not have any overt effect on behavior—if a person who is simulator sick drives exactly like someone who is not—then why are they such pressing issues in the design and use of VR setups? Since they do affect behavior, why not actually measure the behavior directly, rather than indirectly?

8.2.3 General Guidelines

Given the variety of possible real-world tasks, it is nearly impossible to outline guidelines which will hold for all of them beyond the general guidelines that were outlined in Chapters 2 and 3. Since many real-world tasks are camouflaged counterparts of the tasks described in Chapters 4–7, then the guidelines presented there should likely also be valid for the real-world task.

8.3 Specific Variants

A few tasks have been used so often that they have become somewhat standardized. For more detail on them, see the relevant experimental literature. To provide some insight into how even seemingly simple real-world tasks have been studied and refined, we discuss one general method and one specific one.

8.3.1 Method of Adjustment

Arguably the first closed-loop task is one of the three original psychophysical procedures introduced by Fechner (1860): the *method of adjustment*. Many of the real-world tasks that have been subsequently developed can be seen as variants of this. In a method of adjustment task, a direct connection between the input device and the stimulus is provided so that participants can choose the level themselves. An example is perhaps the clearest way to describe the method of adjustment. Imagine that we wish to measure the absolute brightness threshold (just as we did in Chapter 6). The experiment is set up precisely as it was when we used a 2-AFC, with one exception—participants are given a dial. This dial is essentially a dimmer switch and it determines the intensity of the light coming from the stimulus. The participants' task is to set the light to the point where they can just barely see the stimulus. The technique is amazingly effective and efficient. In addition to measuring the final intensity level, there is a considerable amount of information present in the settings the participant used before settling on the final result. Since this is a standard in psychophysics, we will not discuss it any further here. More information can be found in any psychophysical text, such as Gescheider (1997); Falmagne (1985).

8.3.2 Pointing

Many studies focus on spatial location, even if it is not obvious that they do so. In fact, to some degree, nearly every visual perception experiment that uses realistic stimuli has a spatial component. This is due to the simple fact that every object in a scene has a location and no two objects can share the same location. Location is a unique feature for everything in a scene. Thus, if we wish to know if someone detected a stimulus, such as in a 2-AFC task or a visual search task, we could ask the participants to tell us if they think they saw the target. If they have seen it, however, they know also know where it is and thus we could just as easily have them point to the target. This has the added advantage that we can double check to see if the item they felt was the target was in fact the correct item. It has the disadvantage that it adds time and a few additional processes (such as response selection: see the discussion on the go/no go task in Chapter 7 for more on this) to the experiment. Pointing is used heavily in two research fields, as described below.

8.3.2.1 Spatial Orientation Research

People usually have a very good idea of where they are in relation to specific landmarks in their current environment. The odds are high that if you close your eyes right now and try to point to the entrance to the room you are in, you will be pretty accurate. Thus, to test how well people know where they are or where specific landmarks are, we can have them simply point to them with their eyes closed. For more on pointing in spatial orientation, see Riecke et al. (2005, 2006).

8.3.2.2 Intersensory Integration Research

Any given event in the real world affects many different physical media. On the one hand, this means that we receive information about objects and events simultaneously through a number of different sensory modalities, each of which is slightly different than the others. The location, time of occurrence, and other aspects of an event can be seen, heard, smelled, tasted, felt, and sensed in a variety of other ways. In order to be used together, the different sensory systems must be calibrated against each other. If I wish to pick up a cup, I need to be able to move my hand from its current *felt* location to the the cup's *visible* location. That is, I must be able to translate visual coordinates into proprioceptive or kinetic coordinates (and then into a motor sequence). The modality-specific temporal and spatial coordinates of a single event must be comparable with each other. Moreover, I need to be able to compare the visual and felt locations of my hand to ensure that the actual path it takes towards the cup is the one I intended it to use. If errors occur (for example, if there were weights on my hand that I did not know about), then I will need to be able to correct for them online.

The integration of different sensory and motor systems is made harder by the fact that human perceptual systems change over small time scales (e.g., putting on new glasses or reaching for something in the water would involve a time scale of only seconds) as well as longer time scales (e.g., maturation and growth take place over months to decades). Thus, there must be a mechanism for "recalibrating" the systems; von Helmholtz (1962/1867) and Stratton (1897) were among the first to notice that if people put on glasses that shift or rotate the visual field, all spatial behavior (e.g., reaching, point, catching, walking, skiing) is initially impaired, but people adapt rapidly (for more information, see Bedford, 1993b; Harris, 1965, 1980; Redding and Wallace, 1997; Welch, 1978). Consider, for example, when the visual field is shifted to the left with prism goggles. When participants are asked to point at or reach towards a target without being able to see their hand, they initially reach towards the visual location of the object (to the left of the actual location) and fail to touch the object. A few minutes of acting with the new visuomotor relationship (and—very importantly—while all the time receiving feedback about the correct location of the object) returns behavior to nearly normal performance levels (i.e., they can point at the object accurately again). Subsequent removal of the prism goggles leads to a renewed disruption of performance, which is in the opposite direction of the initial error. The size of this *negative aftereffect* is one of the most common measurements of the strength of adaptation (see, e.g., Welch, 1978). The effects of prism experience are assessed either by simply testing for the occurrence of adaptation (e.g., Held and Hein, 1958; Moulden, 1971) or, more commonly, by comparing adaptation at locations used during the prism training and locations not previously used (e.g., Baily, 1972; Bedford, 1989, 1993a; Cohen, 1966).

There is some evidence that when the objects are moving, adaptation takes a fundamentally different form (Field et al., 1999). To interact with moving objects requires the additional ability of being able to predict the future. To catch a flying ball, we need to be able to see it approaching us and to feel where our hand is currently located. The visual system must determine the exact location in space and time of the ball, along with its change in those dimensions. Then, the sensorimotor system needs to figure out where the ball will be in the future, determine where the hand is now and move the hand to the proper location to intercept the ball.

8.3.2.3 Guidelines

When using pointing as a task, we need to think about several issues.

Timing. How much time do participants have to point once the stimulus has been displayed? Can they take as long as they like or should they respond more or less instantly? The first alternative encourages cognitive strategies to play a role, the second tries to prevent them. Naturally, one can get very large differences, not only in accuracy but also in the general trend of behavior. Some actions are fundamentally changed—and not always for the better—once we have a chance to stop and think.

Feedback. When do people receive information about where their hand is? There are three options. The first is an open loop. Here, the participants cannot see their hand and might not receive information about how well they did on the pointing task. This is useful in the pre- and post-test phases of a negative aftereffect experiment. Note that simple ballistic actions such as pointing can be done without visual feedback, while more precise actions such as grasping cannot. Eliminating feedback will not only prevent people from correcting inadvertent mistakes in the action, but will also prevent them from learning new sensorimotor mappings. The second option is terminal feedback: during the pointing or reaching action, participants cannot see their hand. After each trial, however, they can see their hand, including feedback about how well they performed on the task. Finally, there is full feedback or a fully closed-loop task. Each of the three options has advantages and disadvantages.

8.4 Conclusions

Many of the research questions that will interest us will involve real-world actions, such as driving, pointing, and making tea. In these cases, the most direct way to answer the question is to have participants actually perform the action we are interested in. Doing so, however, requires considerable care to make sure not only that the research environment is as close as possible to the real world (one of the most common criticisms of VR research is that VR is not close enough to the real world for the results to be informative of behavior in the real world) but also that every element and factor that might affect the task is controlled for. Finally, real-world tasks are not always appropriate. Some research questions will involve subjective phenomena, such as color perception or risk assessment. Even the research questions, however, will need to be phrased so that the participant actually performs some task during the experiment. It is worth trying to find a task that fits a real-world behavior.

Chapter 9

Physiology

In the chapters up to this point, we have so far assumed the basic experimental setup of traditional "open-loop" monitor psychophysics: a participant sits in front of a monitor, stimuli are presented on this monitor according to an experimental protocol, and the participant's responses are recorded by means of a keyboard, mouse, or other input device. The experimental protocols discussed in Chapters 4 through 7 try to ensure that the experimental question at hand is measured as precisely and objectively as possible.

For many types of questions, monitor psychophysics is all that is needed. There are contexts, however, in which this setup is not sufficient. If one wishes to investigate the effectiveness of computer-generated facial animations, for example, one can measure how well the animated facial expressions are recognized. Recognizability is not the only dimension of effectiveness, however. In general, one might be interested in how well observers *believe* the expressions—a believable avatar will be much more effective in communication than an avatar whose statements are recognizable but are consistently judged as fake or posed. To measure believability is not an easy task when using psychophysical methods, as too many factors will influence the results. Indeed, this is true for any higher-level concept that goes beyond clearly objectifiable absence or presence of a signal. Nevertheless, there are several kinds of measures one could imagine in such a situation: questionnaires about different perceived qualities of the avatar (direct measurement) or maybe about a story it told (indirect measurement), simple forced-choice scenarios in which two or more different avatars are pitted against each other, and many more.

In a way, however, all of these measures are still more indirect. To illustrate this point, imagine an audience listening to an exciting story told by a master-storyteller. Listening to this performance will undoubtedly not only engage the attention and other, cognitive processes in the audience, but it will also cause bodily changes: listeners will most likely adopt a body posture that signals their attention and will focus their gaze on the storyteller, sad stories might cause some people to cry, people might hold their breath with hearts beating faster and palms starting to sweat at a particularly suspenseful point in the story, etc.

These bodily reactions can be recorded and analyzed in addition to the standard experimental measures that were mentioned above. The analysis will then not only pay attention to perceptual and cognitive measures that are to various degrees still amenable to subjective influences, but it will also take into account underlying and—one hopes—less cognitively penetrable physiological changes.

Another, more basic context in which monitor psychophysics is not sufficient, is if one wants to probe deeper. In the early 1980s David Marr wrote a now-classic book called *Vision* that was built on the premise that vision should be understood and analyzed as an inherently *computational* process (Marr, 1982). Consequently, he posited that the study of how vision works could be approached at the three levels listed below.

- *Functional level.* What kinds of problem must the organism solve, and why?
- *Algorithmic level.* How does the organism solve these problems? What kinds of computations does it use?
- *Implementational level.* How are the computations realized in "hardware" (for a biological organism, this would mean the anatomical structure of the brain with its neural networks)?

The aforementioned monitor psychophysics treats the participant as a blackbox with properties that can only ever reveal the functional and the algorithmic level. If one wants to know about the implementational level as well, one needs to open the blackbox further by looking into the physiological functions themselves. Here, the scope ranges from arousal of the whole body to selective activation of certain brain areas down to recording of single neurons in the brain. Although the algorithmic level and especially the functional level should be independent of the implementational level,[1] when taking the human as the subject of study, the detailed investigation of the physiology can, for example, provide important constraints about what kinds of algorithms and functions are possible at all.

This chapter is going to provide an overview of the most common techniques in measuring changes in the participants' bodies in response to stimuli. This includes a number of measurements, ranging from heart rate to brainwaves. A brief description and common features of several of the most widely-used measures are included in this chapter. A list of these follows:

- Psychophysiological measurements (pulse measurement, galvanic skin response (GSR), and pupil dilation)

[1]Some philosophers—perhaps most prominently John Searle—would probably object to this assumption. Here, we chose to follow Marr's proposition.

- Eye tracking
- Electroencephalography (EEG)
- Functional magnetic resonance imaging (fMRI)

9.1 Psychophysiological Measurements

Psychophysiology is a research area that is concerned with measuring bodily changes in response to psychological states. Thus, as described in the introduction to this chapter, if one listens to a suspenseful story, the suspense felt will induce changes in the heart rate (as the heart beats faster to secure ample blood supply in case of fight-or-flight reactions), the pupil diameter (the diameter of the pupil is controlled by the nervous system in response to states of arousal), and the skin conductance (as the person sweats more, the resistance to an electric current across the skin is lowered). In addition, there are of course a host of other potentially measurable changes that occur in the body (for example, hormonal secretion, blood pressure, heart beat frequency, brain wave patterns, etc.). Here we have chosen to focus on pupillometry and the galvanic skin response, as these can be integrated relatively easily into existing behavioral experiments and are relatively accessible. In addition, both methods are minimally invasive and are known to provide stable markers of psychophysiological changes. For other measures used in psychophysiology, the interested reader is referred to the excellent book by Cacioppo et al. (2007).

9.1.1 Pupillometry

As the maxim "The eyes are the windows to the soul" illustrates, eyes are seen as conveying important messages about a person. Here, we are concerned with what the eyes can tell about mental states of a person in general and their emotions in particular. More specifically, we will focus on the dynamics of the pupil, which among other things is known to contract and expand as a function of emotion processing.

9.1.1.1 What It Measures

Pupillometry measures the degree of dilation and contraction of the pupil over time.

9.1.1.2 Measurement Techniques

There are special devices for measuring pupil dilation and contraction (pupillometers), although for most practical purposes in the context of perceptual and cognitive experiments, video-based eye trackers will be the method of choice (see Section 9.2 for a description of typical eye-tracking setups). As the pupil diameter varies quite a lot on its own, it is important to establish both a solid baseline in a well-rested participant and to use many repeated measurements of similar stimuli for later averaging.

9.1.1.3 Typical Questions

- Do people react emotionally to a stimulus?
- How much information processing is being triggered by a stimulus?

9.1.1.4 Data Analysis

It is important to filter the data right after the blink as the lid-close reflex causes a contraction followed by a dilation in the pupil which is independent of the stimulus. Similarly, data collected right after stimulus onset and stimulus offset, which both usually change the brightness level on the screen needs to be discarded. As the interesting response takes around 0.2–1.0 s after stimulus onset, an event-related analysis needs to be made based on careful stimulus timing, that is, changes in pupil size need to be correlated with stimulus onset. Usually, data needs to be averaged over multiple trials and compared against a baseline condition. Any changes (in the positive or negative direction) compared to this baseline constitute the effect of the stimulus on the pupil size.

9.1.1.5 Background

The pupil is the "camera aperture" of the eye, an opening in the iris through which light enters the eye, hits the lens, and is refracted onto the retina. Two sets of muscles set around the pupil control pupil diameter, most importantly to control the amount of light that passes through, making the pupil contract in bright light conditions and dilate in low light conditions. The overall variation in pupil diameter is from around 1.5–8.5 mm and takes a minimum of 0.2 s to occur, with usual response peaks at around 0.5–1.0 s. There are two sets of muscles that control pupil diameter. The radial muscles are used to dilate the pupil. They are controlled by the sympathetic part of the nervous system with signals originating in the hypothalamus. The sphincter muscles are used to contract the pupil. They are controlled by the parasympathetic part of the nervous system with signals originating in a structure in the midbrain. One potential explanation for this particular innervation is that in dangerous situations the sympathetic nervous system prepares the body for "fight or flight," thus causing the pupils to dilate to maximize the amount of light and therefore to maximize the amount of visual information for further processing. It is this particular innervation that makes pupil diameter an interesting measure beyond its use in controlling the amount of light that reaches the retina.

Physiologists had already observed the control of pupil diameter in patients as early as the eighteenth century, with Claude Bernard adding more data on how the sympathetic nervous system caused the pupil to dilate and contract.

In his book *The Expression of Emotions in Man and Animals*, Charles Darwin was cautious as to the exact causes of pupil dilation, although his treatment of physiological response to emotions is very concise in many respects. However, later research has shown that the pupils dilate, for example, as part of the startle reflex after a sudden, strong outside stimulus. This dilation persists even when light is being shone onto the pupil, indicating that the startle reflex is able to override the usual light-controlling response for a short while. In line with this, the pupil also was found to dilate in response to pain (Chapman et al., 1999).

Measurement of pupil diameter was popular in the 1960s and 1970s, and the psychologist Eckhard Hess was the most influential figure in establishing pupillometry for use in perceptual and cognitive experiments. Hess and co-workers found in early studies, for example, that pupil dilation was influenced by picture interest (pupils of female observers dilated when they viewed male bodies and vice versa Hess and Polt, 1960). Although many studies focused on the effect of emotional stimuli on pupil diameter, results remained ambiguous: initially it was suggested that variation in pupil size was correlated with emotional valence, that is, that dilation corresponded to positive stimuli and constriction to negative stimuli (Hess, 1972). Other studies indicated that pupil diameter was related to the emotional intensity of the stimuli (Janisse, 1973) or simply to their novelty (Aboyoun and Dabbs, 1998). As the pupil diameter critically depends on stimulus properties and well-controlled environments, however, results need to be viewed with caution. In a recent study by Partala and Surakka (2003), pupil size variation was therefore measured carefully in response to previously calibrated and annotated, affective auditory stimuli. This study clearly showed that the pupil dilated significantly after presentation of *both* positive and negative sounds, indicating its usefulness as a measure of affective processing in general.

The other large body of work in pupillometry relates to using pupil size variation as a measure of processing load with the pupil dilating in a variety of tasks ranging from perceptual (pitch discrimination Kahneman and Beatty, 1966) to cognitive (mental multiplication Hess and Polt, 1960) to memory (digit recall Peavler, 1974). As such, pupil dilation can be seen as a measure of information processing both in terms of capacity and difficulty (Kahneman and Beatty, 1966; Beatty, 1982). Interestingly, a recent study on memory-recall performance of neutral and emotional terms found that not only did the pupil dilate more to correct responses ("hits"), but that this dilation was much reduced for emotional words, indicating that memory and emotional processing might interact in this measure (Vo et al., 2008).

9.1.1.6 Guidelines

The strongest factor that influences control of the pupil diameter is changes in brightness level. Therefore, stimulus brightness levels should be normalized in order to rule out systematic biases. This is not an issue of course, for non-visual stimulus presentation. As pupil diameter decreases with fatigue, care should be taken to control for fatigue levels (both at the start of the experiment, but also throughout the experiment) with well-rested, fully awake individuals being the best choice. As with accommodation, the flexibility of the two muscles controlling the pupil also decreases with age, leading in turn to greater variability with increasing age, which speaks in favor of within-participant designs when measuring pupil diameter. As with any experiment involving prolonged exposure to many potentially similar stimuli, care should be taken to avoid habituation, which will decrease the response to stimuli overall.

9.1.2 Galvanic Skin Response (GSR)

Besides facial expressions and body postures, changes in the skin are one of the most visible communicators of internal mental and emotions states. People can blush in response to a shameful feeling, and frightening or suspenseful situations can cause a host of changes ranging from goose bumps to turning white as blood rushes from the face to breaking into a sweat. Especially the latter reaction is usually the basis for measuring arousal, as sweating changes the resistance of the skin to conducting currents. This principle is called *galvanic skin response* (GSR, sometimes also called "electrodermal activity" (EDA)) and has found widespread use from basic research to medical and clinical applications. It has also been (ab)used in so-called lie detectors, although its signals are far from easy to measure and rarely correlate with the truthfulness of a statement.

9.1.2.1 What It Measures

GSR systems measure the conductance of the skin as a correlate of arousal.

9.1.2.2 Measurement Techniques

Usually, two electrodes are fastened to the non-dominant hand (either to the tips of two fingers, or to two points on the palm). Contact gel can be used to increase the electric contact between the skin and the electrodes. A small voltage is then applied between the two electrodes, and an amplifier measures the change in current that flows through the closed circuit as the skin conductance changes. Given that conductance is measured in siemens (with the unit symbol of "S," which equals the inverse of resistance $S = \frac{1}{\Omega}$), typical outputs of skin-conductance measurements yield values of μS.

9.1.2.3 Data Analysis

Usually, the data is split into two components: the slowly changing (tonic) skin-conductance level (SCL), and the rapidly changing (phasic) skin-conductance response (SCR). The former changes over the course of minutes and has levels of several μS, whereas the latter has short response latencies (typically 1–4 seconds after stimulus presentation) and has levels of $0.1 \simeq 1.0$ μS. As SCRs can also occur spontaneously, one needs to filter the experimental data to extract the specific SCRs that are correlated with stimulus presentations. Usually, changes in magnitude of both SCL and SCR components of the signal are recorded to assess the effect of experimental manipulations on the arousal state of the participant. Absolute SCLs vary from individual to individual, so when comparing two groups of individuals, normalizing the SCL (and SCR) to relative units can be beneficial. This can be done in a prior calibration phase, in which both components are measured at rest and during a well-defined exposure to suitable stimuli (for example, short, loud tones in the case of SCR calibration).

9.1.2.4 Typical Questions

- Do people react emotionally to a stimulus?
- Does a stimulus evoke sudden or slow changes in arousal?
- Does reaction to the stimulus depend on the sympathetic nervous system?

9.1.2.5 Background

Already in the nineteenth century, skin conductance was used as one of the first psychophysiological measurement methods. One of the more anecdotal, early examples of GSR research comes from the Swiss psychologist C. G. Jung, who used a galvanometer to measure changes in skin conductance in response to emotionally charged words that were read to the patient. Any words which produced a higher than usual reading on the galvanometer were thought to be linked to the underlying psychological problems of the patient. As the device did not use an amplifier, however, measurements proved to be imprecise. In 1888, the French physiologist Fere published a study in which he reported that when applying a weak current across two electrodes on the skin surface, sudden decreases in the skin resistance could be observed in response to different sensory and emotional stimuli. He concluded that for a short time after a (strong) sensory stimulus, the skin becomes a better conductor, which established the core principle behind GSR measurements. As soon as devices with amplifiers became available, the GSR was used as the basis of "lie detectors" (indeed, the so-called E-meter used by the Scientologists is basically a GSR device). Despite several decades of research on this application domain, results have proven to be mixed: GSR performance has been found to vary significantly among individuals, and to be influenced by many internal and external causes, such as habituation to stimuli, temperature changes in the environment, movement artifacts, etc.

Despite its widespread use, the neural and anatomical mechanisms behind the emergence of skin conductance have only recently begun to be fully understood. It seems that the skin's electrical resistance stems from the outermost of the three layers of the epidermis, which is a relatively good insulator compared to the other layers. This high resistance means that when voltage is applied to this layer, only a small current flows through it. This resistance can be modified by the presence of conducting materials in the skin's layers, such as when salty sweat is excreted from the pores and begins to rise to the surface. This decrease in resistance (or, more appropriately for modern recording devices, this increase in skin conductance) is then measured by the GSR device.

It is commonly assumed that the GSR correlates with arousal or strong emotional states, as these states lead to increased perspiration on the palms, the soles, and the forehead (as opposed to temperature-controlled perspiration, which affects the whole body). This particular kind of perspiration is mostly controlled by the sympathetic part of the autonomous nervous system, which prepares the body for a fight-or-flight situation in states of high arousal. This specific control makes the GSR response a good candidate if one is interested in sympathetic body responses (as opposed to examining, for example, heart rate, which is controlled by both sympathetic and parasympathetic functions).

9.1.2.6 Use in Perceptual Research

Due to its strong connection to the sympathetic nervous system's arousal-signaling, the GSR has been used mostly in connection with strong, emotional stimuli in research. The GSR has been instrumental in the research of emotions and facial expressions, with early studies in the 1950s showing how GSR was able to encode the extent of activation present in facial expressions (Schlosberg, 1954). In a classic study, responses to conditioning or habituation with respect to emotional stimuli were investigated: when shown frightening pictures of snakes and spiders that were paired with a light electric shock as a conditioning stimulus, the GSR of participants was found to habituate much slower than when they were shown non-frightening pictures (such as houses, flowers, landscapes, etc. Fredrikson and Öhman, 1979). This was seen as a confirmation of the long-lasting arousal potential of the frightening pictures.

In a series of studies, GSR has been used as an autonomic, covert marker marker for highlighting significant stimuli. In the first study (Tranel et al., 1985), it was established that SCRs were much higher for highly familiar than for unfamiliar faces. This is interesting, as it represents a covert response to the faces in that participants merely had to look at the faces without having been instructed to do a task, that is, they did not display an overt reaction to the stimuli. In a follow-up experiment (Tranel and Damasio, 1985), this response was found to be still present for prosopagnosic patients who were not able to consciously recognize the (previously familiar) faces, but whose other sensory channels seemed to be still able to process the stimuli (see also Bauer, 1984; Tranel et al., 1995). These findings (among others) have led to influential theories that stress the importance of covert subconscious and emotional processing for our overt reasoning (Damasio et al., 1998).

Nowadays, skin-conductance measurements are rarely used on their own, but rather as a secondary, robust measurement that controls for, or indicates the state of, arousal in participants. Two recent exceptions are the studies of Khalfa et al. (2002) and Armel and Ramachandran (2003). In the first study (Khalfa et al., 2002), participants were exposed to different musical stimuli intended to elicit either fearful, happy, sad, or peaceful emotions. Interestingly, skin-conductance responses correlated rather well with the general level of emotional intensity, in that responses were larger for the more intense emotional music pieces (that featured fear and happiness) than for the comparatively weak pieces (that featured sadness and peacefulness). This finding shows that the skin-conductance responses can be selectively modulated also by the strength and emotional intensity of musical stimuli. The second study (Armel and Ramachandran, 2003) used GSR methods to establish the now-famous "rubber-hand" illusion. In such a setup, a participant holds an arm underneath a table. Roughly aligned with where their arm would be normally, a rubber hand is placed. The experimenter now repeatedly stimulates (strokes) both the rubber hand and the participant's hidden hand. After a while, a sudden touch of the rubber hand alone will elicit a reaction from the participant, as measured by both subjective accounts and, most importantly, by strong changes in skin conductance. Interestingly, the study showed that the change in GSR was significantly less when the hand remained in view, or when the hands were stroked asynchronously, indicating that it was the body-image of the participants that was affected by the illusion.

In a combined fMRI/GSR study by Bartels and Zeki (2000), participants who were deeply in love viewed pictures of their partners in order to investigate the potential brain regions involved in romantic love. Interestingly, in the control study involving GSR, the responses to pictures of loved ones were still significantly higher than responses to pictures of friends—despite the fact that the skin-conductance response is unspecific with regard to the valence of the stimuli. A good example of using GSR as a direct indicator of arousal was given in Williams et al. (2001), in which the GSR was recorded simultaneously with an fMRI scanner while fearful and neutral faces where shown. Whenever there was a significant skin-conductance response, the experimenters knew that the autonomic system had reacted to the face, and could therefore target their neurophysiological investigation specifically towards these instances. Indeed, they found a clear dissociation between amygdala-related activity (that is, deeply emotional reactions) versus hippocampus-related activity (that is, fact-related processing) in the brain depending on the GSR signal.

9.1.2.7 Guidelines

The skin-conductance response is prone to a number of limitations that one needs to be aware of; these are listed below.

- It only measures arousal, and generally lacks distinction possibilities along other, important emotional dimensions, such as valence.
- The response varies across individuals, with around 10–20% of the population exhibiting no useful signal at all. The individual variation therefore requires calibration trials before the experiment to check for presence, strength, and baseline of the signal.
- Skin conductance is usually measured on the hand and can therefore be easily contaminated by movement artifacts, which need to be controlled for in the experimental setup. In addition, care should be taken that room temperature and humidity are kept constant in order to avoid changes in body temperature and thus in sweating.

9.2 Eye Tracking

What can eye movements tell us about the processes and thoughts of the observer? In what has become the most reproduced image in the context of eye movement studies, Alfred Yarbus analyzed the scan paths of an observer viewing I. E. Repin's painting entitled *They Did Not Expect Him* (Yarbus, 1967). What he found is that observers were focusing their eyes on specific parts of the painting (such as persons, important objects, etc.) rather than randomly scanning the image. In addition, he found that the scan path, i.e., the trajectories of eye movements, critically depended on the task assigned to participants or context with which the image was viewed: if observers were asked to look for people, their gaze focused only on people, whereas if they were asked to simply look at the painting, many more fixations in other areas occurred. Ever since this study, researchers have used eye tracking to investigate what kind of visual information is focused on, depending on the task.

9.2.1 Elements

9.2.1.1 What It Measures

Depending on the type of eye tracker, the signal that is measured is the rotation of one or both eyes relative to a reference system. Most eye trackers only measure rotations on the vertical (looking left or right) and horizontal (looking up or down) axes, although some can also measure the torsion of the eyes, that is, rotations around the axis going through the eye (these eye movements occur, for example, when the whole body is rotated around this axis). In addition, some eye trackers can also measure pupil dilation.

9.2.1.2 Measurement Techniques

There are three broad categories of eye tracker systems.

- *Magnetic search coils.* A small coil is fixed into a contact lens worn by the observer. In the most simple setup, two perpendicular, oscillating magnetic fields induce a changing current in the search coil as the eye rotates. The rotation of the eye around two axes can be deduced from the change in current.

- *Electrooculogram (EOG).* The eye has a certain resting potential, which can be modeled as a simple dipole with its main axis oriented from the retina towards the pupil. Two electrodes are then fixed either above and below, or to the left and right, of the eye. As the eye moves, the dipole will induce a potential difference in the two electrodes, which can be converted to a rotation signal.

- *Video-based techniques.* The most common technique is to use an infrared light source that creates several reflections in the eye. By using one or two of these brightly visible reflections together with the position of the (dark) pupil, the eye rotation can be determined. Other systems also use two cameras to triangulate the position of the eye in space, which are based on detecting the image structure of the pupil and the eye in the two images. These systems can also be used to measure pupil dilation and constriction (see pupillometry in Section 9.1.1).

For all setups, if one is interested in gaze data (such as which parts of a stimulus observers fixated on), the data needs to be converted into the stimulus reference-frame (for example, the x, y coordinates in an image). In case of a standard monitor setup the first step of any eye-tracking experiment therefore consists of a calibration phase, in which the observer fixates on a defined set of points that span a monitor's screen area. From this set of fixations, the eye-tracking software can then determine which eye positions correspond to which screen coordinates. In addition, most eye-tracking systems can track only the eye positions, meaning that movements of the head made during data acquisition will result in erroneous data. For this reason, in most experiments involving eye trackers, the observer's head is fixed in place for the duration of the experiment (this also ensures a defined distance to the monitor and therefore controls for the viewing angle of the stimulus presentation). This is usually done with a chin rest, or in cases requiring even higher accuracy, with a so-called bite bar, which the observer bites into so that his or her head remains still.

9.2.1.3 Data Analysis

The raw eye-tracking data first is filtered to detect blinks and also sometimes smoothed to filter out noise. In the second step, algorithms try to detect saccades and fixations in the data based using eye-movement statistics. This can be done by looking at the velocity profile of the eye movements, or by determining the points in time at which eye gaze remained in a small visual angle. In the final step, fixation clusters (where people look) and dwell-time statistics (how long people look) can be determined and correlated with properties of the stimulus.

9.2.1.4 Typical Questions

The most common experimental question in which eye tracking is useful is, what part of the stimulus is important? More specifically, one might ask,

- What part of the stimulus generates many fixations or long dwell times?
- In what order is the stimulus scanned?
- How much does the viewing pattern change when the observer is given a different task?

9.2.1.5 Background

Our eyes are constantly in motion, from large saccadic movements that cover a large visual distance in a few milliseconds to microsaccadic movements that introduce a minute jitter. One of the main reasons for eye movements is to gain information about a specific region of interest in the environment. As the amount of information that can be processed by the brain is limited, evolution has devised a trick for the brain to save bandwidth: the spatially varying resolution of the retina. The retina contains two types of light detectors: *cones*, which come in three kinds (sensitive to red, green, and blue wavelengths) and which process color information at bright light levels, and *rods*, which process monochromatic information at low light levels. The density of the cones across the retina, however, is not constant; at the periphery the density is very low, while the center of the retina contains the highest density of photoreceptors (incidentally, this density is almost optimally matched to exploit the full optical resolution of the eye's lens system). This point of highest resolution on the retina is called the fovea, and this is what moves around the scene in order to target an object of interest with the maximal amount of spatial resolution. This is illustrated in Figure 9.1, which shows a rough reconstruction of what the retinal representation of two different fixations on a face would look like. Fixation 1 falls between the eyes, and the resulting reconstruction also has this region at a high resolution, with other parts quickly falling off in resolution. Fixation 2 is on the mouth, accordingly highlighting this region at high resolution. These figures were created by a so-called log-polar mapping, that is by a grid of sensors that are spaced densely with small receptive fields in the middle, and are spaced more widely apart with larger receptive fields on the periphery. The

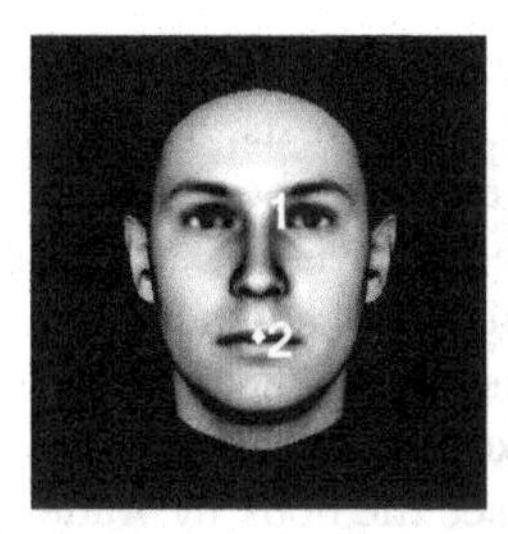

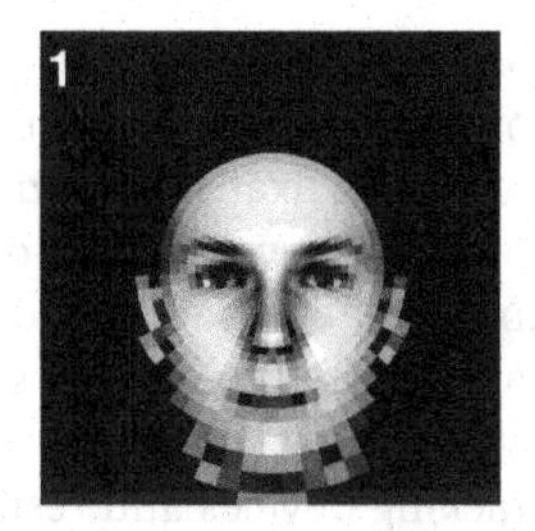

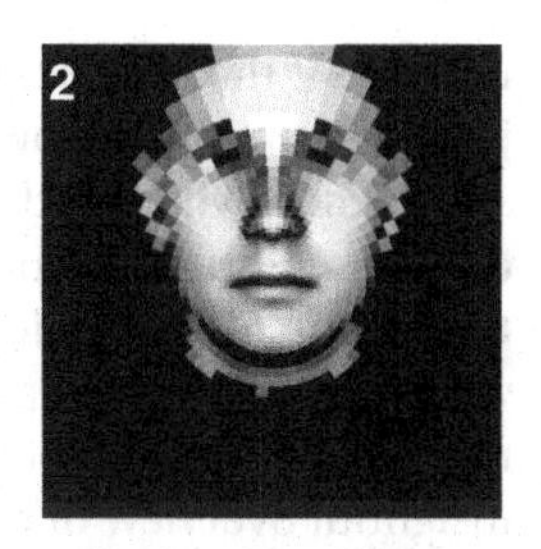

Figure 9.1. Left: A face with two fixation locations: fixation 1 is between the eyes, and fixation 2 is on the mouth. The middle and right images show a simulated reconstruction of the retinal representation of the two fixations. Note the spatially varying resolution in the center, that is, in the fovea.

size of the receptive fields decreases logarithmically from the periphery to the center, and sensors lie on concentric circles.

Our visual system integrates this constantly changing, high-resolution spotlight surrounded by low-resolution peripheral data into one, coherent, seemingly high-resolution mental representation. The fact that we have high resolution only where we fixate can be exploited by so-called gaze-contingent displays, which couple an eye-tracker with visualization software that displays the fixation location with large details, thereby saving precious rendering time for complex environments or large-scale visualization data. In the context of this chapter, however, we are interested in exactly where the fixations land—as these fixations are far from random in terms of their image content, presumably, they will be able to tell us something of their importance for perceptual and cognitive processing.

When viewing a typical natural scene, the human eye makes around 2–3 saccades per second with an average fixation duration of 250 ms and a spacing between fixations of around 7 visual degrees (Kienzle et al., 2009). The first saccade is usually initiated around 100 ms after stimulus onset (note that these first 100 ms seems to be enough time for the visual system to form a rough concept of the viewed scene—the so-called scene gist). A common observation is that during saccades, visual memory is almost not present, that is, during eye movements, visual information will not be processed, a phenomenon called *saccadic suppression* (Matin, 1974). In addition, fixations will always have a minimum spacing—although sometimes corrective saccades are made within a close distance, the same location is rarely fixated on twice within a given (short) time span.

9.2.1.6 Use in Perceptual Research

Investigation into the characteristics of eye movements seems to have begun with experiments on reading by Louis Laval in 1879 (cited in Huey, 1908) who noted that the eyes do not proceed smoothly over the text, but rather execute jerky movements followed by lasting pauses in which the gaze rests. The desire to know what words the eyes had fixated on and what parts of the sentence drew more attention prompted the development of the first eye-tracker systems. One of the earliest (invasive) eye trackers consisted of a "contact lens" with a hole for the pupil that was fixed to a small pointer with which the eye movements could be observed ((Huey, 1908); for an excellent historical and insightful overview of eye-tracking devices and results, see the book by Wade and Tatler (2005)). The first investigation into visual processing using a non-invasive eye tracker was conducted in the early 1930s (Buswell, 1935). The apparatus shone light onto the cornea by means of mirrors. The reflected light of the cornea was focused onto film that moved horizontally and vertically by means of focusing lenses and wedge prisms. As the beam of light was interrupted by a moving fan 30 times each second, sharp dots appeared on the moving films that allowed Buswell to determine the position of fixations accurately. In one of the studies, participants viewed photographs of various types of artworks and the resulting scan paths were analyzed (Buswell, 1935). Interestingly, the study found no differences in fixation patterns between trained and untrained artists. The study also found that no two scan paths were exactly identical, thus demonstrating individual differences in gaze behavior, although global patterns could be derived that defined two overall search strategies of global and local scan paths. Participants also tended to focus on high-contrast regions in the foreground (including faces and people).

In perhaps the most well-known study, Yarbus analyzed the scan paths of an observer viewing the painting *They Did Not Expect Him* (Yarbus, 1967). One of the core findings was that eye movement patterns changed dramatically depending on viewing instructions: for each type of instruction, Yarbus found that the most informative regions of the image were scanned and processed. Similarly, in experiments by Molnar (1981), participants who were told that they were going to be asked about the aesthetic qualities of a painting had longer fixations than participants who were going to asked about the contents. Recent experiments on the role of task context on eye movements have found that fixations fall primarily on task-relevant objects—studies of fixations made during the making of a cup of tea, for example, revealed that during pouring of the tea, participants fixated on the spout of the teapot (Land et al., 1999).

It seems, then, that fixation locations are driven by two types of information: bottom-up, low-level features (such as color, luminance edges, and high-contrast regions) and top-down, higher-level information (such as task influence). The interpretation of fixation locations as loci of attention is even more complicated by the fact that there are also two types of attention: overt attention, or directing our eyes (and the head) towards an object of interest, and covert attention, or the indirect, mental focus on an object of interest. These two processes can be illustrated by the observation that we can look to the right while mentally focusing on the left side of the visual field. In order to get the maximum out of eye-tracking experiments, covert attention should be controlled as much as possible—for example, through the use of short presentation times, or through explicit instructions.

Physical properties like high luminance-contrast and high edge-density (Parkhurst and Niebur, 2004; Einhäuser and König, 2003; Mannan et al., 1996; Tatler et al., 2005) have been found to be prominent target locations for eye fixations. This research has led to models predicting human fixation patterns (Itti and Koch, 2001; Kienzle et al., 2009). Fixations driven by physical properties of local features are usually being observed only for the initial fixations on an image. As the presentation time increases further, however, most existing models gradually fail to predict fixation locations (Itti, 2005), as higher-level processes of image interpretation set in.

9.2.1.7 Guidelines

The underlying assumption behind many eye-tracking studies is that the spatio-temporal pattern of fixations on the stimulus is more or less equivalent to the information flow processed by the brain. This is not always true, however—we can perceive quite a bit of structure in a scene that is rapidly flashed even before the eye is able to initiate a saccade. A series of scene categorization experiments (starting with the one reported in Thorpe et al., 1996) has found, for example, that humans are able to tell correctly which of two presented images contains an animal in as little as 100 ms after stimulus onset by making a quick voluntary eye movement to the animal (Kirchner and Thorpe, 2006). Note that this eye movement is made only *after* the information in the two images has been processed by the brain.

In addition, the famous study by Yarbus already showed that eye movements are critically influenced by the (visual) task at hand. With that in mind, it is always a good idea to split the analysis of the eye-tracking data into the first (or first few) fixations, and the later fixations. If possible, several different visual tasks should also be run to cross-validate these first fixations and to provide a better measure of the inter- and intra-individual variance in fixation patterns for both early and late fixations. If one particular task is of interest (such as, for example, efficient navigation through a complex webpage, or fast detection of a certain stimulus property), explicit instructions might help to reduce variance.

9.2.1.8 Example

Figure 9.2 shows data from an eye-tracking experiment on how people look at paintings (Wallraven et al., 2009). More specifically, participants, whose eye movements were being recorded, viewed paintings and judged either their visual complexity or their aesthetic appeal. The study found that there were some differences between those two tasks: although both tasks were expected

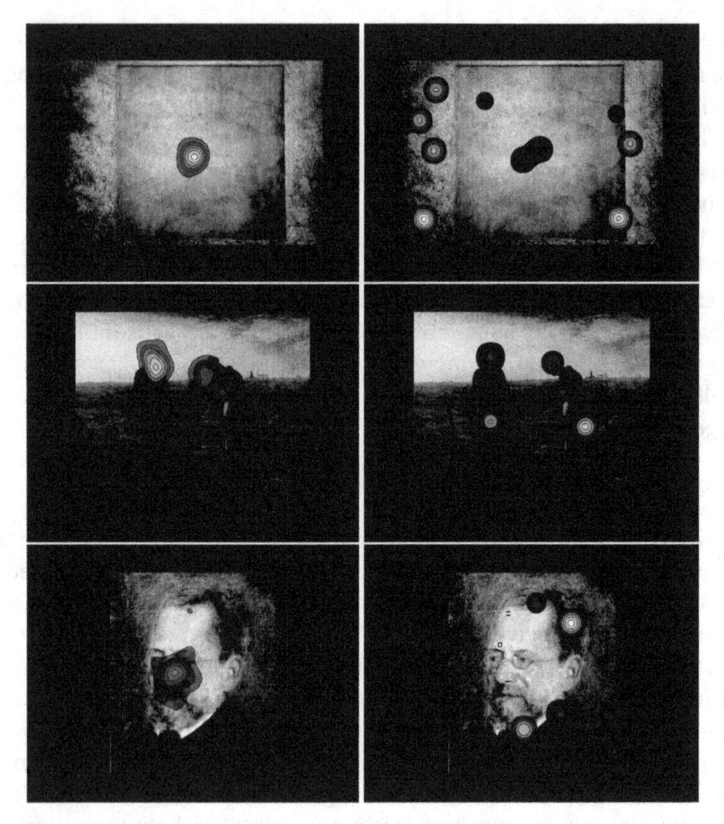

Figure 9.2. Comparison of human eye-tracking patterns (left column) and computational predictions (right column) for three types of images.

to highlight global-processing strategies, the gaze of participants in the aesthetics task wandered farther on the image and took in more details. In addition, processing of the meaning of images was extremely quick, such that by the second fixation, viewers had often already focused on a critical part of an image, such as the visual jokes played by Magritte or crucial figures in the painting. These results confirm that the task plays an important role in defining eye-movement patterns. In addition, visual processing of complex scenes is astonishingly fast, attesting to the speed with which the brain is able to decode and analyze visual input.

Another goal of the study was to compare the human eye-movement data with the predictions of a computational model. The right column of Figure 9.2 shows computational predictions from the well-known Itti and Koch model of visual saliency (Itti and Koch, 2001). The model is based on three different kinds of saliency maps that were computed from an image. Images were first resampled into a multi-resolution model. Each image was then separated into an intensity channel and two color-opponent channels for red-green and blue-yellow contrast, respectively. The intensity image is then filtered to obtain gradient orientation maps at four orientations. Information across two resolutions is then integrated into so-called feature maps. Finally, these feature maps are summed over all resolutions and integrated into one saliency map. This computational model is based on findings from low-level physiology and psychophysics and seems like a good candidate to model bottom-up saliency (Itti and Koch, 2001; Walther and Koch, 2006).

Comparing the two fixation maps in the first row of Figure 9.2, one can see that the computational model picks out the center part of the image, which contains a hole—a rather salient part, which is also fixated by the human observers. The remaining fixations, however, seem strewn all over the image, whereas the human observers stay focused on that one interesting part of the image. Moving on to the second row, we see that the computational model does an excellent job of predicting salient fixations on the two heads of the farmers. However, the reason why these parts are salient for the computational model is because they have a high contrast in relation to the bright background—*not* because they are faces. This is demonstrated in the third row, which compares fixation patterns for a portrait. The computational model highlights many different high-contrast points, none of which come even close to the usual, eye-centered fixation pattern that faces elicit for human observers. In summary, while certain fixations can, indeed, be explained by low-level image properties (and these also highlight important cues that artists use to guide our gaze around the canvas), the vast majority of fixations seems to be guided very early on by scene-interpretation. The computer will need to learn this skill before it can proceed to become an art critic!

9.3 Electroencephalography (EEG)

Ever since the beginning of physiological investigations into the structure of the brain's processing units, the electrical properties of neurons have been under close scrutiny. Early investigations into the electrical activity of the whole brain, however, were performed already in the late nineteenth century, with the first recording of electrical activations of a mammalian brain being published in Russia in 1913. In the 1920s Hans Berger, a German physiologist, published his first electrical recordings from a human participant and gave the recording its current name: *electroencephalogram* (see, e.g., Berger, 1929). He is also credited with observing particular brain rhythms, or waves, in normal and abnormal patients, as well as the characteristic changes in EEG in epileptic patients (see Gibbs et al. (1935) for one of the first published studies). EEG was particularly important for sleep research (Pivik, 2007); in the 1950s, EEG recordings uncovered different sleep phases and in particular helped to distinguish REM (rapid eye movement) from non-REM sleep. EEG remains a popular tool in clinical psychiatry and in perceptual and cognitive neuroscience due to its high temporal resolution and—compared to fMRI—relative inexpensiveness and ease of use. In recent times, EEG has attracted more interest due to its use in biofeedback and, in particular, in brain-computer-interface applications, where a few electrodes can be used to steer a cursor (see Wolpaw et al. (1991) for one of the earliest studies and Wolpaw and McFarland (2004) for a recent one) or even allow for simple game control (at the time of writing in 2010, at least two companies have released or licensed game controllers based on EEG technology with prices of around a few hundred US dollars).

9.3.1 Elements

9.3.1.1 What It Measures

EEG measures the coordinated electrical activity of a large ensemble of neurons close to the brain surface.

9.3.1.2 Measurement Techniques

In most recording scenarios, a large number of electrodes need to be fitted to the head of the participant at specific locations. This is usually done with a skull cap that has all the electrodes fitted already and needs only to be pulled over the head. Contact gel needs to be applied to the electrodes to ensure electric contact between electrode and scalp (although so-called dry electrodes exist). Each electrode receives the combined signal of neurons in a cortical region, which is sampled at a high rate and yields time-series data measured in μV.

9.3.1.3 Data Analysis

After filtering to remove artifacts from amplification and sampling, the signal is processed depending on the application. For measurements of spontaneous brain activity, the signal is commonly split into different frequency bands that are presumed to correspond to different functional states and that characterize changes occurring at specific time periods. Another technique focuses on measuring activity related to stimulus presentations, and thus looks for specific, time-locked events at specific electrode locations. As with the analysis of SCR, this requires robust detections of peaks in signals.

9.3.1.4 Typical Questions

- In which frequency bands does activation occur and what does this tell us about the participants' arousal or alertness level?
- Is a stimulus covertly or overtly processed?

- Which brain regions are involved in stimulus processing and what is the time course of their activation?
- Can I use specific brain signals for interfacing with external devices?

9.3.1.5 Background

A single neuron generates electric activity in the form of action potentials, which travel down its axon, usually in the form of sharp spikes. The spike rate is a measure of activation for a given neuron. At the end of the axon, the signal is converted in the synapse to a neurotransmitter, which cross over to the neighboring neuron. The neurotransmitters bind to receptors in this (post-synaptic) neuron and in turn release a current. The combination of many thousands of these post-synaptic currents then causes the neuron to form an action potential when a certain threshold is passed.

An EEG electrode records this post-synaptic activity from a large ensemble of neurons in the cortex (that is, the outer, most recently evolved part of the brain). In order to generate a measurable signal, the neurons have to respond synchronously and similarly, otherwise the signal will be averaged out at the electrode.

Activity averaged over large scales in the brain manifests itself in different rhythms that have well-defined frequency bands as a result of specific, functionally defined brain activation. These frequency bands are named after Greek letters and range from low frequency, slowly changing rhythms to high-frequency, fast changing activity. The lower-frequency bands are usually related to larger-scale, synchronized activity in the brain and thus manifest as larger amplitudes in the EEG-signal, whereas the higher-frequency bands are more desynchronized and therefore have smaller amplitudes. As one might expect, slower rhythms also correlate with less activity overall, whereas faster rhythms are observed during heightened brain activity. The following list gives the names of the bands, the corresponding frequency band in which activations occur, and their assumed functions:

- *Delta band (1–4Hz).* Activation in this band is often found during sleep. In addition, during early infancy children have more activation in this band than in others.
- *Theta band (4–8Hz).* Again, mostly observed during sleep; during waking states, Theta activation signals reduced alertness levels.
- *Alpha band (8–13Hz).* Activation in this band is observed when participants are relaxed with their eyes closed. As this frequency range lies between sleeping stages and awake stages, the exact functional roles are still under debate, but seem to range from a relaxed "idling" mode to specific, task-modulated activity.

- *Beta band (13–30Hz).* This rhythm indicates increased awareness and cognitive involvement—the pre-stage to a Gamma rhythm.
- *Gamma band (36–44Hz).* The fastest rhythm—not surprisingly—indicates full attentional processing, stimulus awareness, or top-down modulation, that is, full alertness. This band most likely stems from the concerted, synchronized activities of many specific brain regions, which undergo a specific task.

In addition to the recording of *spontaneous* brain activity (such as in sleep research, or patient monitoring), there are two types of experimental techniques in use in perceptual research that measure *stimulus-related* activity: evoked potentials and event-related potentials (ERPs).

The evoked-potentials technique focuses on measuring the response to a simple physical stimulus (this can be a pulse of light, or a tone, or a touch to the finger) using a selected set of (a few) electrodes. As the signal strength of these evoked potentials is much smaller than the baseline EEG signal, averaging over multiple trials needs to be done. The second technique measures the response to more complex stimuli (such as a face, or an emotional stimulus) and seeks to investigate higher-level processing in the brain. The elicited, event-related potentials are usually referred to by their polarity (that is, whether the signal deviated into the negative (N) or positive (P) direction from the baseline) and their timing after stimulus onset. One prominent example is the N170, a robust ERP which is obtained for faces, but not for non-face objects. This degree of invariance against low-level stimulus properties is one of the core properties of ERPs. Later ERPs usually also indicate higher-level processing stages.

9.3.1.6 Use in Perceptual Research

EEG—similarly to the psychophysiological measures—has many properties that signal the alertness or awakeness of the brain, as well as the attentional demands of a situation. Indeed, EEG can be used to demonstrate that attention modulates sensory processing, that is, to indicate top-down influences. One of the most canonical examples of this is a study by Hillyard et al. (1973), in which tones (both higher-pitched target and lower-pitched distractor tones) were presented to both ears of listeners who were instructed to pay attention to only one ear. Sometimes, a higher-pitched tone was also presented in the unattended ear, in which case participants were instructed not to respond. The results showed clearly that very shortly after stimulus onset, two ERPs, the N1 and the P1, were larger for the attended ear than for the unattended ear. The times of the ERPs and their effects were seen as an indicator of attentional modulation of early auditory processing. A similar finding was made for visual processing in which attentional modulation was observed for parts of the early visual cortex (Rugg et al., 1987).

A very robust ERP—the P300—is a good indicator of novelty, or selective attention to a stimulus (see Sutton et al. (1965) for the original study on this ERP). Indeed, this particular ERP has been found to occur in oddball paradigms, independent of the eliciting modality (whether visual or auditory), and related to the subjective importance of the stimulus as well as the cognitive demands placed on the participant in the task (e.g, Cacioppo et al., 2007). Because the P300 is such a robust phenomenon and can be readily and reliably detected, it can be used, for example, in brain-computer interface applications such as cursor control (Wolpaw and McFarland, 2004) that allow spelling of words by brain power alone.

EEG has been used to indicate covert processing or recognition of stimuli, that is, EEG signals have been found that carry stimulus-specific information without the participant being aware of the fact and without any behavioral consequences (see also the section about skin-conductance response). In the first study to use EEG in this manner, a prosopagnostic patient (a patient who has serious deficits in face recognition) was looking at a series of faces which mostly consisted of unfamiliar people with the odd familiar, famous face being thrown in (again, an oddball paradigm). One of the ERPs—the P300—was consistently higher for famous faces than for unfamiliar faces, despite the patient's inability to recognize the familiar faces (Renault et al., 1989).

Finally, EEG signals have also been found to be related to memory and cognition performance—albeit in a non-linear way. As the review by Klimesch (1999) shows, it is possible to derive suitable correlates from an EEG signal, if it is split into alpha and theta bands, and if both slow (so-called tonic) and fast (so-called phasic, or event-related) response characteristics are measured and analyzed. The particular behavior of the EEG data in those two frequency bands can be used to gain insight, for example, into long-term memory performance, or the ability to take in new information. In this case, these two types of memory processes and their corresponding EEG signal can actually be linked to specific processing networks in the brain whose activation will generate the signal.

9.3.1.7 Guidelines

In most recording scenarios, a large number of electrodes need to be fitted to the participant for recording. Nowadays, this is usually done with a skull cap, which already has all the electrodes fitted and only needs to be pulled over the head. Contact gel needs to be applied to the electrodes to ensure electric contact between electrode and scalp.

EEG is sensitive to the currents generated by eye-movements, so these artifacts need to be controlled and corrected for later. In addition, many higher-level (or cognitively related) ERPs are heavily influenced by top-down expectations, or lapses of attention, which means that the experimenter needs to take care in instructing participants about the task and monitoring their performance closely.

Especially for the recording of event-related potentials, many trials need to be averaged for adequate signal power, resulting in longer experimentation times. Prolonged experimentation will increase the risk of fatigue and therefore lead to attentional effects. In addition, the EEG electrodes need to be closely fitted at all times, for longer experiments the contact points need to be checked carefully.

Finally, whereas one can buy rather cheap devices—sold in the context of brain-assisted gaming, or also for simple brain-computer interface solutions—one should be aware that a lot of these devices have very few sensors, leading to very coarse data. One potential confound for such devices might, for example, be the fact that sensors are attached to the side of head—however, without double-checking the location of the sensor, it might actually measure eye-muscle or face-muscle activation rather than brain activity. Of course, in many cases, concentrated efforts in order to control a cursor actually might lead to measurable muscle activity and can therefore also be used as a measure of attentional resources, but this might not relate to the actual brain activity at that moment (for example, simply making grimaces might also have the same effect).

9.4 Functional Magnetic Resonance Imaging (fMRI)

fMRI is a development of standard magnetic-resonance imaging techniques that were pioneered in the 1970s, with the first experiments being run in the early 1990s. It is a non-invasive technique that now has become the most widespread method for functional imaging, that is, for investigating physiological processes in the brain as they relate to changes in the metabolism. At the time of writing in 2010, a search on the publication database PubMed reveals that more than 250,000 papers include the term "fMRI" in their title or abstract. This popularity is due to the non-invasiveness of the technique (requiring, for example, no injections of radio-active substances such as in positron emission tomography (PET)), to its ready availability (most university hospitals and many other research institutes have at least one fMRI machine at their disposal), and perhaps most importantly, due to the absence of radiation exposure (such as is present in conventional tomographs based on x-rays, for example). Many fMRI studies have received widespread media coverage—resulting in sometimes wildly exaggerated claims about, for example, mind-reading, and research areas such as neuro-theology. Accordingly, many have criticized fMRI as a new kind of phrenology[2] that aims to seduce the masses with pretty pic-

[2]Phrenology was developed in the late nineteenth century and claimed to be able to derive every individual character trait from way the skull was formed.

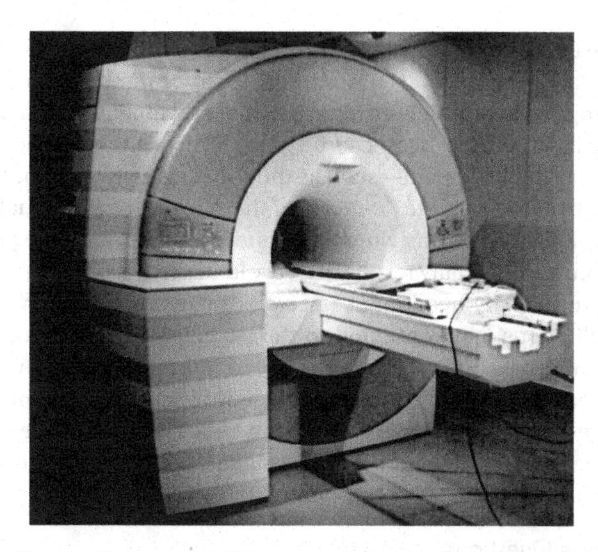

Figure 9.3. Picture of a typical fMRI scanner for research purposes. The participant lies prone on the sledge, which is then slowly driven into the main magnetic field in the bore.

tures of the brain. Nevertheless, many important contributions to understanding the functioning and structure of the brain have been made using fMRI, especially when used in combination with other perceptual and physiological measures. Therefore, as long as the user is aware of the limitations and assumptions underlying the specific techniques used in fMRI, it can be seen as a powerful tool for analysis of the brain.

9.4.1 Elements

9.4.1.1 What It Measures

fMRI measures the change in oxygenation of the blood (also called blood oxygen level dependent, or BOLD) in a specified recording volume in the brain. The BOLD response is due to the blood circulation that supplies brain areas with energy in the form of oxygen. A region that has been active has used up a lot of oxygen, resulting in a large decrease in oxygenation of the blood, which in turn changes its magnetic properties and gives rise to a change in signal that becomes the basis of the fMRI analysis.

9.4.1.2 Measurement Techniques

A participant lies prone on a sled, which is slowly drawn into the fMRI bore (see Figure 9.3). Most fMRI experiments take at least 30 minutes, and some much longer, during which the participant needs to lie still in this (narrow) bore. In order to register different people's brains for later averaging of the data across participants, an anatomical (standard MRI) scan of the brain must be made first, recording the anatomical structure of the individual brain. In addition, a second, so-called localizer scan can be made which is used to define *regions of interest* for each participant—such as motion-sensitive areas, or brain areas that are selective for faces, etc. For each of these sessions, the actual scan consists of several "slices," during which a magnet rotates around the person. The switching of currents usually results in a very loud, rhythmic noise, which often needs to be masked by headphones in experiments requiring concentration. Visual stimuli are usually presented to the participant via a mirror, although goggles that are fed by optic cables also have become popular lately. Recent developments have included fMRI compatible joysticks and other input devices, as well as eye-trackers and even EEG's. The typical output of the experiment consists of the measurement of the BOLD-signal for each voxel in every slice as a function of time.

9.4.1.3 Typical Questions

- Does a specific brain area serve a well-defined perceptual or cognitive function?
- What is the network of brain areas that are active for a given task?

9.4.1.4 Background

The full quantum-mechanical treatment of the physical principles underlying the fMRI signal is outside of the scope of this book. What follows is a basic introduction that serves to highlight the important concepts.

First of all, the physics behind the standard MR signal is based on a physical property of nucleons (protons or neutrons) called *spin*. An atom with an odd number of nucleons has a non-zero spin, which then gives the atom a magnetic moment—this is similar conceptually to the magnetic moment generated by an electric charge that is moving in a circle. Atoms with such a magnetic moment (for example, hydrogen, or the oxygen-carrying hemoglobin) will spin around and gradually start to align with an externally applied magnetic field. The time constant with which this happens is called T_1.

The frequency f_{spin} with which the atoms spin around the external magnetic field depends on its strength and the nuclear structure. If now a second, *oscillating* magnetic field is applied perpendicularly to the direction of the static field, the precession of the atoms is perturbed, when the field's oscillation frequency f is close to f_{spin} (it creates a resonance—hence magnetic resonance imaging). This perturbation in turn creates a feedback signal in the field coil that decays with another time constant T_2 after the oscillating field has been turned off.

In order to locate the signal at a certain point in the brain, a spatially varying gradient field is superimposed on the static magnetic field. This gradient causes f_{spin} to become spatially dependent as well, allowing one to locate the current signal precisely.

The important insight that led to the development of *functional* MRI came when Ogawa discovered that as the ratio of oxygenated to deoxygenated hemoglobin increased, the MR signal also increased (Ogawa et al., 1990). Soon after it was found that brain activation also affected the same signal, most likely due to the changes in blood oxygenation in the surrounding tissue (Kwong et al., 1992). In a seminal study combining multi-cell recording and fMRI, this BOLD response was found to correlate well with the changes in local field potentials, establishing a deeper link between the two signals (Logothetis et al., 2001). However, which exact processes contribute to the BOLD response remains a current field of research.

The actual change in MR signal is fairly weak, being on the order of a few percent at most. Careful statistical analyses and experimental designs must therefore be chosen in order to reliably interpret the signal. The change in signal is dependent on the strength of the static magnetic field, which is why the development in fMRI scanners has led to ever-increasing field strengths. The latest models in use for human experiments have field strengths of 7 T (tesla), with MR scanners in typical clinical applications being 1.5 T or at most 3 T. The spatial resolution of typical scans is measured in voxel size and is now on the order of 0.5 cubic-millimeters.

Finally, the strongest component of the BOLD response itself is due to an increase in bloodflow that happens only a few seconds after stimulus onset. This time period (roughly 5–6 seconds) is the reason for the limited temporal resolution of fMRI experiments, although various experimental and statistical techniques have been proposed to overcome this limitation, and real-time fMRI scanning is possible at the expense of spatial resolution.

Due to the strong magnetic field, safety precautions have to be enforced during scans (a small piece of metal can wreak havoc in such a field).

9.4.1.5 Data Analysis

There are a host of special programs that help with analyzing fMRI data—one of the most common free software packages is SPM (Friston et al., 2006), which implements a typical analysis pipeline in a Matlab toolbox. Usually, the raw data is heavily preprocessed (spatially and temporally filtered, as well as averaged) to increase the signal-to-noise ratio. Since the data arrives in different slices, these slices need to be aligned and corrected for motion of the participant during the scan. Finally, the fMRI signal in each voxel is correlated with either results from a perceptual task or compared against a control condition using a variety of statistical techniques. In addition, when interested in average results across participants, each scanned brain needs to be spatially aligned to correct for individual differences in brain anatomy. It is important to become aware of the assumptions underlying the respective statistical analyses, as the datasets contain many thousands of voxels and are therefore well outside the scope of eye-balling. There are a number of different techniques that not only can determine where and when a particular brain area was (more) active (usually based on general linear models (Friston et al., 1995), which are a kind of ANOVA, see Section 12.7.7), but also how the different brain areas are linked together; that is, a more network-oriented analysis (Friston et al., 2003, 2004).

9.4.1.6 Use in Perceptual Research

fMRI has become a ubiquitous technique with a host of studies appearing each year that deal with every topic in the perceptual and cognitive sciences.

Perhaps one of the interesting questions that can be addressed with fMRI is whether purely mental processes, such as the conscious imagination of visual stimuli, affects the brain in the same way as when the stimuli are actually shown. In an early study, it was shown that participants who engage in mental rotation tasks use brain areas similar to the ones that are active during direct perception (Cohen et al., 1996). In a study with patients with a dissociation between visual perception and sensory input (the so-called Charles Bonnet syndrome), fMRI was used to demonstrate that when patients had hallucinations of colors, faces, and objects, the same areas were active as for normal, direct perception (Ffytche et al., 1998). Hence, those brain areas have been found to be involved in the *conscious* percept. Finally, when participants were asked to imagine faces and places, areas in the brain were observed that were also active for direct perception of faces and places (O'Craven and Kanwisher, 2000).

In this context, it is important to stress that despite several years of concentrated research, just how objects and entities are represented and processed in the brain remains a topic of hot discussion. Proponents of the *modular account* (such as Kanwisher et al. (1997) or Kourtzi and Kanwisher (2000)) see the brain's recognition modules as very stimulus-specific, that is, each important object category such as faces, body parts, and places has its own well-defined, well-localized processing module in the processing stream. Proponents of the *distributed approach* hold that the activation pattern that can be used to identify and categorize objects becomes much more diagnostic when neuronal responses across many different areas are taken into account; thus they see a much less modularized organization of visual processing (see, for example, Haxby et al., 2001). It remains to be seen—perhaps with the development of higher-resolution fMRI scanning—which model will be able to explain the data better.

Finally, two current, hot topics in fMRI research should be mentioned. The first topic concerns social or cognitive neuroscience, that is, the investigation of brain functions relating to higher cognitive functions such as the emotional reaction to social rejection, or moral judgment. In the first study (Eisenberger et al., 2003), the brain areas active when pain was experienced based on social rejection, or those active during physical stimulation were compared and were found to be largely overlapping. This result indicates that the pain felt when an indirect, non-physical, but socially important stimulus might be as real as that felt from direct physical contact. In the second study (Greene et al., 2004), fMRI was used to prove that moral dilemmas (such as choosing which person to save from an impending accident) do not only engage people's reasoning, but rather drive and are driven by their emotional involvement (as indicated by activation of central emotional areas in the brain), thus indicating a clear role for emotion in the context of moral judgments.

The second topic is that of "mind reading" using fMRI—a rather big term, which has been popularized lately in the media. The reality so far is limited to decoding certain, rather low-level states of perception from brain activity (see review by Haynes and Rees, 2006). The current state of the field has advanced quite rapidly, however, with early methods capable of, for example, decoding orientation, position, or object categories from activity in the visual cortex (Haynes and Rees, 2006). Recent studies have shown that it is possible—using a larger set of input voxels together with heavy algorithms from machine learning—to predict which out of many images a person has seen (Kay et al., 2008), and even to reconstruct the actual visual input from brain activation for simple, high-contrast stimuli (Miyawaki et al., 2008). It remains to be seen where the limits of fMRI temporal and spatial resolution are, and how good the quality of the reconstructions will become in future years.

9.4.1.7 Guidelines

For all the great many findings that fMRI has brought to our understanding of the brain, if one asks what fMRI can do in the scope of perceptual or cognitive experiments, a few issues need to be kept in mind, which are highlighted in the list that follows. The interested reader is also referred to the following papers: Henson (2005) discusses the use of fMRI in the context of psychological questions, Logothetis (2008) mentions fundamental limitations to fMRI-analysis, and Vul et al. (2009) highlights the need for more rigorous statistical analyses in the field.

- fMRI places restrictions on the kinds of experiments one can run—not only because of the physical restrictions and the potential discomfort of long experiment times, but also because of particular designs and averaging protocols that need to be observed. Not every experimental design is compatible with fMRI.

- fMRI data is only as useful as the perceptual or cognitive data that the underlying task produces. This means that the experimenter needs to take as much care about the design and analysis of the actual perceptual and cognitive experiment as with the later imaging data. The fallacy of "showing some stimuli and seeing what lights up" was particularly problematic in the early days of fMRI. Nevertheless, all good fMRI studies should involve a careful experimental design with pilot studies outside of the scanner, etc.

- Analysis of fMRI data is a highly complex topic with whole journals devoted to reporting progress in this field. As with statistical tests in perceptual and cognitive experiments, it pays to be aware of the underlying assumptions behind the battery of tests that need to be run on fMRI data. In particular, there are various pitfalls when conducting a multi-staged analysis of voxel activations, such as using a localizer to identify a certain brain area and then using that activation in the later experiment.

- Fundamentally, fMRI highlights regions in the brain that are correlated with experimental data. When one is not interested in the implementational level, that is, how a network of particular brain areas solves the task, or how certain objects and events are represented in the brain, fMRI will add little to the experimental insight gained by running a carefully designed study outside of the scanner. Too often the final conclusion of the study will read only: "a network of connected, cortical regions was found to underlie the activation in task X."

9.4.1.8 Example

For an example of how the end result of a typical fMRI experiment looks, refer to Figure 9.4. This figure shows results from a part of an experiment that has focused on face recognition. In the experiment, several visual stimuli were shown to participants, and brain activation was compared to a baseline, fixation condition. The stimuli consisted of four classes: faces, scrambled faces, objects, and scrambled objects. The scrambling was done in order to rule out very simple low-level activations of visual areas—if activation to faces in an area were *the same* for whole faces as for scrambled faces, one would not deem an area to be involved in higher-level face perception; rather this area might be sensitive to some low-level contrasts or textures in the face that are shared for intact and scrambled stimuli. The result of the contrast between faces and non-face objects is displayed in the top left part of the figure, which shows a view of the brain from below with the front part pointing upwards—note that the normally heavily folded surface of the brain has been unfolded so activation can be clearly seen for all parts of the brain. As can be seen, face-selective activation occurs in both hemispheres of the brain at two anatomically similar locations, which are labeled FFA (fusiform face area) and OFA (occipital face area). These names are derived from many experiments, which have indicated very robust activation for faces over other types of objects in these specific locations. In addition, activation also seems a little stronger in the right hemisphere—a result that has also been found in other studies.

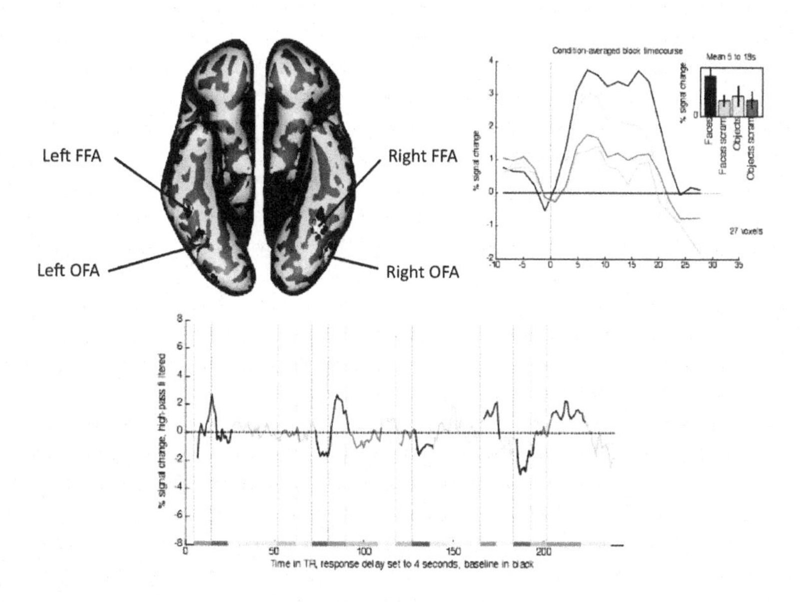

Figure 9.4. An example of fMRI data from an experiment on face recognition (data and analysis by Johannes Schultz).

The plot in the upper right shows the time-course of the BOLD signal in the right FFA. The time-scale on the x-axis is measured in seconds with the vertical line indicating stimulus onset, whereas the y-axis shows signal-change percentage. Note that the signal change overall is small (maximum of 4%) and that the typical BOLD response is clearly visible with an onset time of around 2–3 seconds, a prolonged peak and a subsequent fall-off. The different curves correspond to the different stimuli (faces, scrambled faces, objects, and scrambled objects) and show that the highest activation is, indeed, present for faces, followed by objects, followed by the scrambled stimuli. It should be noted that these curves would actually indicate that FFA is also selective for objects, not only for faces—however, when looking at data averaged across trials (shown as the inset in the top right figure), the variability inherent in the object trials washes out and face activation remains different from the other three conditions.

Finally, the graph at the bottom of the figure shows the raw experimental data as it was recorded with the different stimuli shown after each other—including the baseline trials. One can see that the face trials, indeed, have the highest peaks—in addition, the baseline trials often result in negative values, that is, the area in question is inhibited during those trials.

III

Stimuli

Chapter 10

Choosing Stimuli

There is a central issue in stimulus selection that has to do with the experimental context: is the experiment going to be a user study of an existing set of data (such as whether a particular webpage is effective in attracting the viewer's gaze to specific parts on the screen), or is the experiment going to test a non-specific experimental question (such as which elements in general attract the viewer's gaze when browsing)? The issue is therefore whether the possible test set is constrained from the outset or not. If it is constrained, then the matter reduces to which elements of the whole dataset become part of the test set in the experiment (for example, it might be unfeasible to test all individual pages contained in the webpage that one is interested in); that is, the question is how to sample properly. If the test set is unconstrained, the problem is more general, since then the researcher has to come up with a suitable set.

In both cases, stimulus selection requires the balancing of two diametrically opposed requirements: experimental specificity and generalizability. On the one side, an experiment requires a set of images that will allow the researcher to obtain a unique, interpretable answer to the research question. Thus, the differences between the elements in the stimulus set should be solely due to the question of interest. On the other hand, if the elements are too specific, the results might not generalize beyond the exact stimuli used, and will not help to answer the question in general. Hence, the test set must somehow also be a representative sample of the space of possible stimuli that are relevant to the question.

Over the past decades, researchers in perception and cognition have amassed a wealth of information about the fundamental information-processing mechanisms that humans possess. As with any good experimental science, part of the success comes from being able to reproduce experiments in order to validate published results and to generate a foundation of accepted findings from which to proceed. With the advent of the computer and, later, of the Internet, the ease with which well-controlled stimuli can be created, tested, and distributed has vastly increased. In computer vision, which seeks to develop computational algorithms for scene understanding and object recognition from images, the publishing of databases has become a highly prestigious undertaking, and many of the most highly cited papers in this field have introduced novel datasets to the academic community. In neuroscience, another fast-moving field, researchers have begun to realize how important it is to be able to reproduce and validate findings about brain anatomy and brain function—which is especially true when looking at the studies involving functional magnetic resonance imaging (fMRI), which produce massive amounts of data per trial run. In this chapter, we will take a look at a few databases that might be useful for running perceptual experiments. It should be noted that the choice of databases is far from exhaustive as new developments are constantly being introduced, and that while we have tried to span a broader range of interesting stimuli, the selection is biased towards visual experiments.

10.1 Control

It has been mentioned repeatedly throughout this book that one of the central issues in experimental design is control; that is, the experimenter should take care to have all relevant parameters of the experiment under control so that external, non-relevant factors can be ruled out. In reality, of course, not all factors in the experiment can be controlled and therefore one usually settles for checking that the most obvious and relevant factors are constrained well enough. Partial justification for this is that as long as the other, noncontrolled factors are random, they will simply add to the noise in the experiment. The only problem will be when these factors introduce systematic biases into the experiment.

As an example, consider face recognition—arguably one of the most well-studied problems in perceptual and cognitive science. If the researcher is interested in how well people can recognize unfamiliar faces, then first a suitable test set needs to be created (see also Section 10.2.3 for a list of face databases available for research). For this test set, there are a number of issues to be considered (probably a nonexhaustive list).

- *Size of the stimulus set.* How many faces will people be tested on?
- *Population.* What should the population of faces encompass in terms of age, ethnicity, social background, sex, etc.?
- *Resolution.* How many pixels must each face have?
- *Pose.* Should only frontal faces be considered, or should other views also be considered? If other views are to be included, these should be well-defined.

- *Illumination.* Will the faces be photographed under normal, varying illumination conditions or taken in a controlled setting, perhaps even under different illuminants?
- *Expressions.* Do the faces need to be in a neutral expression or contain varying expressions?
- *Accessories.* Some people wear glasses or make-up. Will this be included in the database or not?
- *Dynamics.* Is the researcher interested in dynamic aspects of face recognition? If so, video data should be recorded.

As can be seen, there are a number of important factors that all have the potential to be important for face recognition and therefore will critically influence the outcome of the experiment. Of course, this list already shows that it will be impossible to design a stimulus set that encompasses all possible factors with enough representative samples. Let us assume that we want to sample a moderate number of expressions and environmental conditions by taking 7 expressions × 7 poses × 5 illuminants × 2 accessory conditions. This means that 490 pictures would need to be captured *per face.* That number needs to be multiplied further by the population factors, taking a moderately dense sample of 2 sexes × 5 age groups × 5 ethnic backgrounds × 20 members per group. This means that 1,000 people would need to be recruited, invited, recorded, paid, and post-processed for a total of 490,000 images—a truly Herculean task.

A full factorial exploration of all possible environmental and population factors thus seems to be far from attainable in most cases (not to mention the fact that this would also mean a full factorial exploration in terms of the experimental design!). While it would be nice to have access to such a resource (and some databases actually approach this complexity, at least for the case of sampling faces—see Section 10.2.3), it is in most cases desirable to limit oneself to the important experimental factors in question and carefully design a complete stimulus dataset in this context. One would hope that the full picture—that is, the influence of all important factors—will emerge over time, as more and more pieces of the puzzle are published and all specific factor combinations are slowly explored across different studies.

10.2 Stimulus Databases

Before giving examples of several stimulus databases below, there are a few important points that need to be mentioned.

10.2.1 Guidelines

Stimulus availability. While it would be good to test on existing, well-defined stimulus databases, as this provides a good way of cross-checking your results, greatly reducing the preparation time for the experiment, it is of course not possible to obtain stimuli for all possible experimental questions. Therefore, the stimuli will need to be created by the experimenters themselves. If this is done, it is advisable in many cases to make the stimuli available if possible (either publicly, or by request only)—this provides a great service to the scientific community and facilitates data exchange.

Source bias. The Internet is a rather useful resource in terms of providing easy access to millions of images, videos, and music files. For experiments on object recognition, a quick Google image search with the object categories of interest might turn up valuable results. However, apart from copyright issues that need to be kept in mind, the idea of controlled experimentation (see Chapter 2) requires a more rigorous approach, which should consider the myriad of factors that could influence the result considerably.

Quality. Connected to the previous point, stimuli used for proper psychophysical experiments should offer features like unique definability, quantifiability, and replicability. For example, grating stimuli, which commonly are used for experiments on low-level properties of visual processing, are physically uniquely defined by their spatial frequency. Such a mathematical description is at the basis of psychophysical methodology and enables the researchers to derive *functional relationships* between stimulus and behavior. In contrast to this, for research in the area of face perception (such as face recognition and facial expressions) stimulus properties are less well-defined, as a mathematical parametrization of faces and facial motions is lacking still.[1] This means necessarily that a stimulus database suitable for higher-level experiments will never be able to capture all relevant dimensions of the research question fully—so it is up to experimenters to decide which risks to take and which factors to ignore in their experimental designs.

10.2.2 Resources

It should be acknowledged that the review of stimulus databases in the rest of the chapter is far from complete and most likely rather biased to the authors' wider research context. For a good resource on databases and stimulus material in general, as well as for novel analysis methods, that is, for an overview of useful research tools, the journal *Behavioral Research and Methods* is recommended to the interested reader. The journal has been around since 1968 and

[1]Note, however, the recent developments by Blanz and Vetter (1999); Breidt et al. (2003) for parametrizations of faces and facial expressions, respectively.

is currently published by Springer; it states that it "covers methods, techniques and instrumentation used in experimental psychology research" with a "particular focus on the use of computer technology in psychology research." As such it includes not only tutorials on statistical techniques, but also descriptions of specialized hardware and software packages that are designed for specific tasks, as well as detailed descriptions of many useful stimulus databases and experimental resources.

With these introductory remarks, let us turn to the list of stimulus databases and resources.

10.2.3 Face Databases

Following up on the previous example, the list below highlights a few important resources for face recognition. For more information on face recognition in the scope of computer vision, take a look at the excellent book (Li and Jain, 2005) (for which a second edition is in the works at the time of writing). The book also includes a good overview of findings in perception research (for another overview, see Schwaninger et al., 2006).

10.2.3.1 FERET and FRGC Databases

In the late 1990s computer-vision algorithms were developed that were promising enough to allow face recognition under natural conditions. The first, government-sponsored program to independently evaluate algorithms from both commercial and academic resources was the FERET program. One of the first tasks for this program was to collect a database of face images that would allow for detailed testing of face recognition under different, controlled environmental conditions. The FERET database (Phillips et al., 1998) contains face images from over 1,000 individuals in different poses, indoor versus outdoor lightings, and also multiple sessions per individual allowing for the assessment of recognition performance over small time changes. An improved version of the database can be obtained from http://face.nist.gov/colorferet/.

The database was later extended for a second iteration of the evaluation—the face recognition grand challenge (FRGC Phillips et al., 2005)—which focused not only on face recognition in small still images, but also in higher resolution images, as well as in three-dimensional data. The dataset for this challenge consisted of both training and test sets for over 400 individuals of varying ethnic backgrounds. For some people, many replicate sessions were available. Interestingly, the results of this test showed that some computational algorithms seemed to fare better in face recognition than their human counterparts (O'Toole et al., 2007; Phillips et al., 2010). The database is available at http://face.nist.gov/FRGC/.

10.2.3.2 CMU Pose, Illumination, and Expression Database

The CMU Pose, Illumination, and Expression Database (Sim et al., 2003) mainly contains static data, but a small amount of videos is also available. A total of over 40,000 facial images from 68 persons were collected. Each person was recorded in 13 different poses, under 43 different illumination conditions, and with four different facial expressions. For the different illumination conditions, the images were captured with and without background lighting and with a flash system. The database is ideal for investigating human and algorithmic robustness to changes in pose and lighting, making it into one of the standard benchmark databases in the computer-vision domain. The facial expressions in the database, however, are limited to neutral, smiling, blinking, and talking.

Recently, another project at Carnegie Mellon University (the CMU Multi-PIE face database) has been put online. It contains images of over 300 people recorded in multiple sessions over the span of five months. Similar to the standard CMU PIE database, recordings were taken from 15 viewpoints and 19 illumination conditions. In addition, however, high resolution frontal images were acquired as well, and a larger range of facial expressions was recorded in all conditions. The database is available from http://www.multipie.org/.

10.2.3.3 Labeled Faces in the Wild

In most cases, humans (and to a lesser degree, also artificial face recognition systems) do not encounter faces in clean, well-lit environments from a frontal pose. The Labeled Faces in the Wild database (Huang et al. (2007), http://vis-www.cs.umass.edu/lfw/) is interesting in that it contains photographs of faces collected from the web, fully acknowledging the fact that these images were taken in uncontrolled conditions. In a way, it might therefore offer more realistic training and testing data. The database consists of over 13,000 images of faces that have been culled from the web and run through a face detector to highlight the location of the person in the picture. The faces contain mostly well-known individuals and have been labeled with the name of the person pictured.

10.2.3.4 Three-Dimensional Face Databases

As faces are a very homogenous class of objects, automatic algorithms can be developed that allow for morphing between them. This morphing can even be done in three dimensions, which can be used to construct a statistical model of the shape- and appearance-dimensions of faces. The first so-called morphable model was developed by Blanz and Vetter (1999), and uses data from 100 male and female faces scanned in a with a three-dimensional laser scanner. One of the nice things about this model is that it becomes possible to control shape and appearance of faces on a high level; for example, one can take a face and manipulate its sex, age, expression, and even its perceived attractiveness with one slider. In addition, because the morphable model represents an extremely good model of what faces look like, one can use it to reconstruct the three-dimensional face from a two-dimensional picture (Blanz and Vetter, 1999). The database in its basic form (without the morphable model) is available from http://faces.kyb.tuebingen.mpg.de and is based on the three-dimensional scans collected by Troje and Bülthoff (1996).

Recently, another morphable model was released, which was collected at Basel University (Paysan et al., 2009). The researchers have made the actual morphable model available that contains the statistical description of the three-dimensional face space—again, based on 100 male and 100 female three-dimensional scans. It can be downloaded at http://faces.cs.unibas.ch.

10.2.4 Facial-Expression Databases

As with research on face recognition, thorough studies on the perception of facial expressions have been made possible only by having access to high-quality databases. One of the critical issues with such databases is the need to take the naturalness of such stimuli into consideration. Wallbott and Scherer (1986) realized that researchers in this field are confronted with "... simulated emotional expressions [that] are clearly not natural enough" and that "the natural expressions of emotion obtained in most studies to date have not been emotional enough."

How can one then reliably induce a particular natural facial expression in order to design and record a database of facial expressions? As one of the most extreme examples (which clearly violates all ethical standards of today's scientific conduct), Landis (1924) investigated the facial expression of disgust. In order to elicit this particular facial expression, he—among other things—drew lines on volunteers' faces with burnt cork before asking them to smell ammonia or place their hands in a bucket of frogs. Subsequently, each volunteer was asked to decapitate a white rat—during this particular ordeal, the volunteers were photographed, and the photographs were then later annotated. This rather extreme example illustrates one of the basic trade-offs with facial-expression databases—namely that between control and naturalness. It also illustrates of course the potential ethical implications of some elicitation protocols.

As many of the example experiments in this book feature research on facial expressions, this section includes a sample of the databases available, highlighting the difference in approaches in order to capture representative data about expressions.

10.2.4.1 Pictures of Facial Affect

Paul Ekman and Wallace Friesen, both eminent researchers in the area of facial expressions, developed the Pictures of Facial Affect (POFA), which contains perhaps the most well-known static pictures of facial expressions (Ekman and Friesen, 1976). The database has been widely used in cross-cultural, psychophysical and neurophysiological research. It consists of 110 frontal, grayscale photographs of six so-called basic or universal facial expressions (happiness, sadness, fear, anger, disgust, and surprise). These were elicited through a careful protocol focusing on single muscle activations to produce well-defined displays of emotion. Neither intensity nor facial configuration were controlled. The database can be ordered at http://face.paulekman.com/productdetail.aspx?pid=1.

Paul Ekman and colleagues also developed the Japanese and Caucasian Facial Expression of Emotion (JACFEE), as well as the Japanese and Caucasian Neutral Faces (JACFeuF) (Matsumoto and Ekman, 1988). These two databases can be ordered from http://www.humintell.com/humintell-products/.

10.2.4.2 AR Face Database

The AR Face Database was created by Martínez and Benavente (1998). The AR Database contains over 4,000 frontal-view, colored images of different facial expressions; 126 different individuals (70 men and 56 women) were recorded. Furthermore, each facial expression is available in four conditions: facial expression without modified features, illumination condition, occlude eyes through sunglasses, and occlude lower half of the face through scarf. The latter conditions were added to test the robustness of computer-vision algorithms against occlusion of facial features. The database is available for download at http://www2.ece.ohio-state.edu/~aleix/ARdatabase.html.

10.2.4.3 Japanese Female Facial Expression Database

The Japanese Female Facial Expression (JAFFE) Database by Lyons et al. (1998) is another static facial-expression database; it can be downloaded from http://www.kasrl.org/jaffe.html. The JAFFE Database includes 213 images of ten Japanese female models showing the seven standard facial expressions (the six basic facial expressions and one neutral expression). The photographs were taken in front of a semireflective mirror; the models controlled the camera trigger.

10.2.4.4 MMI Facial Expression Database

The MMI Database, created by Pantic et al. (2005), includes more than 1,500 samples of static images and image sequences in frontal and profile view. Various facial expressions of emotion as well as single facial action units (AUs) and multiple AU-activations make up the database. The length of the sequences varies, with each video containing a sequence of one or more neutral-expressive-neutral facial expression patterns. The database includes 19 different individuals (44% female) who range from 19 to 92 years of age and who have different ethnic backgrounds (European, Asian, and South American). Recently the database was augmented to include *induced* disgust, happiness, and surprised expressions elicited by suitable video stimuli. The database is available from http://www.mmifacedb.com/ and has been updated continually over the years.

10.2.4.5 DaFEX: Database for Facial Expressions

The DaFEX is a database of posed human facial expressions recorded by Battocchi and Pianesi (2004) belonging to the category of dynamic databases. It consists of 1,008 short videos that are based on the basic expressions plus a neutral expression. All expressions were posed by professional actors. Each expression was recorded in three intensities (low, medium, and high) and in two different conditions: an utterance condition in which actors posed facial emotions while uttering a phonetically rich and visemically balanced sentence, and a non utterance condition in which actors posed facial emotions while being silent. To ensure that all four actors induced emotions in a similar fashion, cue stories for each expression were given before starting the recordings. More information is available at http://i3.fbk.eu/en/dafex.

10.2.4.6 Cohn-Kanade AU-Coded Facial Expression Database

The Cohn-Kanade AU-coded Facial Expression Database is also known as the CMU Pittsburgh AU-coded Face Expression Image Database (Kanade et al., 2000). Action units (AUs) are defined by single muscle movements in the face and have been developed and described in great detail by Ekman and Friesen (1978) in their Facial Action Coding System (FACS). This database has been used widely in computer vision for automatic extraction of the FACS units, as these are fully annotated in the database.

The Cohn-Kanade Database contains 23 facial motions including single action units as analyzed by certified FACS coders. The image sequences contain neutral to apex facial expression and are available in three lighting conditions: ambient lighting, a single high-intensity lamp, and two high-intensity lamps with reflective umbrellas. Out of the 100 participants, 65 were female; 15 participants had an African-American background and 3 had an Asian/Hispanic background. The participants' ages ranged from 18 to 30 years. Recordings were done by two synchronized video cameras, one of which was located directly in front of the participant, and the other of which was positioned 30° to the participant's right. The database was recently augmented to include posed and non-posed, fully coded emotional expressions, which allows for both analysis of FACS units and emotional labels—see http://vasc.ri.cmu.edu/idb/html/face/facial_expression/ for more details.

10.2.4.7 MPI Face Video Databases

The MPI Face Video Database is a dynamic facial expression database containing over 40 facial motions that follow the Facial Action Coding System by Ekman and Friesen, as well as a few compound expressions from one well-trained actor. The uncompressed video data of the Video Database was acquired with a custom camera system using six synchronized cameras arranged in a semicircle around the actor. In order to facilitate the recovery of rigid head motion, the subject wore a headplate with six green markers. For a detailed description see Kleiner et al. (2004); for downloads go to http://vdb.kyb.tuebingen.mpg.de/.

The MPI facial-expression database (Kaulard et al., 2012) was developed in order to overcome the restriction of most databases to only emotional expressions. For this, more than 50 dynamic expressions were recorded from ten male and ten female participants. This was done using the method-acting protocol, which asks participants to immerse themselves in a specific scenario and then act accordingly—a compromise between full control ("Please do a sad smile now!") and a spontaneous protocol (from annotating generic video footage, for example). The database was recorded with three synchronized cameras, and every expression is available in both low- and high-intensity each with three repetitions.

10.2.4.8 Binghamton University 3D and 4D Facial-Expression Databases

The Binghamton University 3D Facial Expression (BU-3DFE) Database by Yin et al. (2006) is based on a combination of prototypical three-dimensional facial expression models and two-dimensional texture images. The current version contains static three-dimensional data of the six basic expressions. With this three-dimensional database it is possible to investigate the influence of facial geometry on recognition of facial expressions, since it allows for detailed manipulation of the three-dimensional geometry of the face. The database includes 100 participants (56 female and 44 male) of varying racial background who range from 18 to 70 years. The six basic facial expressions were performed in four different intensities in front of a 3D-scanner; a neutral expression has also been included, yielding a total of 2,500 facial-expression models. For each model a texture image was captured in two views (about $\pm$ 45°). The database can be obtained from http://www.cs.binghamton.edu/~lijun/Research/3DFE/3DFE_Analysis.html.

With the advent of improved three-dimensional scanning technology, it has become possible to record three-dimensional scans in real-time, yielding temporal information about the three-dimensional deformation of the face. The Binghamton University 3D Dynamic Facial Expression Database (BU-4DFE, Yin et al., 2008) provides such a resource for facial expressions. Each of the

101 participants again replicated the six standard expressions, resulting in a total of 25 three-dimensional meshes per second, each of which contains 35,000 vertices. The challenge with such a database lies in the post-processing, as the scans themselves are not in correspondence; that is, a given vertex index i, which in frame N specifies the tip of the nose will not be on the tip of the nose anymore in frame $N+1$. The database can also be obtained from http://www.cs.binghamton.edu/~lijun/Research/3DFE/3DFE_Analysis.html.

10.2.5 Object Sets

10.2.5.1 The Colorized Snodgrass and Vanderwart Set

The colorized Snodgrass and Vanderwart set is a set of 260 common objects based on the line drawings that were used in the famous study by Snodgrass and Vanderwart, which have long since been the standard for line drawing–based object experiments. The stimuli are colorized and contain texture information and are described in Rossion and Pourtois (2004). The set can be found at http://www.nefy.ucl.ac.be/facecatlab/stimuli.htm.

10.2.5.2 Diagnostic Color Objects

Many objects have diagnostic colors (especially, of course, fruits such as bananas, oranges, etc.). This set is a collection of images of such objects used in the study by Naor-Raz et al. (2003). Download at http://stims.cnbc.cmu.edu/ImageDatabases/TarrLab/Objects/.

10.2.5.3 The Bank of Standardized Stimuli Set (BOSS)

BOSS is a large set of photographs of everyday objects, carefully processed to include only the foreground object. Many objects are also available from different viewpoints, and some have more than one exemplar in a given category. The full set includes around 1,500 photos, while the normalized set includes 480. The latter set has been normalized (that is, participants have rated the properties) according to name, category, familiarity, visual complexity, object agreement, viewpoint agreement, and manipulability (Brodeur et al., 2010). The images can be accessed at http://www.boss.smugmug.com/.

10.2.5.4 Greebles

Greebles are a now-classic set of novel objects that were used for demonstrations of perceptual-expertise learning for new stimulus classes. The objects consist of a small number of geometric shapes (parts) that are stuck together symmetrically and bear a vague resemblance to a figure or a face. There are many different kinds of greeble families available based on this concept. Go to http://stims.cnbc.cmu.edu/ImageDatabases/TarrLab/NovelObjects/ for more information. The first paper to use these stimuli was Gauthier and Tarr (1997).

10.2.5.5 Shepard and Metzler Objects

In 1971, Roger Shepard and Jacqueline Metzler conducted a famous experiment using possibly the first computer-generated stimuli in experimental psychology: perspective drawings of three-dimensional shapes that were rotated both in depth and in the picture plane (Shepard and Metzler, 1971). In the experiment, two rotated shapes were presented side by side; participants had to recognize whether they were from the same or from different shapes. The data was strikingly linear and did not differ between picture-plane rotation and depth rotation, suggesting a "mental rotation" model in which the shapes are mentally rotated through the required angles. The Shepard and Metzler objects have been remodeled and can be found at http://stims.cnbc.cmu.edu/ImageDatabases/TarrLab/NovelObjects/.

10.2.5.6 BTF Database

The term BTF refers to the *bidirectional texture function,* which in computer graphics describes the interaction of an object made out of a certain material with incident light. The BTF is fairly high-dimensional, as it needs to describe the interaction with light of all wavelengths and from all directions (think about the complex reflectance pattern that wool would, for example, make)—however, once acquired, images of extraordinary realism can be created without the need to model the physical interaction of such complex materials in detail. In order to sample the BTF, a piece of material needs to be photographed from many different directions and lighting angles, resulting in a large matrix of data. An overview report describes methods for acquiring, analyzing, and rendering of BTFs (Müller et al., 2005)—the BTF database itself is available from http://cg.cs.uni-bonn.de/en/projects/btfdbb/.

10.2.5.7 Various Three-Dimensional Objects

There are many web resources that offer links to both free and commercial three-dimensional models such as http://www.turbosquid.com and http://www.the3darchive.com/. Possibly the best, easily accessible web resource for three-dimensional objects is the "3D warehouse" associated with the freely available Google Sketchup three-dimensional modeler (http://sketchup.google.com/3dwarehouse/). Since the models are contributed by users the quality might vary, but many gems of sometimes astonishing fidelity can be found here. The advanced search can filter the queries for models according to complexity, user rating, location, etc.

Another (older) database of three-dimensional models sorted by category is the Princeton Shape Benchmark (http://shape.cs.princeton.edu/benchmark/, see Shilane et al., 2004). This database contains training and test sets for categories such as aircraft, animals, body-parts, books, buildings, etc., and has become one of the standard benchmarks with which to test algorithms on shape retrieval, three-dimensional matching, etc.

10.2.6 Image Collections

10.2.6.1 The Computer Vision Homepage

The somewhat out-of-date repository of the computer vision homepage at http://www.cs.cmu.edu/afs/cs/project/cil/www/v-images.html still has a great set of resources (beware of broken links, however).

10.2.6.2 Caltech Databases

The Caltech Computational Vision group led by Pietro Perona has a large set of well-known databases to offer, many of which have become standard benchmarking datasets for computer-vision algorithms. These include the Caltech-101 and Caltech-256 databases, which contain 101 and 256 object categories, respectively. Each category for the Caltech-256 database contains at least 80 exemplars. Go to http://vision.caltech.edu/archive.html for the full list of databases.

10.2.6.3 ImageNet

ImageNet (http://image-net.org/) is a structured way of browsing a well-defined library of images according to an ontology defined by the linguistic resource WordNet (http://wordnet.princeton.edu/). Each concept in this linguistic resource is described by a number of words, which together are called a *synonym set.* In the current version of WordNet, there are over 100,000 synonym sets (of which most are nouns), which are used to index a repository of images. For example, browsing from the super-category "instrumentation" leads one to the subcategory "furnishing," which includes 210 synonym sets, one of which is "seat" (1,738 pictures). This category in turn contains further sub-categories of "chair" (1,460 pictures), again broken down into "armchair," "wheelchair," etc. The current release of the dataset contains 17,624 synonym sets with a total of over twelve million images.

10.2.6.4 LabelMe

The LabelMe database (http://labelme.csail.mit.edu/) created by Bryan Russell, Antonio Torralba, and William Freeman is an online tool that allows users to segment and annotate images of photographs so that a scene can be broken down into its objects and their boundaries. The data provided by LabelMe has become an important resource for computer-vision researchers as it provides natural image statistics. An extension is available for annotating videos—a much more challenging task (http://labelme.csail.mit.edu/VideoLabelMe/).

10.2.6.5 Viperlib

Viperlib is a noncommercial, nonprofit Internet page that collects images, presentation materials, and even tutorials on various topics in visual perception and makes them available freely to the research community. Images and material can be contributed easily by anyone, and the current number of images as of March 2011 is around 1,800. Viperlib is run by Peter Thompson and Rob Stone from the University of York and can be accessed at http://viperlib.york.ac.uk/.

Viperlib is structured into the following categories:

- Anatomy/Physiology
- Attention & Learning
- Colour
- Depth
- Development
- Faces
- Illusions
- Imaging
- Lightness/Brightness
- Miscellaneous
- Motion
- Spatial Vision
- Viper2go Tutorials
- Vision Scientists
- Visual Abnormalities

Many of the images that have been contributed illustrate illusions or demonstrate other perceptual effects, but overall, the Viperlib is a great community resource that has a lot to offer and is available for free.

10.2.7 Sound and Multisensory Data

10.2.7.1 Sound-Event Database

The Auditory Lab at Carnegie Mellon University hosts a large database of common sounds structured into specific event types (involving impact, deformation, rolling, and air and liquid interactions) created by Laurie Heller. Access at http://stims.cnbc.cmu.edu/SoundDatabases/AuditoryLab/.

10.2.7.2 Avian Sounds Database

The Avian Vocalization Center at the University of Michigan hosts a database of thousands of bird sounds complete with accurate species identifications. Go to http://avocet.zoology.msu.edu/ for more information.

10.2.7.3 Real-World Computing Music Database

This interesting database was compiled for music information processing (such as genre classification, instrument and voice analysis, etc.). In its most recent iteration it contained 100 pieces of popular music, 50 pieces of classical music, 50 pieces of jazz music, 50 instrument samples, as well as 100 pieces covering a variety of genres (popular, rock, dance, jazz, Latin, classical, marches, world, vocals, and traditional Japanese music). The MIDI files (that is, the musical score), as well as song lyrics are available as well. This database was designed by Goto et al. (2003) (see http://staff.aist.go.jp/m.goto/RWC-MDB/) and is freely available and copyright-cleared.

10.2.7.4 Real-World Events Database

The Real-World Events Database by Jean Vettel (also from Carnegie Mellon University) contains videos and accompanying sounds of everyday events, such as splashing water or snapping twigs. This database would be very useful in particular for multimodal integration studies. Access at http://stims.cnbc.cmu.edu/SoundDatabases/Events/.

10.2.8 Human Behavior

10.2.8.1 CMU Graphics Lab Motion Capture Database

This is a web resource hosted by Carnegie Mellon University that is continuously updated and which contains motion-capture data of human actions and interactions (see http://mocap.cs.cmu.edu). Motion capture is a kinematic recording technique used extensively in the motion industry and in sports analysis. There are several different technologies in use which generate motion-capture data, but all of them generate three-dimensional data of a few important points with a very high temporal resolution. Most typically, small reflective markers are placed on the joints of a person, which are then illuminated by infrared light and recorded by several infrared cameras placed in the environment. The three-dimensional position of each reflective marker can be reconstructed from two camera pictures that are in a known configuration (that is, they are calibrated). Motion-capture setups have been used to drive animated characters, such as Gollum in *The Lord of the Rings* movie trilogy, directly from an actor's facial movements.

The CMU database holds more than 2,500 individual motion-capture recordings of human behaviors, which are broken down into human interaction, interaction with the environment, locomotion, physical activities and sports, situations and scenarios, and test motions. It is one of the most extensive, freely available resources for this type of data.

10.2.8.2 The POETICON Corpus

The POETICON enacted-scenario corpus is a database for (inter)action understanding that contains six everyday scenarios taking place in a kitchen/living-room setting. Each scenario was acted out several times by different pairs of actors and contains simple object interactions as well as spoken dialogue. In addition, each scenario was first recorded with several HD cameras and also with motion-capturing of the actors and several key objects. Having access to the motion-capture data allows not only for kinematic analyses, but also for the production of realistic animations where all aspects of the scenario can be fully controlled. Wallraven et al. (2011) has more information, and the corpus can be downloaded from http://poeticoncorpus.kyb.tuebingen.mpg.de.

Chapter 11

Presenting Stimuli: The Psychtoolbox

After having chosen a suitable set of stimuli, any researcher who is interested in running perceptual experiments will need to worry about how to display or render the stimuli and how to record the answers of the participants. For the end user, the perfect stimulus-presentation software would allow for easy specification of various experimental designs and would interface with various hardware devices both on the input and output side. In addition, it would offer precise and accurate timing under all experimental conditions, as well as play back or render a number of different sensory stimuli—ranging from text in UNICODE format to images to movies in various formats, up to 3D-Audio—preferably while keeping audio and video synchronized. Finally, it should save the acquired data in an easy-to-parse format, and gracefully handle missing data values, as well as unexpected experimental situations.

Of course, such software does not exist; instead the experimenter has to review his or her constraints carefully and decide which software solution will provide the ideal compromise between flexibility and ease of use. Here, we have chosen to focus on one of the most popular open-source stimulus presentation software packages—the Psychtoolbox. There are of course numerous other solutions for stimulus presentation, both commercial (the most popular package is perhaps E-Prime by Psychology Software Tools, Inc.) and noncommercial. For a more complete overview, the interested reader is referred to the metalist provided at http://www.psychology.org/links/Resources/Software/.

The Psychtoolbox is a software toolkit designed for use with the Matlab scientific programming environment (or its free counterpart Octave). It is multiplatform in that it was developed under Mac OS X, but also will run fine under the newer versions of Windows. Among other features, it encompasses a large set of helper functions that enable time-accurate and synchronized presentation of visual and auditory stimuli, very precise keyboard access, an interface to OpenGL-accelerated two-dimensional and three-dimensional graphics, extensive code for determining and fitting psychometric functions, colorimetric calibration data, interfaces to popular eyetracking hardware, and much more. After a long hiatus, the Psychtoolbox is now at Version 3 and has been completely rewritten to take advantage of new developments in computing hardware such as GPU acceleration. One of the major advantages of the Psychtoolbox is its large, very active, and helpful online community where answers and support for many detailed questions are provided by several people free of charge. This community offsets the sometimes out-of-date help files and the lack of a detailed tutorial or a comprehensively edited documentation.

11.1 History

The first version of the Psychtoolbox was released 1995 for Macintosh by David Brainard. The main reason for choosing Macintosh at the time was the popularity of the Macintosh operating system in psychology departments. The operating system also had a built-in interrupt function that was tied to the screen framerate, which made it possible to provide easy timing functions for the Macintosh—something that was much harder to program for operating systems working with an unconstrained set of hardware devices, such as MS-Windows. Version 2 of the Psychtoolbox was released in 1996 for the Mac OS by David Brainard and Denis Pelli, and had many improvements (Pelli, 1997; Brainard, 1997). Its popularity generated demand for a MS-Windows version, which was released in 2,000 by Allen Ingling and others. This version resulted in over 20,000 downloads for Windows and around 9,000 downloads for the Mac OS, which, despite the relatively small numbers, made it into a very popular software toolkit.

With the release of Apple's Mac OS X and the subsequent discontinuation of the OS-9 operating system, many of the C plugins written for the old operating system were not well suited for the completely overhauled Mac OS X programming interfaces. In addition, since the C source code was not modularized, this made it difficult for new developers to contribute—and therefore the project was not able to benefit from the growing popularity of the open-source movement. Finally, the Psychtoolbox 2 did not take advantage of modern graphics hardware, which slowly revolutionized the way people were approaching display and rendering on the computer. The Psychtoolbox 3 (PTB-3) was developed as an answer to those problems and resulted in a complete redesign and code restructuring.

PTB-3 (Kleiner et al., 2007) was released in 2004 for Mac OS X by Allen Ingling and in 2006 for the MS-Windows operating systems, as well as for the various flavors of the popular open-source Linux distributions. PTB-3 contains many Matlab-script M-Functions adapted from PTB-2. Most importantly, however, all C plugins were rewritten from scratch for increased modularity, better maintainability and portability. Wherever possible, it was decided to switch to common, well-known, and platform-independent programming APIs as well as open-standards/open-source toolkits such as OpenGL and OpenAL. For many basic functions, such as drawing images and graphics primitives on the screen, as well as producing three-dimensional sound effects, PTB-3 takes advantage of modern graphics and sound hardware. With the new release, the Psychtoolbox also has switched to the GNU Public License (GPL), which facilitates more third party code contributions and provides users with a well-defined licensing scheme, and easy code-sharing and reusing.

As the PTB-3 has converted to an open-source contribution scheme, a modern versioning system was put in place. The popular Subversion system was used as code-management system, which makes it possible to document, archive, and track all code changes, as well as to resolve concurrent code-editing conflicts. It also provides a nice software update mechanism to users—on average, the PTB-3 is updated roughly once a month (sometimes even faster for major bug fixes). At the time of this writing, PTB-3 has about 25,000 known installations.

11.2 Contents

The typical installation is handled by a Matlab script DownloadPsychtoolbox.m that can be obtained from http://www.psychtoolbox.org. After successful installation, the Psychtoolbox resides in a self-contained folder on your computer's hard drive. If you type in `help Psychtoolbox`, you will get a listing of this folder's contents, which you can then browse in more detail in Matlab. The quality of the documentation is usually rather good for recently added functions, but can become very technical in places. Another documentation effort has been set up on the Psychtoolbox wiki at http://docs.psychtoolbox.org/Psychtoolbox, which is also worth looking at. The following list gives a commented overview of the folder structure and the functionality included with the Psychtoolbox.

- *PsychAlpha.* Not quite ready code for alpha-testing.
- *PsychAlphaBlending.* Set of tools for correct, OpenGL-assisted transparency operations on pixel data.
- *PsychBasic.* Main code repository for core functions of the Psychtoolbox; includes important scripts and mexfiles for commands such as Screen, KbWait, etc.
- *PsychCal.* Set of functions to calibrate a monitor (most importantly, its color characteristics). Note that this and other monitor functions might

be specific to the old CRT-type monitors. Have a look at the slides of the included presentation in PsychDocumentation about the issues with LCD monitors and timing.

- *PsychCalDemoData.* Example calibration data for typical monitors.
- *PsychColorimetric.* Extensive library of color-vision related functions. Includes conversion routines between different color spaces, as well as for converting into eye-receptor-specific formats.
- *PsychColorimetricData.* Sample data containing many useful eye characteristics (such as eye length, cone density, etc.).
- *PsychContributed.* Repository for third-party code, that does not necessarily underlie the GPL. Includes, for example, an interface to the video capture library of the popular ARToolkit used in many augmented reality applications.
- *PsychDemos.* The most important directory for beginners. It contains a large number of examples, illustrating the various aspects of the Psychtoolbox.
- *PsychDocumentation.* Sadly, not yet a full-fledged documentation. Includes slides from a presentation of the Psychtoolbox, as well as various other specialized topics.
- *PsychFiles.* Some useful Matlab scripts for dealing with files in folder hierarchies.
- *PsychGamma.* Code for fitting gamma-functions of monitors, which specifies the luminance output for a given input voltage.
- *PsychGLImageProcessing.* OpenGL-accelerated image-processing functions implemented as GLSL shaders—examples include alpha blending, convolution, etc.
- *PsychHardware.* Various interfaces to input devices, sensors, and eye trackers—mouse and keyboard are dealt with in PsychBasic.
- *PsychInitialize.* Bookkeeping code for the Psychtoolbox.
- *PsychJava.* Helper functions based on Java code, which in the Psychtoolbox currently refers mostly to keyboard functions.
- *PsychMatlabTests.* Legacy examples for bugs in earlier versions of Matlab on Mac OS 8 and 9.

- *PsychObsolete.* Contains legacy code needed to keep older versions of the Psychtoolbox functional.
- *Psychometric.* Set of tools that fit and estimate parameters of psychometric functions.
- *PsychOneliners.* Functions that require little code, but come in handy for various tasks—have a look with `help PsychOneliners`.
- *PsychOpenGL.* Essential functions for interfacing the Psychtoolbox with the advanced potential of OpenGL and with that of modern-day GPUs. Each OpenGL command has a Matlab wrapper which enables standard OpenGL code to work flawlessly and efficiently from within the Psychtoolbox.
- *PsychOptics.* Functions for calculating specific properties of the human eye, such as its Modulation Transfer Function.
- *PsychPriority.* Tools for manipulating the priority queues of operating systems to ensure that the Psychtoolbox has the maximum resources for time-critical tasks.
- *PsychProbability.* Various handy tools for randomizing variables and experimental factors, as well as for calculating the χ^2-distribution and the Bernoulli-distribution.
- *PsychRects.* Functions for handling and arranging rectangles—extremely useful for arranging items on a screen in well-defined patterns.
- *PsychSignal.* Contains Low-, High-, and Band-pass filters for images, as well as some min/max matrix operations.
- *PsychSound.* Interface to OpenAL, which is an advanced, free sound API that can even deal with spatialized (three-dimensional) sound.
- *PsychTests.* Simple tests for timing different functionalities of the Psychtoolbox, useful for benchmarking and debugging.
- *PsychVideoCapture.* Still somewhat experimental interface to provide real-time, low-latency access to video-capture hardware, such as webcams.
- *Quest.* Toolbox for implementing a QUEST procedure that estimates detection thresholds in psychophysical experiments efficiently.
- *Contents.m.* Helpfile for describing the contents of the Psychtoolbox.
- *DownloadPsychtoolbox.m.* Script for downloading the Psychtoolbox.

- *License.txt.* A copy of the GPL license.
- *PsychtoolboxPostInstallRoutine.m.* Tidying-up functions that are run after the Psychtoolbox has been installed.
- *ptbflavorinfo.txt.* Specifies whether the Psychtoolbox version is stable, or beta.
- *SetupPsychtoolbox.m.* Setup environment variables.
- *UpdatePsychtoolbox.m.* Script for updating the Psychtoolbox—call frequently to ensure that the Psychtoolbox always stays updated and that the latest bug fixes are applied!

11.3 Getting Started

First off, as mentioned above, the help function of the Psychtoolbox can be accessed via `help Psychtoolbox`, which gives an overview of all its various parts. Typing `help PsychBasic` will give a good overview of the core functionality, as it lists the most basic commands. A good place to get started is by typing `help PsychDemos`, which will list all of the Psychtoolbox's many included demo files that cover and demo all of its functionality at varying levels of detail. A short introduction to the most basic commands follows, which should help to show how things are set up in the Psychtoolbox.

One of the most important—and also most complex—commands is the `Screen` command. It provides an interface to the screens that are managed by Psychtoolbox (and usually, under the hood, by OpenGL). An overview of the functionality of this command can be obtained by typing `help Screen`, or simply `Screen`. This will yield a large list of subcommands that themselves take additional parameters to control the behavior of the display in more detail. The subcommand categories that you will see in the help, together with some relevant examples, are given in the list that follows.

- Open or close a window for the Psychtoolbox: This includes the `OpenWindow` command, which creates a window for the Psychtoolbox to draw on. Note that the Psychtoolbox usually takes over the full screen. The `Close` command closes this window and returns control to Matlab. Most code actually uses `CloseAll` at the end as this takes care of all memory reserved by the Psychtoolbox.
- Generate a texture from image data: Since the Psychtoolbox relies on OpenGL, image data from Matlab needs to be converted into an OpenGL-compatible format. The command `MakeTexture` does just that given a standard Matlab image array.

- Drawing lines and solids: Optimized commands that draw lines, ovals, polygons, etc.
- OpenGL functions for drawing of primitives (OS X only, look at the extended OpenGL functionality of the Psychtoolbox below for commands that give access to all OpenGL commands).
- Definition/Drawing of text: This includes commands that select the font and its properties, as well as commands for drawing text on the screen. Note that a more convenient form of drawing text is offered by the separate `DrawFormattedText` command. The Psychtoolbox can make use of Unicode fonts and has no problem using all of the operating system's fonts with nice looking anti-aliasing enabled.
- Copy images from/to the screen: Includes `DrawTexture`, which is used to draw a texture created with `MakeTexture` into the back-buffer.
- Synchronize with the screen: None of the previously mentioned drawing commands actually display results on the screen, until a `Flip` command is issued, which causes the graphics card to switch its back-buffer and its front-buffer—this is needed for highly accurate stimulus timing and smooth animations/video playback.
- Get information about the color lookup table of the screen.
- Get and set basic properties of the screen: Includes commands for getting the frame rate of the display, the window size or resolution, the number of attached screens, etc.
- Get and set basic properties of the computer/operating system environment.
- Movie commands: A set of commands to play synchronized and well-behaved videos in the Psychtoolbox. This works best under OS X as Quicktime codecs are built into the operating system itself. However, by now Windows also offers similar functionality if Quicktime is installed. For a demo, have a look at the demo `SimpleMovieDemo.m` included in the Psychtoolbox.
- Video-capture commands: A set of commands to record and play back data from video-capture devices. For a closer look, start `VideoCaptureDemo.m`.
- Full access to OpenGL commands: `Screen('Preference', 'Enable 3DGraphics')` enables access to OpenGL 3D graphics, and OpenGL commands are wrapped in `Screen('BeginOpenGL', windowPtr)` and `Screen('EndOpenGL', windowPtr)` similarly to the C-code versions.

11.4 Example 1: A Simple Experiment

The following Matlab script illustrates the basic use of a few core commands in the Psychtoolbox. It first opens a full screen window for the Psychtoolbox, then loads two images and converts them into textures. The actual display code first displays a small fixation spot, followed by a short (so-called priming) image presentation and another, longer (target) image presentation. The code is shown in standard Matlab format with extensive comments (preceded by a % sign) explaining each command line. The comments should be enough to guide a person who is somewhat experienced in Matlab syntax through the code.

A special feature of this piece of code is that it demonstrates rather precise timing—on a standard laptop, results are accurate up to the frame rate of the monitor (16 ms for the usual 60 Hz of the LC-display). For even more precise timing, it is recommended to use a CRT monitor with a high frame rate (100 Hz or higher), or to use a 120 Hz LC-display.

```
% simpleScreenDemo.m
%
% Simple demo that opens a screen, then draws a fixation point which
% the participant should look at for 500ms, then draws a priming-image for
% 100ms, followed by the target image for 500ms. It uses precise timing as
% per the recommendations of the Psychtoolbox documentation.
%
% This demo uses two sample images that are shipped with matlab, called
% 'durer' (the engraving Melancholia I by Albrecht Duerer) and 'detail'
% (a detail from the same engraving showing a magic square / matrix)
%
% 02/28/11 Created by Christian Wallraven based on demo code from
% Mario Kleiner

% standard preamble clearing the workspace
clc;
clear all;
close all;

% First, get a screen to display our stimuli on. This tells the
% Psychtoolbox to open a double-buffered, full-screen window.
% '0' is the main screen with the menu bar, other screen numbers could be
% chosen in a dual-monitor setup. The 'OpenWindow" subcommand returns two
% important parameters: 'win' is the handle used to direct all drawing
% commands to that window - the Psychtoolbox name for the window.
% 'winRect' is a rectangle defining the size of the window - in this case
% this corresponds to the screen resolution of the main window (see also
% "help PsychRects" for help on such rectangles and useful helper
% functions). The default background color for this window is white.
[win, winRect]=Screen('OpenWindow',0);

% The following command is required for precise timing. It measures the
% average time that the operating system and graphic card require to
% execute this switching from backbuffer to frontbuffer. For a more
```

```
% detailed explanation see the file 'Psychtoolbox3-Slides.pdf' in the
% PsychDocumentation folder, or type 'Screen GetFlipInterval?'
slack = Screen('GetFlipInterval', win)/2;

% This loads the full engraving - note that after loading, the image data
% is in the Matlab variable 'X', which we rename to save for later use.
load durer
ti = X;
% We then convert this image into an OpenGL texture that is used for
% optimized drawing on screen using the 'MakeTexture' subcommand. Note,
% that this command returns a handle to that texture which is later needed
% for drawing.
targetImage = Screen('MakeTexture', win, ti);

% This loads the engraving detail - note that after loading, the image data
% is in the Matlab variable 'X', which we rename to save for later use.
load detail
pi = X;
% We then convert this image into an OpenGL texture that is used for
% optimized drawing on screen using the 'MakeTexture' subcommand. Note,
% that this command returns a handle to that texture which is later needed
% for drawing.
primeImage = Screen('MakeTexture', win, pi);

% Draw fixation spot to backbuffer. At this point, nothing is shown to the
% user - we need to issue a 'Flip' command to actually show the stimulus on
% the screen.
% We use the default draw routines that draw a filled oval (in our case, as
% a circle) with a black color. Also note the use of the CenterRectInRect
% function supplied by the Psychtoolbox, which makes it much easier to draw
% a 20 x 20 circle centered on the screen.
Screen('FillOval', win, 0, CenterRect([0 0 20 20], winRect));

% The 'Flip' subcommand waits until the monitor's next refresh cycle and
% displays the image immediately. It returns the time at which it did so,
% which gives the most accurate representation of when the stimulus
% actually was shown on screen.
tfixation_onset = Screen('Flip', win);

% Now, draw the **prime** stimulus image into the backbuffer:
Screen('DrawTexture', win, primeImage);

% Use the previous timestamp to show the prime exactly 500 msecs after
% onset of the fixation spot, making sure to gather the next timestamp for
% this picture onset. This takes into account the system overhead for
% actually performing the flip operation as measured by the slack variable.
tprime_onset = Screen('Flip', win, tfixation_onset + 0.500 - slack);

% Now, draw the **target** stimulus image into the backbuffer:
Screen('DrawTexture', win, targetImage);

% Show target exactly 100 msecs after onset of prime image, again saving
% the stimulus onset time
ttarget_onset = Screen('Flip', win, tprime_onset + 0.100 - slack);
```

```
% Finally, show this target exactly for 200 msecs, then blank screen.
ttarget_offset = Screen('Flip', win, ttarget_onset + 0.200 - slack);

% Print results
% On a standard MacBook Laptop running MacOS X, this yields the
% following output showing very precise timing:
%
% time from fixation to prime onset: 0.501092 secs
% time from fixation to prime onset: 0.100188 secs
% time target shown: 0.200402 secs
%
fprintf('\n\n\ntime from fixation to prime onset: %f secs\n',...
    (tprime_onset-tfixation_onset));
fprintf('time from fixation to prime onset: %f secs\n',...
    (ttarget_onset-tprime_onset));
fprintf('time target shown: %f secs\n\n\n',...
    (ttarget_offset-ttarget_onset));

% Exit the Psychtoolbox screen and return to Matlab.
Screen('CloseAll');
```

11.5 Example 2: A More Involved Experiment

The following Matlab function gives a fully commented example of a ready-to-run experiment. It is based largely on a sample experiment shipped with the Psychtoolbox, but with added functionality and expanded comments for explaining the code. In order to run, the code requires two additional files, called `studylist.txt` and `testlist.txt`, examples of which are given below the Matlab code. Finally, the last file gives a sample output file for a short experiment run with the given input text files. Both text files, as well as the required input stimuli, are shipped with the Psychtoolbox and can be found under `Psychtoolbox/PsychDemos/PsychExampleExperiments/OldNewRecognition`.

```
function OldNewRecogExpTrans(parNo,hand,transparency)
% OldNewRecogExpTrans(parNo,hand,<transparency>);
%
% Example of an old/new recognition experiment.
%
% This is an example of a simple old/new recognition experiment. It is split
% into two phases, a study phase and a test phase:
%
% Study phase:
%
% In the study phase, the participant is presented with 3 images of objects
% in randomized presentation order and has to learn/memorize them. Each
% image is presented for a 'duration' of 2 seconds, then the image
% disappears and the participant has to advance to the next presentation by
```

```
% pressing the 'n' key on the keyboard. The list of 'study' objects is read
% from the file 'studylist.txt'.
%
% Test phase:
%
% In the test phase, the participant is presented with test images defined
% in the file 'testlist.txt', again in randomized order. Test images are
% presented for a 'duration' of 0.5 seconds or until the participants
% responds with a keypress. The participant has to press one of two keys,
% telling if the test image is an "old" image - previously presented in the
% study phase, or a "new" image - not seen in the study phase. The keys used
% are 'c' and 'm', the mapping of keys to response ("old" or "new") is
% selected by the input argument "hand" — allowing to balance for
% handedness of participants / response bias.
%
% In addition, during the test phase the images can be presented in a
% transparent fashion, with reduced contrast. This is controlled by the
% optional, third input argument "transparency", which is given as a
% percentage.
%
% At the end of a session (when all images in the 'testlist.txt' have been
% presented and tested), the results - both response of the participant and
% its reaction time - are stored to a file 'OldNewRecogExp_xx.dat', where
% 'xx' is the participant number given as input argument "subNo" to this
% script.
%
% Input parameters:
%
% parNo    participant number; use parNo>99 to skip check for existing file
% hand     response mapping for test phase
%
% e.g.: OldNewRecogExp(99,1);
%
% Example DESIGN:
%
% STUDY PHASE: study 3 objects
% TEST  PHASE: shown 6 objects, decide whether the object is old or new
%
% This script demonstrates:
%
%    - reading from files to get condition on each trial
%    - randomizing conditions (for the study and test phase)
%    - showing image/collecting response (response time, accuracy)
%    - writing data to file "OldNewRecogExp_<parNo>.dat
%
% History:
%
% 05/24/05 Quoc Vuong, PhD, University of NewCastle wrote and contributed
% it, as an example for usage of Psychtoolbox-3, Version 1.0.6.
% 03/01/08 Mario Kleiner modified the code to make use of new functionality
% added in Psychtoolbox 3.0.8.
% 02/28/11 Christian Wallraven extended the comments and added some new
% functionality
```

```
%%%%%%%%%%%%%%%%%%%%%%%%%
% any preliminary stuff
%%%%%%%%%%%%%%%%%%%%%%%%%

% Clear Matlab/Octave window:
clc;

% check for Opengl compatibility, abort otherwise:
AssertOpenGL;

% Check if all needed parameters given:
if nargin < 2
    error('Must provide required input parameters "parNo" and "hand"!');
end

% if we only have two arguments, set transparency level for images to
% fully opaque
if nargin==2
    transparency = 100;
end


% This will stop Matlab to listen for keypresses in the command window -
% use with care, as it will also require the user to perhaps press CTRL + C
% when Matlab quits unexpectedly
ListenChar(2);

% Reseed the random-number generator for each experiment
% This is an important Matlab procedure as otherwise each randomization
% will start with the SAME numbers after each Matlab restart
rand('state',sum(100*clock));

% Make sure keyboard mapping is the same on all supported operating systems
% Apple MacOS/X, MS-Windows and GNU/Linux:
KbName('UnifyKeyNames');

% Init keyboard responses (caps doesn't matter)
advancestudytrial=KbName('n');

% Use input variable "hand" to determine response mapping for this session.
% This is good practice as it avoides "handed-ness bias", that is, half of
% the participants get one keyboard layout with the key for the
% old-response on one side, and for the new-response on the other
if (hand==1)
    oldresp=KbName('c'); % "old" response via key 'c'
    newresp=KbName('m'); % "new" response via key 'm'
else
    oldresp=KbName('m'); % Keys are switched in this case.
    newresp=KbName('c');
end

%%%%%%%%%%%%%%%%%%%%%%%
% file handling
%%%%%%%%%%%%%%%%%%%%%%%
```

```
% Define filenames of input files and result file:
% name of data file to write to
datafilename = strcat('OldNewRecogExpTrans_',num2str(parNo),'.dat');
% study list
studyfilename = 'studylist.txt';
% test list
testfilename  = 'testlist.txt';
% folder, in which all stimulus images reside
stimfolder = 'stims/';

% Check for existing result file to prevent accidentally overwriting
% files from a previous participant/session
% This behavior can be avoided by using participant numbers > 999, which
% becomes useful for quick debugging, or for testing out the code
if parNo<999 & fopen(datafilename, 'rt')~=-1
    fclose('all');
    error('Result file already exists! Choose different participant number
        .');
else
    datafilepointer = fopen(datafilename,'wt'); % open ASCII file for
        writing
end

%%%%%%%%%%%%%%%%%%%%%%%%%
% experiment
%%%%%%%%%%%%%%%%%%%%%%%%%

% Embed core of code in try ... catch statement. If anything goes wrong
% inside the 'try' block (Matlab error), the 'catch' block is executed to
% clean up, save results, close the onscreen window etc.
try
    % Get screenNumber of stimulation display. We choose the display with
    % the maximum index, which is usually the right one, e.g., the external
    % display on a Laptop:
    screens=Screen('Screens');
    screenNumber=max(screens);

    % Hide the mouse cursor:
    HideCursor;

    % Returns as default the mean gray value of screen:
    gray=GrayIndex(screenNumber);

    % Open a double buffered fullscreen window on the stimulation screen
    % 'screenNumber' and choose/draw a gray background. 'w' is the handle
    % used to direct all drawing commands to that window - the "Name" of
    % the window. 'wRect' is a rectangle defining the size of the window.
    % See "help PsychRects" for help on such rectangles and useful helper
    % functions:
    [w, wRect]=Screen('OpenWindow',screenNumber, gray);

    % Set text size (Most Screen functions must be called after
    % opening an onscreen window, as they only take window handles 'w' as
```

```
% input):
Screen('TextSize', w, 32);

% Do dummy calls to GetSecs, WaitSecs, KbCheck to make sure
% they are loaded and ready when we need them - without delays
% in the wrong moment:
KbCheck;
WaitSecs(0.1);
GetSecs;

% Set priority for script execution to realtime priority:
priorityLevel=MaxPriority(w);
Priority(priorityLevel);


% run through study and test phase
for phase=1:2 % 1 is study phase, 2 is test phase

    % Setup experiment variables etc. depending on phase:
    if phase==1 % study phase

        % define variables for current phase
        phaselabel='study';
        duration=2.000; % Duration of study image presentation in secs.
        trialfilename=studyfilename;

        % create string with instructions for participant
        message = ['study phase ...\nstudy each picture ...  '...
            'press _n_ when it disappears ...\n'...
            '... press mouse button to begin ...'];
    else        % test phase

        % define variables
        phaselabel='test';
        duration=0.500;  % Duration of test image presentation in secs.
        trialfilename=testfilename;

        % create string with instructions to participant
        str=sprintf('Press _%s_ for OLD and _%s_ for NEW\n',...
            KbName(oldresp),KbName(newresp));
        message = ['test phase ...\n' str ...
            '... press mouse button to begin ...'];
    end

    % Write instruction message for participant, nicely centered in the
    % middle of the display, in white color. As usual, the special
    % character '\n' introduces a line-break:
    DrawFormattedText(w, message, 'center', 'center', WhiteIndex(w));

    % Update the display to show the instruction text:
    Screen('Flip', w);

    % Wait for mouse click:
    GetClicks(w);
```

```
% Alternatively, one could use KbWait to wait for any keypress as
% in the following command:
%
% KbWait(-1,2);


% Clear screen to background color (our 'gray' as set at the
% beginning):
Screen('Flip', w);

% Wait a second before starting trial
WaitSecs(1.000);

% Read list of conditions/stimulus images.
% Note that textread() is a matlab function
%
% objnumber   arbitrary number of stimulus
% objname     stimulus filename
% objtype     1=old stimulus, 2=new stimulus
%             for study list, stimulus coded as "old"
[ objnumber, objname, objtype ] = textread(trialfilename,'%d %s %d')
    ;

% Randomize order of list
ntrials=length(objnumber);          % get number of trials
randomorder=randperm(ntrials);      % randperm() is a matlab function
objnumber=objnumber(randomorder);   % need to randomize each list!
objname=objname(randomorder);       %
objtype=objtype(randomorder);       %

% loop through trials
for trial=1:ntrials

    % wait a bit between trials
    WaitSecs(0.500);

    % initialize KbCheck and variables to make sure they're
    % properly initialized/allocted by Matlab - this is done to
    % time delays in the critical reaction time measurement part of
    % the script:
    [KeyIsDown, endrt, KeyCode]=KbCheck;

    % read stimulus image into matlab matrix 'imdata'
    stimfilename=[stimfolder char(objname(trial))];
    imdata=imread(char(stimfilename));

    % make image transparent
    if(phase==2)
        if (isempty(size(imdata,3)))
            % if we have a grayscale image, augment by making array
            % three-dimensional
            imdata(:,:,2)=round(255*transparency/100);
        else
```

```
        % if we have a color image, augment color array
        % by another component
        imdata(:,:,4)=round(255*transparency/100);
    end
end
% blending function - OpenGL accelerated
Screen('BlendFunction', w, GL_SRC_ALPHA,GL_ONE_MINUS_SRC_ALPHA);


% make texture image out of image matrix 'imdata'
tex=Screen('MakeTexture', w, imdata);

% Draw texture image to backbuffer. It will be automatically
% centered in the middle of the display if you don't specify a
% different destination:
Screen('DrawTexture', w, tex);

% Show stimulus on screen at next possible display refresh cycle
    ,
% and record stimulus onset time in 'startrt':
[VBLTimestamp startrt]=Screen('Flip', w);

% while loop to show stimulus until participant's response or
% until "duration" seconds elapsed.
while (GetSecs - startrt)<=duration
    % poll for a response using KbCheck
    % during test phase, participants can respond
    % BEFORE stimulus terminates
    if ( phase==2 ) % if test phase
        if ( KeyCode(oldresp)==1 | KeyCode(newresp)==1 )
            break;
        end
        [KeyIsDown, endrt, KeyCode]=KbCheck;
    end

    % Wait 1 ms before checking the keyboard again to prevent
    % overload of the machine at elevated Priority():
    WaitSecs(0.001);
end

% Clear screen to background color after fixed 'duration'
% or after participants response (on test phase)
Screen('Flip', w);

% loop until valid key is pressed
% If a response is made already, then this loop will be skipped
% as the the KbCheck is made in the loop above.
% If you do not want this behavior, call
% FlushEvents('keyDown');

if ( phase==1 )
    % study phase: go to next stimulus with proper key
    while (KeyCode(advancestudytrial)==0)
        [KeyIsDown, endrt, KeyCode]=KbCheck;
```

```
                WaitSecs(0.001);
            end
        end

        if ( phase==2 )
            % test phase: wait for one of the two valid keypresses
            while ( KeyCode(oldresp)==0 & KeyCode(newresp)==0 )
                [KeyIsDown, endrt, KeyCode]=KbCheck;
                 WaitSecs(0.001);
            end
        end

        % compute response time in milliseconds
        rt=round(1000*(endrt-startrt));

        % compute accuracy
        if (phase==1 ) % study phase: no accuracy needed
            ac=1;
        else            % test phase: accuracy needs to be determined
            ac=0;
            % code correct if old-response with old stimulus,
            % new-response with new stimulus
            if ( (KeyCode(oldresp)==1 & objtype(trial)==1) ...
                    | (KeyCode(newresp)==1 & objtype(trial)==2) )
                ac=1;
            end
        end

        resp=KbName(KeyCode); % get key pressed by participant

        % Write trial result to file as a formatted string
        % Note, that this is done for each trial, rather than after the
        % experiment in one go - in case of a sudden program failure
        % this will at least ensure that the data is saved up to that
        % point.
        fprintf(datafilepointer,'%i %i %s %i %s %i %i %s %i %i %i\n',
            ...
            parNo, ...
            hand, ...
            phaselabel, ...
            trial, ...
            resp, ...
            objnumber(trial), ...
            transparency, ...
            char(objname(trial)), ...
            objtype(trial), ...
            ac, ...
            rt);

    end % for trial loop
end % phase loop

% Cleanup at end of experiment - re-enable keyboard listening, close
% window, show mouse cursor, close result file, switch Matlab/Octave
```

```
    % back to priority 0 — normal priority:
    ListenChar(0);
    Screen('CloseAll');
    ShowCursor;
    fclose('all');
    Priority(0);

    % End of experiment:
    return;
catch
    % catch error: This is executed in case something goes wrong in the
    % 'try' part due to programming error etc.:

    % re-enable keyboard listening by Matlab
    ListenChar(0);

    % Do same cleanup as at the end of a regular session...
    Screen('CloseAll');
    ShowCursor;
    fclose('all');
    Priority(0);

    % Output the error message that describes the error:
    psychrethrow(psychlasterror);
end % try ... catch %
```

File: studylist.txt The first column codes stimulus number, the second column lists stimulus file name (without stimulus directory), and the third column lists old/new, which for the study phase is by default set to old, that is, to `1`.

```
1   stim1.jpg   1
2   stim2.jpg   1
3   stim3.jpg   1
```

File: testlist.txt The first column codes stimulus number, the second column lists stimulus file name (without stimulus directory), and the third column codes `1` for old and `2` for new stimuli.

```
1   stim1.jpg   1
2   stim2.jpg   1
3   stim3.jpg   1
4   stim4.jpg   2
5   stim5.jpg   2
6   stim6.jpg   2
```

File: OldNewRecogExpTrans_999.dat This sample output file has 11 columns, which code participant number, handedness, name of the block, trial number for each block, key pressed after each stimulus, real stimulus number as specified in studylist.txt and testlist.txt, whether the stimulus was old (`1`) or new (`2`), whether the the participant answered correctly (`1`) or incorrectly (`0`), and the response time in milliseconds. In this sample file, the participant was 100% correct on all test trials.

```
999 1 study 1 n 1 5 stim1.jpg 1 1 2471
999 1 study 2 n 3 5 stim3.jpg 1 1 2410
999 1 study 3 n 2 5 stim2.jpg 1 1 2430
999 1 test  1 c 2 5 stim2.jpg 1 1 802
999 1 test  2 c 1 5 stim1.jpg 1 1 784
999 1 test  3 m 4 5 stim4.jpg 2 1 704
999 1 test  4 m 5 5 stim5.jpg 2 1 708
999 1 test  5 c 3 5 stim3.jpg 1 1 753
999 1 test  6 m 6 5 stim6.jpg 2 1 778
```

IV

Data Analysis

Chapter 12

Statistical Issues

Up to now we have gathered a lot of information on how to design and run experiments. We now find ourselves at the final stage: after having run a good number of participants through a study, how do we analyze the data properly? What kind of statistical tools are appropriate? The answer, of course, depends on the particular type of experiment that was run and the questions behind it.

This final part of the book provides a broad overview of the central factors involved in analyzing the participants' responses. In addition to examining, summarizing, and interpreting the responses, knowledge of which forms of data analysis one can and will do should have a strong impact on the actual experimental design.

This chapter provides an introductory look at statistics. It will include a discussion of the difference between descriptive and inferential statistics, between type I and II errors, and between individual and group analyses. Finally, an overview of the most common statistical tests will be given. Throughout this chapter, we have tried to highlight common pitfalls and problems, and to provide guidelines for how to use the data analysis tools in the context of the experimental design. Most importantly, we try to sensitize the reader to look at not only the statistical but also the practical significance of the result.

12.1 Variables

In any experiment, there are variables that are measured. In the context of this book, these variables usually are the responses of participants as they solve a particular task as part of an experiment. It is important to stress that the measured variables are assumed to be random variables, that is, that participants' responses are *samples* drawn from an underlying *statistical distribution* of possible values. The goal of the experiment is to use this sample to draw some conclusions about this distribution, for example, to confirm a *hypothesis* or a *model* about the data.

Before we explain the concepts of sampling, distribution, and hypothesis testing further, let us focus on what types of variables there might be. In general, an experimenter might be interested in different kinds of measurements. As an example, let us consider an experiment in which the quality of several computer-animated avatars should be measured. The first goal of the experimenter is of course to define what exactly "quality" means, such that the term "quality" can be turned into a measurable variable.

12.1.1 Interval Variables

The experimenter might choose to think about quality in very specific terms and postulate that what makes an avatar good in terms of quality is that its facial expressions are easy to recognize. That is, quality becomes recast in terms of recognition accuracy. The experimenter therefore has the different animations display a set of facial expressions, which participants have to recognize—for example, by entering the corresponding number from a list of given expressions. In such a design there is a right and a wrong answer, as the experimenter knows what expression was shown and can find out whether the participant selected the correct expression from the list. After having gathered recognition performance data from a suitable number of participants, the experiment would yield a quality measure for each animation style, which is the average recognition accuracy for that particular style. Recognition accuracy as a variable behaves very much like any other measurable number, and therefore it is possible to calculate its average, its variance, and other properties.

Another example for such a variable would be response time, or the time it took the participant to make a response after the stimulus has been shown. In the context of this experiment, response time would also be a viable quality measure, as it would surely be important how long participants take to make decisions about the facial expression that is being shown.

Variables like recognition accuracy and response times are called *interval variables*[1] and have the following properties.

- *Ordinal properties.* Interval variables can be ordered, that is, given two variables a and b, it is clear whether one variable is less than, greater than, or equal to the other variable.

- *Interval properties.* Interval variables have the interval property; that is, a given increase (for example, of 10%) in a variable always corresponds to the same conceptual increase. If one participant is 80% accurate, another participant is 70% accurate, and a third participant is 90% accurate, the

[1] To be precise, such a variable is actually a *ratio variable*, that is, in addition to the two properties listed, it also has the notion of a zero point such that a zero value of that variable means the complete absence of the corresponding property. In most perceptual or cognitive experiments, however, distinguishing between these two variable types is usually not necessary.

differences in accuracy of 10% between participant 1 and participant 2, and that of 10% between participant 1 and participant 3, are exactly the same difference in terms of performance.

Most often, interval variables are assumed to follow the normal distribution, which makes it possible to conduct rigorous statistical tests that would, for example, allow us to determine whether one animation was significantly better than another in terms of recognition accuracy and/or response time. In most cases, these statistical tests are based on calculating means and variances of the variables to compare.

12.1.2 Ordinal Variables

Another way of defining quality in the avatar experiment might be to have participants rate the believability of the avatar on a scale of, say, 1 to 7. Again, one needs to define believability in more concrete terms—most importantly, one needs to anchor the lower and higher end of the scale by giving a concrete example of what the ratings of 1 and of 7 should stand for.

After having tested a suitable number of participants, the experimenter now has gathered a set of numerical values for each animation type. Even though at first glance these numbers would seem to behave like real numbers, they actually are not quite like that. There is no guarantee that a rating difference of 1 actually represents an equal conceptual change, as the scales might be used differently by different participants. For example, some participants might avoid using the extreme values of 1 and 7, while others might spread their answers much more across the scale or might make more distinctions for higher-quality animations than for lower-quality ones, etc. This uncertainty results in the loss of the interval-property for rating-scale variables; that is, our variable only retains the so-called ordinal property discussed in Section 12.1.1.

As *ordinal variables* do not have the interval property, there is a different set of statistical tests that needs to be used for these types of variables. Most importantly, rather than calculating averages and variances, such tests will rely on rank-order statistics based on medians and quartiles, as they cannot make use of the normality assumption (the assumption that they will follow a normal distribution). Ordinal variables result every time an ordering or ranking (by time, performance, height, quality, etc.) is made in the measurements.

As a more extreme example of an ordinal variable, the experimenter could choose to define three quality levels, such as, high, medium, and low. By giving a more detailed definition of what these levels might entail, a typical survey-type experiment would now consist of showing each computer animation to participants and asking them to indicate to which of the three quality levels the avatar would belong. Using only these three types of answers, it would probably be a good idea to test many participants, as the data consists of only one answer per participant. Having tested a suitable number of participants, the experimenter can now look at which quality level was chosen by how many people for each animation type. Note that it is possible to put these variables into a mathematical order; that is, high > medium > low. As an alternative, the experimenter could have assigned numbers from 1 to 3 to the three different categories, in which case the variable "animation quality" would still be an ordinal variable.

12.1.3 Nominal Variables

Finally, the experimenter might decide to make three different quality categories that are based on what the avatar could be used for. For example, the categories could be "suitable for a kiosk system," "suitable for the movie industry," and "suitable for computer games." This experiment uses categorical measurement variables, which are also called *nominal variables.* As the term implies, these are named categories that are abstract containers for various concepts. Without any additional assumptions, these three categories cannot be ordered in any meaningful way, it makes no sense to calculate an average value across the three categories. Analysis of nominal variables typically looks at *frequency distributions,* that is, how often a particular category was chosen as in the above example. Such variables occur most often in experimental designs that are based on questionnaires or on free descriptions.

12.1.4 Guidelines

It is important to note that while it is possible to come up with clear mathematical definitions about the three variable categories that were just presented, assigning a given variable *uniquely* to one category might not be such an easy task. The example given in Section 12.1.2 for the rating scale illustrates this problem: when each number between 1 and 7 is anchored properly (by giving a clear definition for each value) and an increase of 1 on the numerical scale always corresponds to a similar conceptual increase in quality, the scale actually becomes an interval scale. Indeed, research on rating scales has shown that in most cases treating them as interval variables will not overly distort the statistical results.

Another exception is when only two categories are formed for an ordinal variable; since the difference (or interval) between two variables is always fixed, it becomes possible to treat this variable as an interval variable. For example, an experiment might only have two values on a rating scale, such as 0 for a non-effective avatar, and 1 for an effective avatar. Here, the difference of 1 between the two conditions specifies the full range of the scale, and as such all properties of the interval variable are retained.

There are two main reasons why the definitions of the different variable types are included here despite their potential conceptual difficulties: first, the three terms occur regularly in statistics books and statistical analyses, and second, the broad distinction between these types helps to illustrate the applicability of different statistical tests and their underlying assumptions. Indeed, the usefulness of these variable types lies not in a strict adherence to the definition, but rather in their utility as benchmarks in determining whether a certain type of scale structure more closely approximates the reality of the experimental task.

12.2 Distributions

In the preceding discussion about scale and variable types, one thing that was possible for all variables was to count how often they occurred. Doing this for all possible values of the variable will result in a frequency distribution. Obviously, the shape of the resulting distribution is not constrained at the outset and can take any number of forms.

Figure 12.1 shows four types of distributions that are often seen in real-world experimental data. The first type is a *constant distribution* in which all variable values are equally likely to occur. The second type is the *Gaussian distribution*, the most important distribution for our purposes as it models a wide range of experimental data reasonably well. If the frequency distribution is normalized such that all values sum up to one, the Gaussian distribution

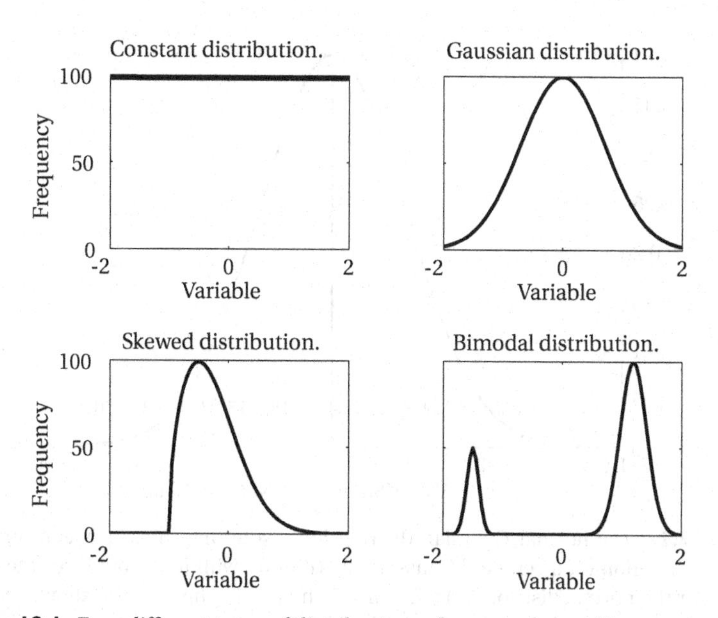

Figure 12.1. Four different types of distributions often encountered in experiments.

is also called the *normal distribution.* Indeed, the normal distribution underlies a whole series of statistical tests which make it possible to compare different datasets for statistical difference based on the assumption that each of the datasets comes from a normally distributed population. The third type is a *skewed distribution* in which certain variable values are more likely to occur than others, but the distribution itself is not symmetric (as is the Gaussian distribution), instead leaning towards a certain range of values (it is skewed). The final type of distribution is a *bimodal distribution,* in which two ranges of values seem to occur quite often, while the other intervals do not occur at all.

12.2.1 The Gaussian or Normal Distribution

The Gaussian distribution is defined as

$$\mathcal{N}(\mu,\sigma) = \frac{1}{\sqrt{2\pi}\sigma} e^{-\frac{(x-\mu)^2}{2\sigma^2}},$$

with μ as the mean of the distribution and σ as its standard deviation. Note the normalization factor, which is added to ensure that the integral $\int_{-\infty}^{\infty} \mathcal{N}(\mu,\sigma) = 1$, as is required with a probability distribution. The normal distribution is obtained by setting $\mu = 0$ and $\sigma = 1$. The graph of the normal distribution is shown in Figure 12.2. The distribution is symmetric around the mean and has an integral of 1. Figure 12.2 also shows the percentage of scores that lie within ± 1.

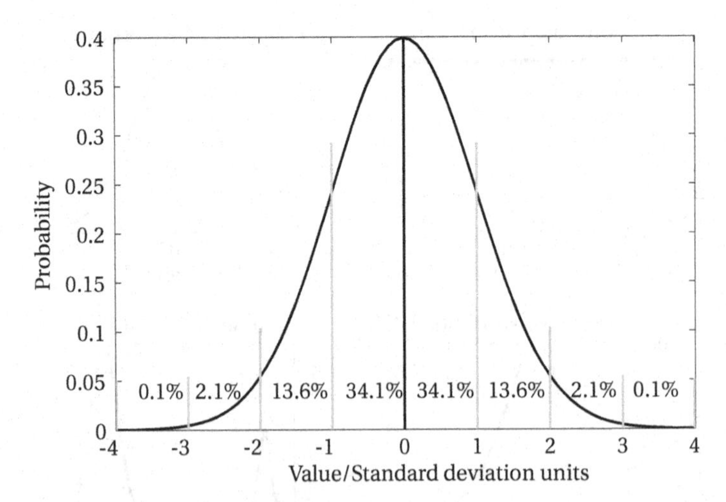

Figure 12.2. The normal/Gaussian distribution. Note that the x-axis is in units of standard deviation for a centered Gaussian distribution with $\mu = 0$, but gives the value directly for the normal distribution when $\mu = 0$ and $\sigma = 1$. The plot also shows the percentages of scores that fall into successive bins of size 1.

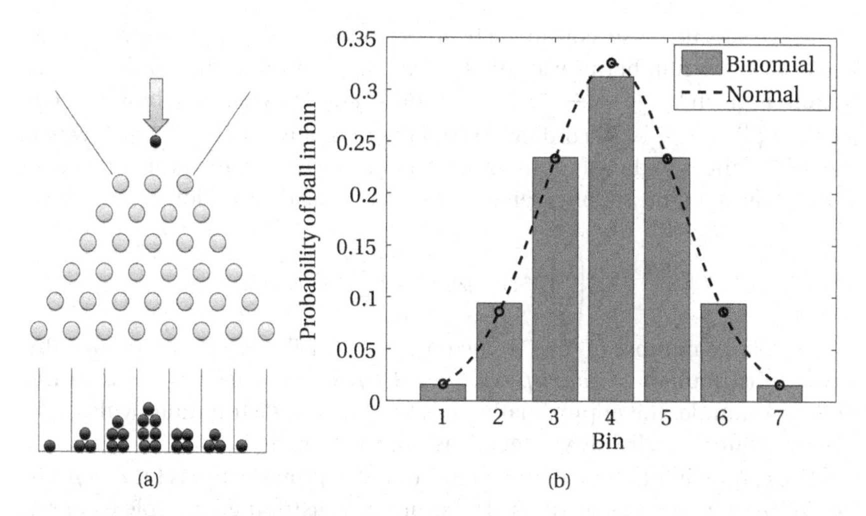

Figure 12.3. (a) Illustration of the "bean machine," in which balls fall in from the top and bounce off rows of pegs, eventually to be collected in bins at the bottom. (b) The resulting probability distribution of balls is given in a binomial distribution, which resembles a normal distribution.

For a regular Gaussian probability distribution the graph would look exactly the same, but the x-axis would be labeled with units of *standard deviation* (see Section 12.4.2 for an explanation of standard deviation). For example, whenever a statistical process follows a Gaussian distribution, roughly 68.2% of all values lie within ± 1 standard deviation, roughly 95.4% lie within ± 2 standard deviations, and roughly 99.8% lie within ± 3 standard deviations around the mean. These values will come in handy to remember when gauging the behavior of data. Another value that will be important later in hypothesis testing is that exactly 95% of the values lie within 1.96 standard deviations: or in other words, if a variable is normally distributed, only 5% of the values are expected to lie outside of 1.96 standard deviations.

As mentioned above, the normal distribution is the most important distribution for statistical analyses as it forms the basis for a whole series of important statistical tests. The reason for this lies in one of the most important theorems of statistics: the *central limit theorem.* The central limit theorem is beautifully illustrated by the so-called bean-machine invented by Sir Francis Galton. This device is shown in Figure 12.3(a) and consists of rows of pegs and several bins at the bottom, much like the Japanese pachinko machines. A ball is thrown into the machine from the top and bounces off the pegs, making its way down into a bin. If many balls are thrown into the machine and the number of peg rows is large enough, the distribution of balls in the bins will quickly start to approximate the characteristic Gaussian bell shape.

To see why this is so, consider a bean machine consisting of n rows of pegs. Recall that the number of ways that a ball can end up in the kth bin at the bottom is given by $\frac{n(n-1)(n-2)...(n-k+1)}{k!}$. This is also known as the binomial coefficient $\binom{n}{k} = \frac{n!}{k!(n-k)!}$. Accordingly, given the probability of $p = 0.5$ with which the ball bounces to the left or to the right at any given pin, the probability that a ball ends up in the kth bin is given by the binomial distribution

$$\binom{n}{k} p^k (1-p)^{n-k} = \binom{n}{k} 0.5^n.$$

For a large number of rows n, the binomial distribution tends towards the normal distribution $\mathcal{N}\left(\mu = np, \sigma = \sqrt{np(1-p)}\right)$. Even for $n = 6$, as in the present example, the approximation of the binomial distribution by the normal distribution is already very good, as shown in Figure 12.3(b).

The central limit theorem now states that this property holds for *any probability distribution,* that is, sums of an arbitrarily distributed variable will tend to a normal distribution as the number of trials gets larger.

12.2.2 Guidelines

Because of the central role of normal distributions, it is often assumed that the experimental data follows the normal distribution without further checks to see whether this is really the case. Violations of the normality assumption result in more or less accurate statistical results depending on how strong the deviation from normality is. In the case of the bimodal distribution shown in Figure 12.1, for example, the approximation of the two peaks by one weak, average peak would surely result in misinterpretations of the data. With all experimental data, one of the first steps should therefore be plotting the frequency distribution to check for obvious, non-normally distributed data. In addition, there are several statistical methods to test for normality of the data. Another possibility is to transform the data so that normality can be assumed; this is useful, for example, for constructing confidence intervals around means of skewed data, or for making data amenable to tests assuming normality. For more information on these transformations, see Cohen et al. (2003).

In the context of statistics there are several other types of distributions that play an important role, as they relate to specific types of random variables. While it may be unlikely that these distributions are directly observed in experimental data, their value lies more in allowing for certain kinds of statistical tests based on parameters derived from the data. These special distributions are briefly introduced in the context of hypothesis testing in Section 12.6.1.

12.3 Descriptive versus Inferential Statistics

The difference between description and inference is crucial in understanding statistical analyses of experimental data of any kind. Here are two short definitions of descriptive and inferential statistics.

The goal of descriptive statistical methods is to uncover general patterns or trends through analysis of experimental data. These results are always constrained by the data itself—calculating the average value of the data, or plotting the distribution of data values, for example. Descriptive statistics should always be used as the first step in any experimental analysis in order to investigate the data to check for outliers, errors in data acquisition, etc. *Descriptive statistics remain specific and analyze only what has been measured.*

The goal of inferential statistical methods is to draw conclusions and make predictions from the experimental data. Inferential statistics often constitute the second step in experimental analysis: for example, one might want to test specific hypotheses that go beyond the gathered data, such as whether the average value of the experimental results that came from only 20 participants is representative of the general population. *Inferential statistics seek to generalize, and they try to make claims beyond the measured data based on statistical assumptions.*

12.4 Descriptive Statistics

As the definition states, descriptive statistics seek to describe the data, to characterize it in terms of numerical quantities that can be calculated based on the gathered experimental data. Given this scope, it is not surprising that most of the methods used in descriptive statistics are intimately connected to frequency distributions made from the data. The most common methods are frequency counts, ranges (high and low scores), means, modes, median scores, and standard deviations.

To illustrate these different methods, consider the data shown in Table 12.1, which consists of ratings from 24 students on the difficulty level of an exam. The rating was to be given as a number between 1 and 5, and was anchored by providing definitions for the two extreme values (1=very easy, 5=very difficult) and for the middle value (3=neutral). The lecturer has gathered the data and is now interested in getting a feeling for how difficult students found the exam, so that he or she can think about whether to revise questions for future exams, or whether to modify the grading strategy.

12.4.1 Measures of Central Tendency

One of the easiest things to do with the data from Table 12.1 is to count how often each value occurs. This will yield a frequency count, which is shown in

Student	1	2	3	4	5	6	7	8	9	10	11	12
Rating	3	3	4	3	5	5	3	4	3	3	4	5
Student	13	14	15	16	17	18	19	20	21	22	23	24
Rating	3	4	4	3	3	4	5	5	4	5	1	3

Table 12.1. Data showing ratings from 24 students on the difficulty level of an exam (1=very easy, 3=neutral, 5=very difficult).

Figure 12.4(a). It is easy to see that the value 1 occurs one time, that the value 2 does not occur at all, etc. Note also that this plot clearly shows that the distribution of the ratings is far from symmetric; indeed, the data is skewed heavily towards larger rating values above 3.

Going one step further, the lecturer might simply be interested in what value occurs most often—after all, the value that most students chose could be seen as representative for the class. This value is called the *mode* of the frequency distribution, or the value that occurs most often. Looking at Figure 12.4(a), the mode is 3, as it was chosen ten times; the exam would thus be judged as neutral in terms of difficulty.

The question is, whether 3 is, indeed, a good representative value. In this case, Figure 12.4(a) shows that the two higher ratings of difficulty, 4 and 5, also received a fair number of votes (seven and six, respectively). This means that in addition to the ten students who selected a neutral level of difficulty, 13 students actually found the exam to be either moderately difficult, or very difficult. Surely then, judging the overall outcome of the ratings as neutral would neglect the opinion of more than half the students. In order to take everyone's opinion into account, one could calculate the *mean* of all ratings, which is defined as the sum of all elements divided by the number of elements, or

$$\text{mean} = \bar{x} = \mu = \frac{\sum_{i=1}^{n} x_i}{n},$$

where x_i is the ith element, and the sum runs from 1 to all n elements. In our case, the mean of the ratings is 3.71. Using this measure, the exam was rated close to moderately difficult.

As the mean takes all data into account, one might think that this is in general the best measure to use. Considering Figure 12.4(a) again, however, there is one small problem. Whereas 23 students chose values of 3 and higher, only one student chose a value of 1 (judging the exam as very easy). It could be that this student's rating reflected a true judgment of the exam's difficulty, but since every other student selected higher values, the lecturer might wonder whether that was the case. It is also possible that the student had misunderstood the scale, and wrote down 1, when 5 was what he or she really meant. Ultimately, of course, the answer can only be determined by interviewing the student again,

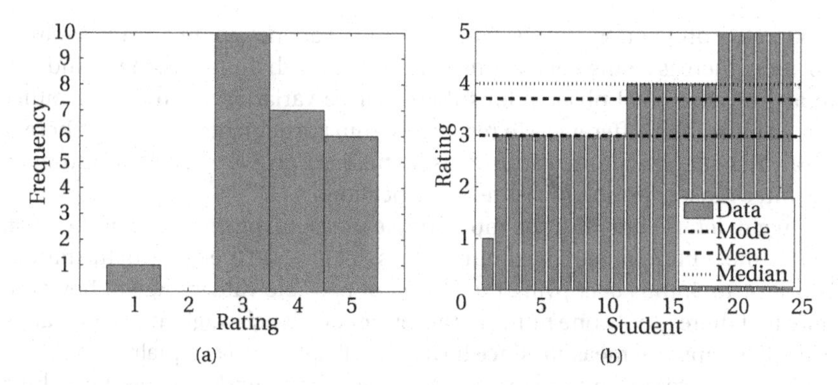

Figure 12.4. (a) Frequency counts for the data from Table 12.1, (b) Full data plot including mode, mean, and median of the data from Table 12.1. Note that the y-axis for (b) is plotted from 0 to 5 in order to make all of the ratings visible, even though the rating scale only ranges from 1 to 5.

and/or looking at the student's exam score, which may also give an additional hint.

Without doing this, however, the overall assessment of the exam's difficulty as given by the mean is very sensitive to this one *outlier.* There is another measure that is less sensitive to such extreme variations; this measure is called the *median* and is defined as the value above which 50% of the data lies. In order to determine the median value, one sorts all of the data in increasing order and then looks for the value above which 50% of the data points fall. In the case of the rating data this value is *4*, which gives an overall moderately difficult level for the exam.

Figure 12.4(b) then shows all three measures (mode, mean, median) together with the full dataset. These three measures are also referred to as measures of *central tendency,* as they all try to characterize the center of the underlying distribution. Note that for symmetric distributions all of these measures will give the same value, but for skewed distributions they will differ quite substantially. A simple, eyeball test to check for normality of the data is to verify that all three measures of central tendency are the same (which they must be for a normal distribution).

12.4.2 Measures of Dispersion

Given that there are measures of central tendency, there also must be other types of tendency measures. Indeed, the other important type of measure in descriptive statistics relates to the *dispersion* (or *spread*) of the distribution. Returning to the exam ratings, the lecturer now has concluded that—using the median as a suitable criterion—students on average found the exam to be moderately difficult. As the discussion for the frequency count already showed, however, there is a range of responses associated with this assessment. Indeed, in any experimental observations there will be variations in the data, either due to individual differences in responses from participants or due to inherent measurement noise from instruments. In this section, we will briefly introduce the three most common measures of dispersion.

Again, let us start with the most simple measure, or the *range* of the data; that is, the minimum and maximum values. In our case the range of the ratings is from 1 to 5, thus occupying the full set of possible values. Again, however, note that there is only one rating at the lower end, which might make the range a less than optimal measure since it weights all data entries equally.

The most common measure of dispersion is the *standard deviation,* which measures the squared difference from the mean for each data point. The squared differences are then summed up and divided by the number of data points;[2] the square root of this final value yields the standard deviation

$$\text{standard deviation} = \sigma = \sqrt{\frac{\sum_{i=1}^{n}(x_i - \bar{x})^2}{n}}. \tag{12.1}$$

Using this definition, the standard deviation of the exam ratings becomes $\sigma = 0.98$. The mean and standard deviation are often reported together, such that one would say that on average, difficulty ratings of the exam were 3.71 ± 0.98. A plot of the rating data showing the mean and the standard deviation is shown in Figure 12.5(a).

Another important measure that is related to the standard deviation is the *variance* of the data. This is simply the squared standard deviation or

$$\text{variance} = \sigma^2 = \frac{\sum_{i=1}^{n}(x_i - \bar{x})^2}{n}. \tag{12.2}$$

As we saw previously, the mean of the ratings might be overly influenced by one outlier. The same holds true for the calculation of the standard deviation, as is shown in Equation 12.1, in order to calculate the standard deviation, the sum of squared differences to the mean is determined. One outlier—that is, a value that is particularly far away from the mean—will therefore influence the standard deviation disproportionately.

In order to correct for this, one can resort to a strategy similar to the one discussed previously for the calculation of the median. The data is first sorted, but instead of determining the point above which 50% of the data lie, the points above which 25% and 75% of the data lie are determined. These two values are called the *lower quartile* (25%) and the *upper quartile* (75%), respectively.

[2]Note that when dividing by the total number of samples, this is the definition of the standard deviation of the sample and therefore is only a good estimate of the true standard deviation when the whole population has been measured (see Section 12.5.3).

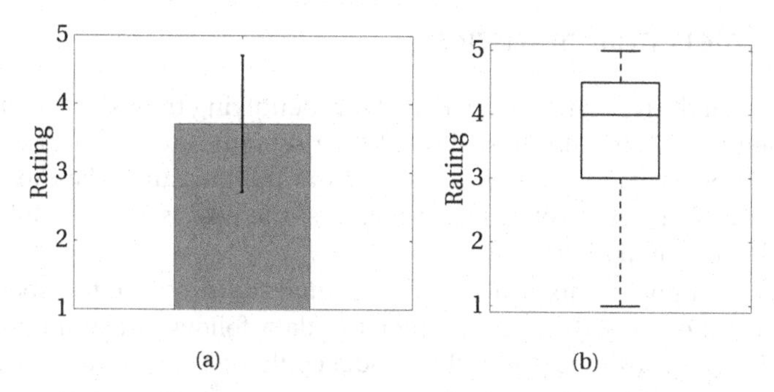

Figure 12.5. (a) Representation of the data in Table 12.1 using a bar plot showing the mean and standard deviation. (b) Box-plot of the data in Table 12.1 showing the median, the upper and lower quartiles, and the range of the data.

The 25%, 50%, and 75% quartiles are sometimes abbreviated as q_1, q_2, and q_3, respectively.

The corresponding measure for dispersion is the difference between the upper and lower quartile, called the *interquartile range.* For symmetric distributions—and most importantly for the normal distribution—these upper and lower quartiles will, of course, also be symmetrical around the mean or median of the data. In our case, however, the distribution is not symmetrical (compare to Figure 12.4(a)), and therefore the lower and upper quartiles will not be symmetrical—the lower quartile is 3.0 and the upper quartile is 4.5, with a median of 4. The interquartile range is therefore 1.5.

The typical way of plotting data with the median as the central measure and the interquartile range as the measure of dispersion is called the *box-plot,* which is shown in Figure 12.5(b). Note that the asymmetry of the distribution is well captured in the asymmetric "box" around the median and that in this case the plot also shows the full range of the rating data as the "whiskers" around the box—strictly speaking, this plot is then a *box-whisker-plot.* The length of the whiskers in this case is calculated as a multiple of the interquartile range: the upper whisker is drawn up to $q_3 + 1.5(q_3 - q_1)$, and the lower whisker is drawn down to $q_1 - 1.5(q_3 - q_1)$.

12.4.3 Guidelines

Descriptive statistics should be the first port of call for any analysis of data. Plot the data in many different ways, calculate the different measures of central tendency and dispersion, try to spot outliers in the data. In most cases, problems with the data can be already identified at this stage—remember that the human brain has evolved to extract patterns from visual data, and we might as well use these skills for our purpose.

12.5 Inferential Statistics

Whereas descriptive statistics are useful for identifying trends in the data and might be used to spot outliers, skewed distributions, etc., in most cases the experimenter wants to *generalize* results from the measured data to a larger scope. That is, given a set of measurements, the goal is to make inferences about a larger population.

Another important application of inferential statistics is to test specific assumptions about the data, or whether the data follows a certain statistical model. In its most simple form, this model could be stated as a yes/no question: is this computer animation better than another one given the measurements? The goal of the analysis could also be to test for correlations between two variables: is an increase in physical realism of the computer animation correlated with an increase in perceptual realism? Finally, the validity of a model can be tested that describes observed measurements as a function of several input variables.

A host of statistical tests have been developed to test the validity of certain models given the measured data. The most common statistical tests will be introduced in Section 12.7. Before we go on to describe the tests, however, we will first give an intuitive introduction to hypothesis testing and then introduce a few key concepts that will provide a better understanding of inferential statistics.

12.5.1 Testing for a Difference in Means

As an example, consider a hypothetical result of the avatar experiment that compared two computer animations differing in realism. Eighteen participants were invited to take part in the experiment, and their task was to identify ten different facial expressions. Nine participants were in group 1, which was shown the low-realism avatar, and nine participants were in group 2, which was shown the high-realism avatar. The data is shown in Table 12.2 and in Figure 12.6. We see first that the means are much higher than the chance level of 10% (the dashed line in Figure 12.6) that would have been obtained if the participants had simply selected an expression at random. Second, we observe that the high-realism avatar resulted in slightly higher recognition accuracy. The question now is: is this difference in some way *statistically significant*?

In order to answer this question, let us look at the potential increase in recognition accuracy. The difference in mean values between the two conditions is 8.9%. If the question is whether this 8.9% is significantly different from 0%, one might ask what is the probability that we obtained the 8.9% difference merely by chance. Equivalently, one might ask whether the two groups originate from the same statistical distribution, or whether they originate from different distributions. If we could somehow quantify this probability, we could

Participant	Low realism	Participant	High realism
1	80	10	90
2	70	11	80
3	80	12	80
4	80	13	90
5	90	14	90
6	70	15	80
7	60	16	80
8	80	17	90
9	80	18	90
μ	76.7	μ	85.6
σ	8.7	σ	5.3

Table 12.2. Table showing hypothetical data from an experiment in which two groups of nine participants had to recognize ten facial expressions that were either displayed by a low-realism avatar or by a high-realism avatar. The data represents percent correct.

answer whether the observed difference is truly meaningful in the sense that if we repeated the experiment with different participants, we would expect a similarly high difference between the two conditions.

In order to calculate this probability we can do a simulation experiment. We have two groups of data values (one each from the low- and high-realism experiments). Let us create two new groups, by randomly reassigning each value from one group to the other and vice versa; that is, we will permute the two groups. We then calculate the mean difference of these two new groups. If this difference is as large or larger than the observed difference of 8.9%, we know that somehow a random permutation of our values might have caused a similar effect. Note that for the two groups shown here, there are only four possible cases in which the difference can actually be larger; namely, when the data

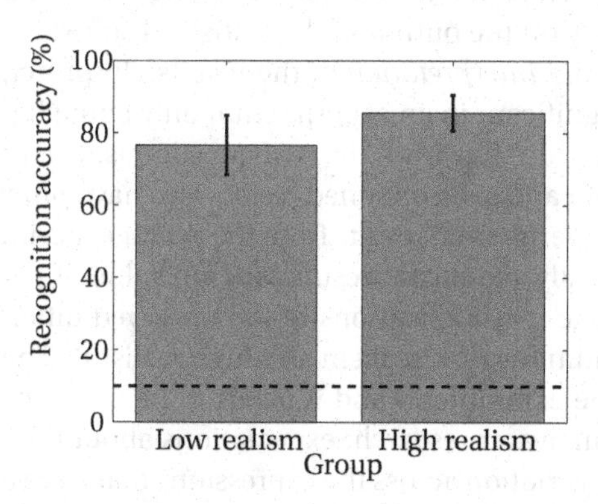

Figure 12.6. Means and standard deviations of the two groups from the data in Table 12.2. The dashed line indicates the chance level for this task

from participant 5 is switched with data from participants 11, 12, 15, and 16. By repeating this *resampling procedure* many times, we can count how often we observe an equal or larger group difference. In this case, the result of the resampling is that roughly 1.73% of the time, we find a value that is equal to or greater than the original 8.9% difference. The value of 1.73% is close to the theoretical value of 1.8% (see Section 12.7.5 on t-tests) and is exactly the number we were looking for; that is, the probability that the observed difference of 8.9% has arisen by chance, or random sampling.

We have now a quantification of the probability that we wanted to determine. This value is usually referred to as the *p value of the significance test.* In addition, 1.8% also seems like a rather low probability, which might lead to us to conclude that there is, indeed, a significant difference between the two groups and that the high-realism avatar, on average, will result in higher recognition performance. It has been common practice to accept differences as statistically significant whenever the probability that they might have occurred by chance is lower then 5%, that is, $p < 0.05$.

It should be noted that this value is only agreed upon by convention and was derived famously by Ronald Fisher in a (mock) experiment involving the question of whether someone could tell the difference between a cup into which milk and then tea had been poured or vice versa. This "arbitrariness" has drawn a lot of criticism as the threshold, for example, determines whether one obtains a significant result and therefore finds a publishable effect, or whether one can advertise a drug to be effective in a medical treatment, etc. Sadly, this "p value-centric" thinking is still rather prevalent in scientific publishing and should be treated with caution—not only because it is only one side of the story (see discussion on effect size in Section 12.5.7 for more detail), but also because in some cases it might be perfectly fine to reject a hypothesis when $p < 0.10$, or to be more conservative when $p < 0.01$, for example.

To be more specific, we would like to strongly recommend the experimenter not only rely on the output of the statistical software package, but pay close attention to the *interpretation* of the results. In principle, it is possible to get *any* effect significant in an experiment if only enough people are tested, or enough trials are run; however, the real question is, whether in addition to the statistical significance, the obtained results also have *practical significance.* Returning to our experimental results from the avatar experiment (in which we obtained a statistically significant result with a p value of $p = 0.018$), the next step will be to gauge the implications of the observed difference of 8.9% between the two conditions (low realism and high realism). This value of 8.9% is the *effect size* of the experiment, and it needs to be put in context with other published results in avatar research, expectations about the effects of realism on recognizability, variation across the expressions that were used in the experiment, etc. The full picture of the implications of the experimental data only emerges by combining and interpreting all of this background information.

12.5.2 Population

Consider again our experiment, in which the goal is to evaluate the quality of a computer animation. One particular task that could be used for participants is to recognize as quickly and as accurately as possible the expressions that the computer animation tries to render. Or, a company might be interested in market research to determine current consumer demands. In this case, an elaborate questionnaire might be designed that gathers information about typical buying habits, previous purchases, and important buying decision factors.

Although the number of participants in these two cases might be very different, it would probably not be feasible to measure the *whole* population for either experiment (in the first experiment, this would be the whole population of potential avatar users; in the second experiment this would be the whole population of potential customers). Nevertheless, intuition tells us that it should not be necessary to measure the whole population to make predictions about whether the avatar would be recognized efficiently, or what kinds of products consumers would like to see—after all, in these experiments, we are not interested in what a particular person does, but rather in the overall, general trend. One important component of experimental design is therefore to chose a representative *sample* of the population, such that the data from this sample will generalize to the whole population. By applying statistical considerations, it is possible to estimate the size of this sample.

A population in its most general definition is a set of potential measurements that are observable in principle. In the context of this book, a population usually refers to a set of people that a particular experiment is targeted to in order to make an observation. It is important to stress that while it would be nice to run experiments with results that always generalize to all of humanity, the resulting *sample size* (see below) would make such experiments prohibitive. Rather, it is advisable to define an experimental population as narrowly as possible, such that the factors of interest are well controllable.

As an example, the computer animation experiment might want to test whether the avatar is recognizable

- for computer graphics experts (who can probably spot detailed errors much more easily),
- for non-experts, but people who are in general computer-literate,
- for non-experts, who are not computer-literate, in case the computer animation might be targeted at a general audience, and
- for a very specific target group defined by the application, such as museum visitors when the avatar is deployed in a kiosk-type system providing guidance in a museum, medical staff in case of an expert system, etc.

Each of these four items involves different populations, and therefore also defines a potential set of people to recruit for the experiment.

12.5.3 Random Sampling

Once a suitable population for an experiment is defined, it is in general not feasible to test this whole population due to limited resources. Therefore, an experiment usually involves a *random sample* of the population. The process of selecting this subpopulation is called sampling. Ideally, the selection process should be fully random, that is, the probability that any member of the population is selected is the same for every member of the population. Whenever the population is sampled in any other way, resulting in a nonequiprobable selection, the sample is nonrandom and potential *sampling biases* might happen.

If the population under study can be fully defined in advance (such as the number of students at a university in the case of a questionnaire about study habits, or the registered customers of a software program), random sampling would simply select a member of the population to become a member of sample with probability $p = \frac{S}{N}$, where S is the sample size, and N is the size of whole population (typically $S << N$).

In many cases, however, the size of the population is less well-defined, and access to the whole population cannot be provided. Typically, psychology experiments in university labs recruit members of the student population for their studies; studies that encompass the whole range of psychological interest ranging from low-level, perceptual issues to high-level, moral judgments. While usually the participant selection process is mentioned, rarely is the fundamental bias of selecting undergraduate psychology students being examined. How representative are these samples? Indeed, if the goal of these experiments is to take a representative sample of the whole population, it can be argued that students form a highly artificial sample—namely a sample from Western, educated, industrialized, rich, and democratic (WEIRD) societies, as a recent article puts nicely (see Henrich et al. (2010a) for a short version, Henrich et al. (2006) for a much more expanded version).

The mentioned article cites, for example, early work by Segall et al. (1966) who gathered data from different cultural backgrounds in the 1960s on a particular perceptual illusion (the Müller-Lyer-illusion, Müller-Lyer, 1889). In this well-known illusion, the length percept of two lines is influenced by the end-segments (see Figure 12.7). Interestingly, the cross-cultural comparison showed that, for example, members of a foraging culture in the Kalahari desert were impervious to the illusion (that is, they saw the two lines as being of equal length), whereas the typical undergraduate population showed by far the largest illusionary effect. This is one of the examples given in the study (Henrich et al., 2006, 2010a), in which critical differences emerge between the typical university sample and other populations. The study lists several such discrepancies, comparing modern industrialized societies with "small-scale" human societies; Western industrialized societies with non-Western industrialized societies; Americans with people from other Western societies; and university-educated Americans with non-university educated Americans. Sometimes very large differences were found between groups that clearly demonstrates that there are rather strict limits on how far results from a "typical" population generalize.

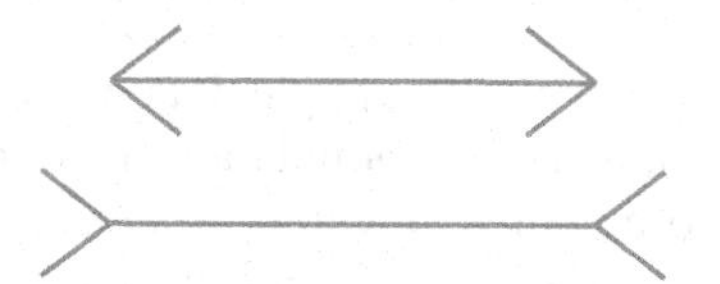

Figure 12.7. Illustration of the Müller-Lyer illusion, in which the two lines depicted here appear to be of unequal length, even though they are of equal length.

Indeed, many more contrasts like this can be found which might limit the generalizability of results to a larger population: age, social background, literacy, media consumption levels, political background, and so on. Depending on the experiment, all of these factors might critically influence the result; at best, by only introducing noise into the experiment, and at worst, by providing a particular, systematic bias in the data. As such sampling biases are, in general, very hard to avoid (mostly due to limited resources and control over the whole population), the experimenter should be aware of the potential dangers and strive to either eliminate the sampling biases as much as possible (which usually means running more control experiments with members of other populations), and/or to limit the conclusion of the experiment accordingly, that is, to estimate how far the results that were obtained can actually generalize from the sample.

12.5.4 Sample Parameters versus Population Parameters

One of the consequences of being able to test only a sample of the whole population is that one has to distinguish carefully between calculations done on the sample and their corresponding population values. Returning to our avatar experiment, we can see that the mean recognition accuracy of group 1—the group that viewed the low-realism avatar—is $\mu = 76.7\%$. The standard deviation of these values is $\sigma = 8.2$. Does this mean that all people viewing this avatar would have the same mean and the same standard deviation? Furthermore, does this also mean that we would find the difference of 8.9% when we tested the whole population of potential avatar users? We certainly would like it to be that way, as this would allow us to generalize our results from the sample of the nine participants to all avatar users.

Strictly speaking, however, we only have access to the *sample parameters*, and *not* to the *population parameters.* Fortunately, however, statistical methods come to the rescue: when assuming, for example, that all observations are normally distributed, we can prove that the sample mean μ is the best estimator of the true population mean μ_{pop}. This means that the more measurements we take, that is, the larger our sample becomes, the better we will approximate the true population mean with the sample mean.

Things are not that easy with the standard deviation, however. As we just saw with the sample data, we would like to make a statement about two unknowns: the population mean and the population standard deviation. The problem is now that, when we calculate the standard deviation for our sample, we would technically need to know the *population mean*:

$$\sigma = \sqrt{\frac{\sum_{i=1}^{n}(x_i - \mu_{pop})^2}{n}}.$$

We do not know the population mean, however, so therefore we can only use the sample mean μ. Because the data is usually closer to the sample mean than to the true population mean (simply due to the fact that we have used the sample to calculate it), this then also means that the estimate of the population standard deviation

$$\sigma_{est} = \sqrt{\frac{\sum_{i=1}^{n}(x_i - 3\mu)^2}{n}}$$

is too small! The estimate therefore needs to be adjusted somehow. This adjustment is provided by a very simple change: instead of dividing by n, one divides by $n-1$, such that

$$\sigma_{est} = \sqrt{\frac{\sum_{i=1}^{n}(x_i - \mu)^2}{n-1}}.$$

This change will obviously play less of a role as the sample size n grows large. In trying to justify this correction one usually encounters another concept in statistical analysis: *degrees of freedom* (df). In this case, the term "degrees of freedom" means that given a sample size of $n = 1$, there is no internal

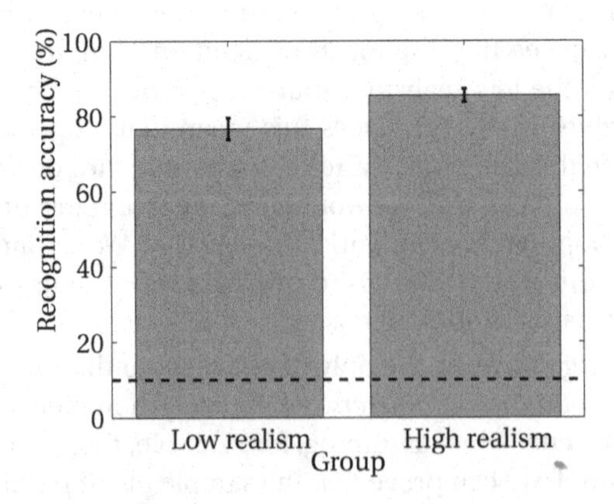

Figure 12.8. Mean values of the two groups from the data in Table 12.2. The dashed line indicates the chance level for this task. Error bars indicate ± 1 standard error of the mean (SEM).

variation in the data, so talking about a sample standard deviation makes no sense. Accordingly, the sample standard deviation is not defined for $n = 1$. One needs at least two data points $n = 2$, before being able to define a measure of the *sample* variation. Another way of thinking about this is that if you know the value of the sample mean, you only need $n - 1$ data values, as the nth value is determined by the sample mean. Hence, in calculating the sample standard deviation using the sample mean, you have decreased the degrees of freedom in the data.[3]

Another issue with sampling is that we would intuitively assume that the more data we acquire, the more certain we would be that the estimate of the sample mean would approach the true population mean. Indeed, it can be proven that this is true. What is more, whenever we have normally distributed data, that is, whenever we know that the population is normally distributed with population mean μ and population variance σ^2 ($\mathcal{N}(\mu, \sigma^2)$), the resulting samples will be distributed according to a normal distribution

$$\mathcal{N}\left(\mu, \frac{\sigma_{est}^2}{N}\right)$$

for N samples being taken. The quantity $\frac{\sigma_{est}^2}{N}$ is referred to as the *standard error of the mean* and captures exactly the fact that the more samples we gather, the more accurate our estimation of the true population mean becomes. Note also that the standard error of the mean is calculated based on the estimated sample standard deviation σ_{est}

$$\text{standard error of the mean} = SEM = \frac{\sigma_{est}^2}{N}.$$

One common use of the standard error of the mean is for plotting mean results. In the case of the avatar experiment the SEM values become $SEM_1 = 2.9$ and $SEM_2 = 1.8$, respectively using the sample standard deviation and $N = 9$ samples. Recall that in Figure 12.6, the mean and standard deviation of the data were plotted, which is rather unusual for plots of experimental data. Most commonly the mean is plotted together with error bars of $\pm$ 1 standard error of the mean. This plot is shown in Figure 12.8. Note that the error bars have become much smaller, indicating the fact that this plot now visualizes the uncertainty inherent in the mean rather than the spread of the sample data that is captured by the sample standard deviation in Figure 12.6. Note also that the error bars of the two conditions *do not overlap*—although not true in general, this can often be taken as an indication that the means of the two groups might be significantly different in the statistical sense. Again, however, we stress that calculating the standard error of the mean does *not* obviate the need for running a proper statistical test (such as the t-test for comparing two group means).

[3]For those still unsatisfied by this explanation, we can recommend the article by Book (1979).

12.5.5 The Null Hypothesis

In many cases, the experimenter has a specific question in mind—a hypothesis—that he or she wishes to put to the test. Owing to a long tradition that was started by one of the godfathers of modern statistics, Ronald Fisher, most statistical inference proceeds by first formulating a so-called *null hypothesis*. This null hypothesis (usually written as H_0) is the premise that whatever cause/effect relationship was formulated as the original hypothesis does *not* occur. The original hypothesis is also referred to as the alternate hypothesis and is written as H_A.

A few examples of alternate and null hypotheses follow.

- You possess a coin that you suspect to be biased, such that heads and tails would occur with unequal probability (H_A). The null hypothesis is that the coin is fair.

- The computer-animation experiment has tested two different avatars with different realism. Your alternate hypothesis is that the avatar with increased realism also results in better recognition accuracy for expressions. The null hypothesis is that the two groups result in equal performance.

- You want to test whether cultural background has an influence on how a particular visual illusion is perceived (H_A). You test participants drawn from ten different cultural backgrounds. The associated null hypothesis is that culture does not influence perception of the visual illusion.

- You have ratings from students in your class about the difficulty of an exam, as well their exam results. You suspect that the more difficult they found the exam, the worse their grade would be. The null hypothesis is that there is no correlation between the two variables.

The goal of the experiment, then, is to employ statistical tests on your experimental data to reject the null hypothesis. As the experimental data is a random variable, however, this statement *cannot be made in a yes/no fashion*. That is, the typical result of a statistical test will be the probability of obtaining the observed data, given that H_0 is true. In other words, testing the null hypothesis asks whether it would be surprising to obtain the observed data given that there is no effect. This degree of surprise is captured in the p value, such that experimental results often might read, "The observed difference in recognition accuracy for the two computer animations was significant at $p < 0.05$."[4] Hence, given that there was no actual difference between the two computer-animation

[4] Note that this is not all of the information that will need to be provided. For more information, see the APA style guide by the American Psychology Association.

types, the probability of obtaining the experimental results is less than 5%. The criterion of 5% is called the *alpha level.*

The catch comes when the test *fails* to reject the null hypothesis. Typically, the criterion for this is when the p value is larger than 0.05; the outcome of the test is then said to be non-significant and the results would read, "The test for differences between the two computer animations failed to reach significance ($p = 0.21$, not significant)." Unfortunately, this does *not* then tell us that the null hypothesis is true (that there really is no effect), but rather leaves us with two types of situations that are related to Type I and Type II errors, respectively, and that are explained in Section 12.5.6.

12.5.6 Type I and Type II Errors

Recall that null-hypothesis testing gives us the probability that we will obtain the observed results, given that the null hypothesis is true. If we fail to reject the null hypothesis, this gives rise to two possibilities. The first possibility is that the null hypothesis was, indeed, true; in which case the failure to reject it has a probability of 95%, given the usual threshold of 5%. This situation is referred to as a *Type I error* and is associated with the threshold, or alpha level. The alpha level therefore specifies how likely the test will result in a *false alarm,* or how often we are willing to reject the null hypothesis even though there was actually no effect. Again, usually hypothesis testing involves an alpha level of 5%, with results being significant whenever the p value is below this threshold.

The second possibility, however, is that the null hypothesis was *not* true and that we have failed to reject it. This can happen, for example, because we did not use enough participants, because there were problems with the experimental design, etc. This error is the so-called *Type II error,* and it is associated with another value, the *beta level.* The beta level is the flip side of statistical testing and relates to the *power* of the test. Power is the probability with which the test itself is able to detect a positive result, that is, to reject the null hypothesis. Usually, power is set to a lower level ($\beta = 80\%$) than the alpha level. Accordingly, here the probability that we failed to reject the null hypothesis is 1 beta level (typically, then, $1 - \beta = 20\%$).

Why is the beta level set to 20% when the alpha level is set to only 5%? From a conceptual standpoint it would be desirable to keep both values as low as possible, both to minimize the risk of false alarms and to keep the power of the test as high as possible. The problem is that in order to satisfy both criteria, the sample size that will be required in order to get a significant result will be prohibitively large. The reason comes from so-called power analysis, a concept explained in the next section. Another way of putting this trade-off is that in hypothesis testing we are being conservative: we would rather not falsely conclude that something exists when it does not (Type I) rather than fail to detect a hypothesis when it is in fact true (Type II).

12.5.7 Power Analysis and Effect Size

Power analysis was introduced into statistics by Jacob Cohen, although similar ideas had been around for a while. Power analysis is important as it focuses on the size of the effect rather than on the p value alone. One of the dangers in interpreting statistical tests is that often the researcher is tempted to only look at the p value, while disregarding the actual effects that the statistical tests have dealt with. Hence, an experiment with a p value of 0.001 is sometimes said to be highly significant, whereas a p value of 0.05 is said to be only significant. This terminology would imply that the effect in the former experiment is somehow bigger than that of the latter. Again, however, the p value only describes one side of the story.

To see why this is so, consider the example given by Cohen in one of his well-known review articles (Cohen, 1990). Researchers had reported finding a "small but significant" correlation between children's height and scores on tests of intelligence and achievement, and so Cohen set out to see just how small this effect could be. Assuming that a p value of 0.001 had been obtained for the significance of the correlation, he calculated that with the selected sample size of 14,000 children, a correlation coefficient (see Section 12.7.15 for the definition of the correlation coefficient) of 0.0278 would have been deemed highly significant. When assuming causality, that is, that height causes intelligence, this would mean that in order to raise a child's IQ from 100 to 130 one would need to increase his or her height by 4.2 meters! Conversely, increasing the height of children by 10 centimeters would require their IQ to be raised by 900 points. Whereas the actual correlation coefficient found in the study was around 0.11 (corresponding to a height increase of 1.1 meters for 30 points of IQ, and a 230 point IQ increase for a 10 centimeter height increase), these calculations demonstrate just how small the effect really was. The main reason for the result was the comparatively large sample size of the study (14,000 children), which made it very likely that even small effects would turn out to be significant.

How can we now capture the idea that we also want to talk about the size of the effect, and not only about its significance? The *effect size*, in general, is simply the magnitude of the effect that is expected to be found in the population. Examples are that you expect to find a correlation coefficient of 0.3 linking height to IQ, or that the difference in recognition accuracy between the two computer animations in the avatar experiment would be 10%. Generally, effect size is not discussed in these absolute terms, but is usually given as a relative proportion that does not depend on the specific variables used. There are two main measures of effect size that are common: Cohen's d, which relates to the raw effect size, and η^2, which relates to the proportion of variance explained by the independent variable(s).

As there is sadly no free lunch in statistics, however, the matter is further complicated by the fact that the calculation of these measures depends on the *type* of statistical test at hand—whether it is a paired t-test, an independent t-test, a repeated measures ANOVA, or a correlation, all require a different calculation of Cohen's d and/or η^2. To illustrate the basic idea, Section 12.5.8 will focus on a case that compares two related groups.

Readers who are interested in a comprehensive, free software package that contains power analyses compatible with a large number of experimental designs should look at G*Power (http://www.psycho.uni-duesseldorf.de/abteilungen/aap/gpower3/). It includes power analyses for the standard family of tests (F-tests, t-tests, χ^2-tests, and z-tests). G*Power is available for both Windows and MacOS X operating systems.

12.5.8 Cohen's *d*

Cohen's d is a relative measure of effect size that can be used when two groups are compared. For a situation in which two independent groups are compared (as in an independent t-test), Cohen's d is defined as

$$d = \frac{\mu_1 - \mu_2}{\sqrt{(var_1 + var_2)/2}},$$

where $\mu_{1,2}$ are the mean values of the two groups and $var_{1,2}$ are the variances of the two groups. The measure d can of course be calculated only *after* the experiment is done, and thus is a *post hoc* measure of effect size.

As an example, consider again the values for the avatar experiment with different realism levels, which are reproduced here in Table 12.3. Applying the formula above, Cohen's d in this case becomes $d = \frac{85.6-76.7}{\sqrt{(75+27.8)/2}} = 0.8$.

Participant	Low realism	Participant	High realism
1	80	10	90
2	70	11	80
3	80	12	80
4	80	13	90
5	90	14	90
6	70	15	80
7	60	16	80
8	80	17	90
9	80	18	90
μ	76.7	μ	85.6
σ	8.7	σ	5.3
var	75	var	27.8

Table 12.3. Table showing hypothetical data from an experiment in which two groups of nine participants had to recognize ten facial expressions that either were displayed by a low-realism avatar or by a high-realism avatar. The data represents percent correct.

Effect Size	Cohen's d
small effect	0.2
medium effect	0.5
large effect	0.8

Table 12.4. Recommendations for effect size values.

Of course, now we are left with the same problem as we had for p values: is $d = 0.8$ a large effect, or a small effect? Based on several meta-analyses, Cohen gave cautious recommendations for the effect size, which are summarized in Table 12.4. As the table shows, our effect size actually is *large*.

12.6 Hypothesis Testing

In Section 12.6.1, the most common statistical distributions related to hypothesis testing will be briefly described. They will be referenced again in the context of the relevant statistical tests. In general, a statistical test for significance will always proceed according to the steps described below.

First, an appropriate statistical test for the problem at hand is chosen. This test also determines the appropriate probability distribution, as the test makes an assumption about how certain measured characteristics of the data are distributed as statistical variables. Next, a significance level is determined that marks the probability that the observed results would have been observed by chance. The significance level is called the p value of the test, and usually is set to $p = 0.05$, although lower values might be chosen for more stringent testing. Recall that the probability distribution determines the likelihood that a certain value of a measure variable is observed. Therefore, the next step is to calculate the value for which the area underneath the probability function is $A = 0.95$. Any experimentally observed value *larger* than this *critical value* will therefore occur by chance less than 5% of the time. In order to calculate the critical value, the probability distribution will need to be integrated to form the *cumulative distribution function* (approximations of these functions are integrated into all modern statistical software packages).

12.6.1 Important Distributions

12.6.1.1 The χ^2, or Chi-Square, Distribution

This distribution is important in hypothesis testing, as it describes the behavior of the sum of squares of independent, normally distributed variables. In other words, whenever you have a set of random, normally distributed variables X_i and you calculate terms like $\sum_i X_i^2$, the resulting distribution will be a χ^2 distribution. To illustrate why this distribution will be important in hypothesis

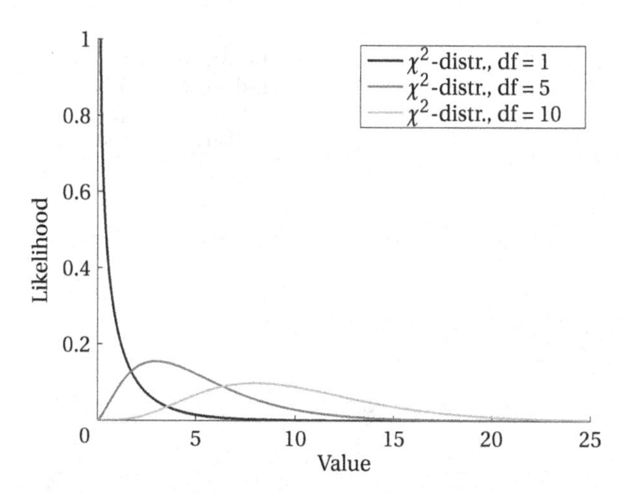

Figure 12.9. The χ^2 distribution for 1, 5, and 10 degrees of freedom.

testing, imagine that you want to test a specific model about the experimental data. Any such model makes X_i' *predictions* about the data. The most common way to decide whether this model provides a good *fit* is to determine the *goodness-of-fit* by summing up the squared errors that the model makes, that is $\sum_i (X_i' - X_i)^2$. This error is then χ^2-distributed, and therefore the distribution can be used to infer how well the model can explain the data in a statistical sense. The χ^2 distribution is plotted in Figure 12.9 for different degrees of freedom.

12.6.1.2 F-Distribution

This distribution arises whenever there is a ratio of two variables $\frac{X}{Y}$, where both X and Y are χ^2-distributed. This distribution is most often encountered in the so-called analysis of variance tests (ANOVA, see Section 12.7.7). Briefly, in an analysis-of-variance test one seeks to determine whether several groups of data might behave differently. The important measure that is used to quantify this difference is the ratio of the variance *between* groups and the variance *within* groups. Each of the variance terms is χ^2-distributed (see the definition of variance in Equation (12.2)), and therefore the resulting ratio is F-distributed. Since the F-distribution is a function of two variables, each of the variables is associated with its own degree-of-freedom value df_X, df_Y. In an ANOVA context, X would refer to the variance that exists between groups and thus $df_X = \#\ \text{groups}-1$, whereas Y would refer to the variance within a group and thus $df_Y = \#\ \text{items per group}-1$, typically $df_X << df_Y$. For sample plots of the F-distribution for different $df_{X,Y}$, see Figure 12.10.

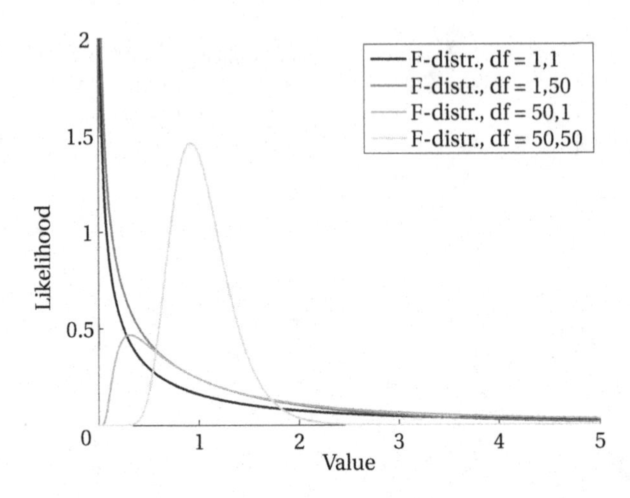

Figure 12.10. The F-distribution for various degrees of freedom; $(\mathrm{df}_1, \mathrm{df}_2) = (1,1), (1,50), (50,1), (50,50)$.

12.6.1.3 (Student's) t-Distribution

"Student" refers to the pen name of Irish chemist William Gosset, who worked on improving the yield of barley crops and on monitoring the quality of different brews of stout beer. The t-distribution is a probability distribution that arises in the context of small sample sizes, and therefore is applicable in the

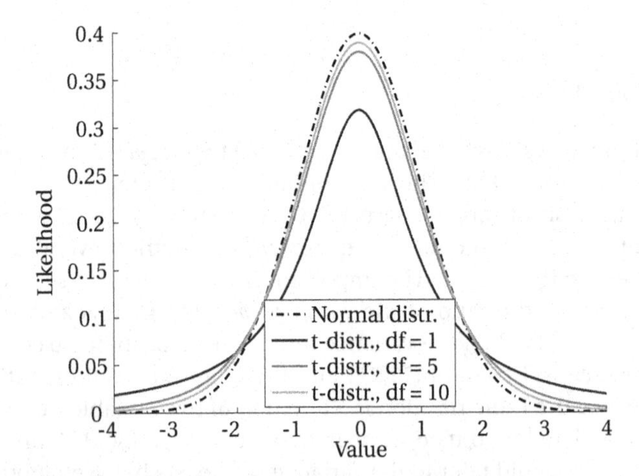

Figure 12.11. The t-distribution for df = 1, 5, 10 compared to the normal distribution (shown in dashed lines). Note how the t-distribution approximates the normal distribution as the degrees of freedom become larger.

context of experimental work, in which the number of participants or trials is on the order of 10 to 20. Accordingly, the t-distribution depends on the degrees of freedom df in the sample and approaches the normal distribution as df increases (see Figure 12.11).

12.6.2 One-Tailed versus Two-Tailed Hypothesis Testing

Any statistical test can be conducted one-tailed or two tailed. To illustrate the difference between the two kinds of tests, let us assume that we have two data samples, x_1, x_2. The difference between the two tests simply refers to the type of question one is interested in.

- If the test is two-tailed, then one might ask whether, for example, two group means are different: $m_1 \neq m_2$
- If the test is one-tailed, then one might ask whether one group mean is either smaller or bigger than the other group mean: $m_1 < m_2$ or $m_1 > m_2$.

The names derive from the fact that in hypothesis testing, the hypothesis is rejected or accepted depending on whether the test measure falls within or outside the tails of the underlying probability distribution (see Figure 12.12). In other words, if $p = 0.05$, then for one-tailed testing of a difference, we know from which side of the distribution the values can come. This then leads to a larger probability of detecting a difference between two means (that is, the test has greater power). In fact, the p value for a one-tailed test is exactly half of that of the two-tailed test.

Because of their greater statistical power, researchers might be tempted to use one-tailed tests for all of their needs. However, it should be noted that there needs to be an a priori justification for their use that needs to come from theoretical grounds. When comparing the average height of two different age groups of children, for example, the use of a one-tailed test is justified, as one would expect changes due to normal growth processes that differentiate the groups. When comparing the heights of two groups of adults, however, one should use a two-tailed test, as the result might come out either way. As a general guideline, the default test in any situation should be a two-tailed test, and only in exceptional, well-justified cases should a one-tailed test be used.

12.7 Common Statistical Tests

The following sections list the most common statistical tests that are used in experimental analysis. Of course, the list is far from exhaustive, and many tests have been proposed for very specific situations. Choosing the right statistical test is an art in itself, as many factors need to be taken into account. The flowchart in Figure 12.13 represents one way of approaching the selection of

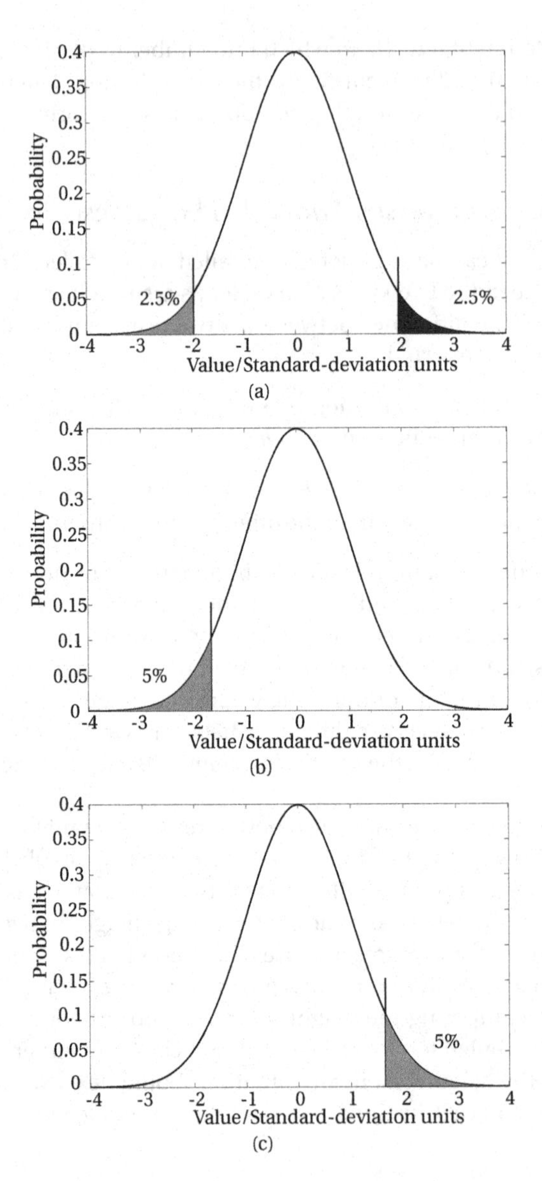

Figure 12.12. Illustration of one-tailed versus two-tailed tests. (a) Normal distribution for two-tailed testing, in which the two tails cut off the first 2.5% and the last 2.5% of area under the curve. The cutoff points are at ± 1.96. (b),(c) Normal distributions for one-tailed testing using the (b) left, or (c) right tail. Each tail contains 5% of the area under the curve with cutoff points at ± 1.65.

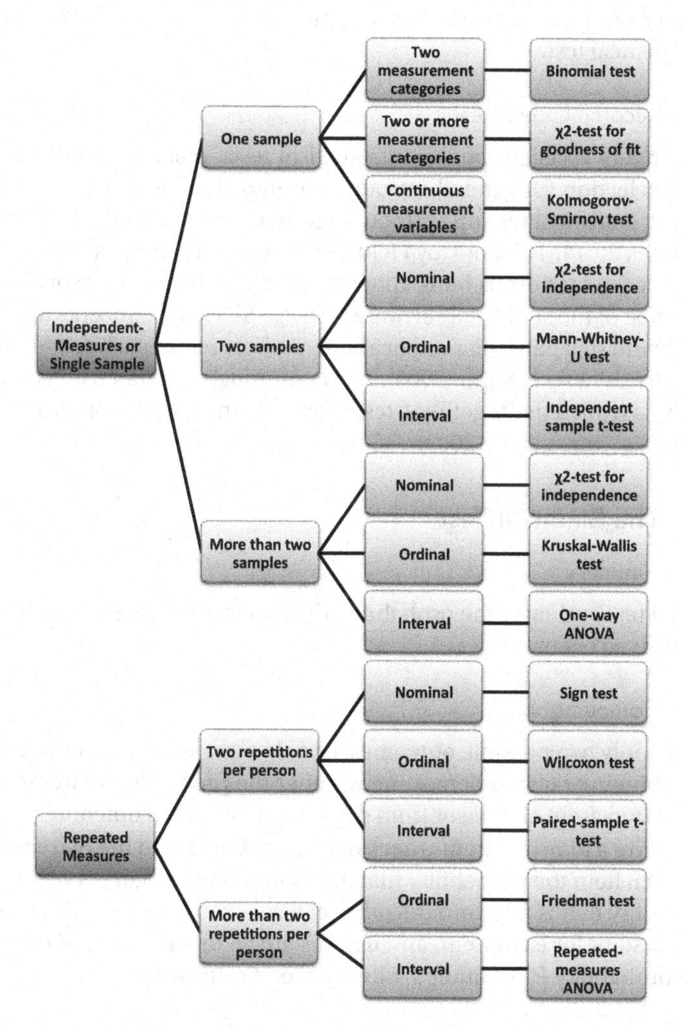

Figure 12.13. Flow chart illustrating how to select the proper statistical test based on properties of the experimental design, the number of samples or conditions, and the variable type.

suitable statistical tests. Based on the properties of the experimental design, the number of samples/conditions, and the variable type, one can select the proper statistical test.

12.7.0.1 Recommended Reading

There are many excellent books on statistical tests, making it hard to narrow down the selection. A general introduction into statistical concepts, as well as the basic statistical testing methodology can be found in the two books by Howell (2009, 2010). The book by Field (2009) is an excellent, well-written, and sometimes even humorous book about statistical testing (even though it is set in the context of the statistical software package SPSS, the concepts are valid of course beyond this particular application). A short reference book for 100 different statistical tests is Kanji (2006). A very thorough book focusing on regression analysis (which ties together t-tests, ANOVA, and correlation in a common framework) is Cohen et al. (2003).

12.7.1 The Binomial Test

12.7.1.1 Summary

A binomial test evaluates the probability that two measurement categories are equally likely to occur.

12.7.1.2 Computation

If there are only two possible outcomes, or only two categories in the data, one can use a binomial test to determine whether the proportion of items in each category differs from chance or from some predetermined outcome.

If you have a total of n items that can be distributed into the two categories, then for each item the probability that it falls into one category is p, and hence the probability that is falls into the other category must be $1 - p$. If this probability is constant for each item, and if items are all independent of each other, then the probability P_k of finding k items in one category is

$$P_k = \frac{p^k(1-p)^{n-k}n!}{k!(n-k)!} = \binom{n}{k} p^k (1-p)^{n-k},$$

as given by the binomial distribution.

If you want to know whether the number of items n_1 in one category is different from the other, you need to add up all the probabilities of finding 1 to n_1 items in this category, hence

$$P = \sum_k^{n_1} P_k.$$

This value is, of course, itself the desired p value. However, note that this is just the one-tailed value—for a two-tailed test, in which you are interested in whether the groups are merely different, you need to multiply this p value by 2.

12.7.1.3 Guidelines

- *Sample size.* The binomial test is usually only applied for small number of items, as the exact formula requires large factorials to be computed which might not be feasible in case of many items. For a larger number of items (say $n > 1000$) the χ^2 test or the normal approximation (see next item) would be better choices.

- *Normal approximation.* If $pn_1 > 10$ and $(1-p)n_2 > 10$, the binomial distribution can be well approximated by a normal distribution, and the z-score can be used to determine the p value with

$$z = \frac{\frac{n_1}{n} - p}{\sqrt{\frac{pq}{n}}}.$$

  See also Figure 12.2.

12.7.1.4 Example

The canonical example for the binomial test is that of a fair coin: you have a coin and suspect that it might not be fair, that is, that the probability for heads is different from that for tails. If we have the data from ten coin tosses, in which six tosses turn out to be heads, and four turn out to be tails, can we say that the coin is not fair?

Evaluating P yields $P = P_1 + P_2 + P_3 + P_4 + P_5 + P_6 = 0.0010 + 0.0098 + 0.0439 + 0.1172 + 0.2051 + 0.2461 + 0.2051 = 0.8281$, therefore there is a $2 * (1-\mathrm{P}) = 34.4\%$ probability that the coin is not fair. With eight heads, we would get $2 * (1-\mathrm{P}) = 2.4\%$, which is less than the usual 5% that we would like to get for a significant result.

12.7.2 The χ^2 Goodness-of-Fit Test

12.7.2.1 Summary

The χ^2 goodness-of-fit test is used to test whether a sample of categorical data comes from a specific distribution.

12.7.2.2 Computation

The χ^2 goodness-of-fit test is useful in situations when one wants to test whether categorical data is derived from a known, fully specified distribution. Although the test is optimized for categorical data, any data can be converted to this format by use of histograms, thereby making the test also applicable for ordinal or interval data in principle.

In order to prepare the data with sample size n for the χ^2 test, the first step consists of preparing a histogram of the data. For this, c value categories are either predefined or created by specifying upper value limits $x_{u,i}$ and lower limits $x_{l,i}$ for each bin. How many items $n_{i,c}$ fall into each category is then counted. Sometimes this count is normalized by n to yield the relative frequencies for each histogram bin or category. In order to determine the test statistic, we need to first calculate the expected, or predicted frequencies for each bin. Given that we know the expected data distribution and with this also the cumulative distribution function (ECDF), we can determine the expected frequency ef_i for each bin for categorical items by

$$ef_i = \mathrm{ECDF(i)} \times n;$$

or, if we have created the histogram from ordinal or interval data by

$$ef_i = (\mathrm{ECDF(x_{u,i})} - \mathrm{ECDF(x_{l,i})}) \times n.$$

With these expected frequencies and the observed frequencies n_i, we can determine the test statistic to be

$$\chi^2 = \sum_{i=1}^{c} \frac{(n_i - ef_i^2)^2}{ef_i},$$

which—as the variable suggests—is a χ^2-distributed variable. In order to determine the statistical significance, we need to fix the degrees of freedom for this test. For c (non-empty) bins in the histogram and k parameters in the known distribution (such as the mean, the variance, etc.), we have $\mathrm{df} = c - k$ for the test.

12.7.2.3 Guidelines

- *Sample size.* Since the test is based on frequency counts in different categories, accurate results require a large number of samples. In addition, as a rule of thumb, the sample size per bin needs to be at least five in order for the test to be valid. If bins contain fewer samples, a different binning strategy should be used (see next item and the example below).

- *Noncategorical data.* Since it is in principle possible to convert any data into categories by binning the data values and creating a frequency histogram from the data, the test can be applied to all measurement variable types. However, the choice of the histogram bins is crucial as it depends on the sample size, and since it determines the degrees of freedom for the χ^2-test.

12.7.2.4 Example

Figure 12.14 shows the histogram of a dataset consisting of 100 random samples drawn from a standard normal distribution with a mean of $\mu = 0$ and a standard deviation of $\sigma = 1$ split into $c = 20$ bins. For comparison, the expected frequencies calculated from the true distribution are shown as the black curve. Note that there are some differences due to the discretization of the bins, but that the general trend seems to be similar. If we calculate the χ^2 statistic, we obtain a value of $\chi^2 = 7.60$. Of the 20 bins, there are two empty bins and the bins at the tail ends of the histogram only contain very few items, so we need to pool further by selecting fewer bins. In order to obtain at least five samples in each bin, we need to select $c = 11$ histogram bins, resulting in $\mathrm{df} = 8 = 11 - 3$ degrees of freedom for the test, and a final p value of $p = 0.47$. Hence, we can conclude that the 100 random numbers are probably from the normal distribution.

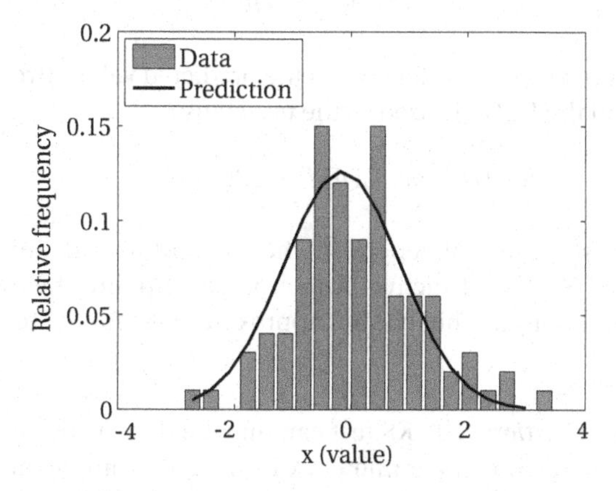

Figure 12.14. Histogram generated from 100 random numbers drawn from a normal distribution with mean $\mu = 0$ and standard deviation of $\sigma = 1$ plotted together with the expected frequencies generated from the true cumulative normal distribution.

12.7.3 The Kolmogorov-Smirnov Goodness-of-Fit Test

12.7.3.1 Summary

The Kolmogorov-Smirnov (KS) goodness-of-fit test is used to test whether a sample of continuous data comes from a specific distribution.

12.7.3.2 Computation

The KS goodness-of-fit test is an exact test to determine whether a sample of continuous data is generated from a known, fully specified distribution. This has important applications, for example, in testing whether data is generated from a normal distribution, a logistic distribution, or any other known data-generation process. If the test turns out to be positive, then one can resort to further parametric tests for testing specific hypotheses about the data, as these are more powerful than their nonparametric counterparts.

The KS goodness-of-fit test uses the observed sample values to create an empirical (cumulative) distribution function (ECDF). This is done by first sorting the n sample data values x_i and then determining for each x_i the number of data points $np(i)$ that are smaller than x_i. From these values $np(i)$ a step function is created which gives the ECDF at each point i as

$$\text{ECDF(i)} = np(i)/n.$$

Obviously, if the data are sorted, this reduces to

$$\text{ECDF(i)} = i/n.$$

The maximum absolute difference in the predicted values (from the known distribution) and the ECDF is used as the test statistic

$$D = \max_i |\text{CDF}(i) - \text{ECDF}(i)|.$$

This test statistic does not depend on the distribution function that is used, which makes the KS test attractive from a practical point of view. Statistical software packages contain the relevant approximations for the test statistic.

12.7.3.3 Guidelines

- *Known distribution.* The KS test can only be done with the distribution fully specified with all parameters known. For example, in order to test for normality, both the mean and the variance of the distribution need to be supplied for the test. If these values are not known, they can be estimated from the data, but then the resulting test statistic D needs to be corrected using Monte-Carlo techniques.

- *Value range.* The data needs to be well distributed along the range of values tested, that is, the data will need to span a large amount of the predictable range.

- *Sample size.* As the ECDF is a step-function, the KS test will become more accurate as more samples are used, which creates a better approximation to the known distribution function.

12.7.3.4 Example

Figure 12.15 shows the ECDF from a toy dataset generated from 100 numbers randomly drawn from a standard normal distribution with mean $\mu = 0$ and

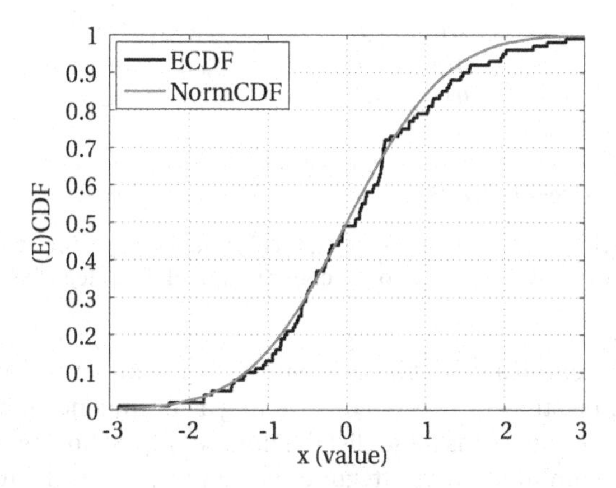

Figure 12.15. Empirical cumulative distribution function generated from 100 random numbers drawn from a normal distribution with mean $\mu = 0$ and standard deviation of $\sigma = 1$ plotted together with the true normal cumulative distribution function generated with the same parameters.

standard deviation of $\sigma = 1$. The plot compares this with the true normal cumulative distribution function generated with the same parameters. Note the close fitting of the two curves in the plot.

The KS test for this data yields a test statistic of $D = 0.0788$ as the maximum difference between the empirical and the predicted functions. This results in a p value of $p = 0.53$ with which we can reject the null hypothesis that the sample comes from a different distribution than the hypothesized normal distribution.

12.7.4 The χ^2 Test for Independence

12.7.4.1 Summary

The χ^2 test for independence evaluates whether two samples of categorical data come from the same distribution—often in the context of a *contingency analysis.*

12.7.4.2 Computation

The same logic that is behind the χ^2 goodness-of-fit test can also be used to test for independence between two categorical variables. This is most often applied in a so-called contingency analysis in which a table is constructed that counts the different frequencies for all combinations of factors. In other words, a sample is being evaluated according to two criteria, all different categories of the criteria are evaluated, and the χ^2 test will evaluate whether one categorical variables is dependent on another.

	y_1	y_2	...	y_{cy}	Overall
x_1	$n_{1,1}$	$n_{1,2}$	...	$n_{1,cy}$	$n_{1,*}$
x_2	$n_{2,1}$	$n_{2,2}$	...	$n_{2,cy}$	$n_{2,*}$
...			...		
x_{cx}	$n_{cx,1}$	$n_{cx,2}$	...	$n_{cx,cy}$	$n_{cx,*}$
Overall	$n_{*,1}$	$n_{*,2}$	...	$n_{*,cy}$	$n_{*,*}$

Table 12.5. Contingency table for two categorical variables x, y with cx, cy categories. Each cell lists how many items fall into the combination of the categories.

Given two categorical variables x, y with cx, cy number of categories, respectively, the first step consists of constructing a contingency table as shown in Table 12.5. Each cell in this tables list the number $n_{i,j}$ of how many items are observed in the combination of categories. Of course, $n_{*,*}$ is the total number of items in the two samples.

The contingency table gives us the observed frequencies for each condition. Next, we need to determine the *expected frequencies* for each cell. As we do not know anything about the distribution, we will simply assume that they are evenly distributed, such that each cell gets

$$e_{i,j} = \frac{n_{i,*} n_{*,j}}{n_{*,*}}$$

expected items, where $n_{i,*}$ is the ith row-sum and $n_{*,j}$ is the jth column-sum.

The test statistic is now based on a χ^2-distributed variable given by

$$\chi^2 = \sum_{i,j} \frac{(n_{i,j} - e_{i,j})^2}{e_{i,j}}.$$

The degrees of freedom is equal to df $= (cx-1)(cy-1)$, which together with χ^2 can be used to determine the critical p value.

12.7.4.3 Guidelines

- *Categorical versus ordinal or interval data.* With the same method as for the χ^2 goodness-of-fit test, the χ^2 test for independence can, of course, also be made applicable for ordinal or interval data. More powerful analysis of group behavior for interval variables that are normally distributed can be accomplished by using ANOVAs.

- *Relative versus absolute.* The χ^2 test for independence can only test whether two variables are dependent or not; it does not yield any information about which of the variables is greater or less in value.

- *Sample size.* As with the χ^2 goodness-of-fit test, the sample count in each cell needs to be $n_{i,j} > 5$ in order for the χ^2 test to be applicable. Smaller cell counts will greatly distort the results.

12.7.4.4 Example 1

We can repeat the canonical coin toss example using the χ^2 test for independence: a coin is tossed ten times, and we record six heads and four tails—is the coin fair? Noting that the sample size for the tails bin is four and therefore technically too small to do the test, we get Table 12.6, which contains both the observed frequency n and the expected frequency e.

With this data, the χ^2 value becomes: $\chi^2 = \frac{(6-5)^2}{10} + \frac{(4-5)^2}{10} = 0.2$. With the associated degrees of freedom of df = 1, we get a corresponding p value of p = 0.65, which speaks highly in favor of a fair coin.

	Heads	Tails	Overall
	6 (5)	4(5)	$n_{1,*} = 10$

Table 12.6. Contingency table for one categorical variable for a simple coin toss. Each cell lists both observed and expected frequencies—the latter in brackets.

12.7.4.5 Example 2

Let us return to one of the favorite examples in this book: the avatar experiment for testing whether a certain computer animation looks realistic or not. For our final study, we want to take a large sample of university students ($n = 1{,}000$) for testing our avatar. Because we suspected that the students' experience with computer games might have an influence, we asked the students to indicate how much experience they had with computer games (none, some, or a lot). In addition, we split the students into half male and half female and recorded their gender. Before proceeding with the actual experiment, we want to test whether gender and computer-game experience might be dependent variables.

Note that the first categorical variable (experience) is actually ordinal, whereas the second variable (gender) is actually categorical. However, for this example, we will treat both variables as categorical. The contingency table for our experiment is shown in Table 12.7.

With this table, the χ^2 value becomes $\chi^2 = 13.56$. With the associated degrees of freedom of df = (2 − 1)(3 − 1) = 2, we get a corresponding p value of p = 0.001, which leads us to clearly reject the hypothesis that gender and computer game experience are independent variables. Note that we would have

	None	Some	A lot	Overall
Male	51 (56.5)	239 (261.5)	210 (182)	$n_{1,*} = 500$
Female	62 (56.5)	284 (261.5)	154 (182)	$n_{2,*} = 500$
Overall	$n_{*,1} = 113$	$n_{*,2} = 523$	$n_{*,3} = 364$	$n_{*,*} = 1000$

Table 12.7. Contingency table for two categorical variables in which students with different levels of experience in computer games choose one of three categories. Each cell lists both observed and expected frequencies (in brackets).

gotten exactly the same χ^2 value if we had swapped male and female genders—hence this test can only tell us that the two variables are somehow correlated, but not how this correlation is laid out. Of course, from the contingency table here, we can see that the male students have a much larger proportion of high experience members, leading us to conclude that actually being male and having a high degree of computer game experience might be correlated.

12.7.5 The Independent-Sample t-Test

12.7.5.1 Summary

Independent-sample t-tests are used to test whether a set of interval measurements is equal to a given value (one-sample t-test), or whether two sets of interval measurements come from the same distribution (two-sample t-test).

12.7.5.2 Computation

There is a whole family of t-tests, all of which calculate a measure that follows Student's t-distribution to check for statistical significance. The most simple t-test is to test whether an observed set of n data points x_i is different from a given value v—the so-called independent one-sample t-test. For this test, one calculates the sample mean $\bar{x} = \frac{\sum_{i=1}^{n} x_i}{n}$ and sample variance $s^2 = \frac{1}{n-1}\sum_{i=1}^{n}(x_i - \bar{x})^2$ and determines the degrees of freedom of the test, which in this case is $n-1$. The t-test measure that is determined is then

$$t = \frac{\bar{x} - v}{s/\sqrt{n}}. \tag{12.3}$$

This value is compared against the value of the t-distribution that is determined for a desired significance level (e.g., $\alpha = 0.05$) and the degrees of freedom of the sample $\mathrm{df} = n - 1$. Equation (12.3) also shows the reason for the difference of the t-distribution compared to the normal distribution: as the denominator features the sample variance instead of the population variance σ^2, for a small number of trials (or degrees of freedom) the sample variance exhibits more variation than the (true) population variance. As the degrees of freedom increase, the sample variance of course approaches the population variance.

In many cases in experimental analysis, one wishes to determine whether two sets of data points come from different populations, that is, whether the two populations have different mean values (note, that the t-test assumes equal variance and equal number of data points in the two datasets; see also Guidelines below). The relevant t-statistic then is based on the difference between the sample means and the average sample variance

$$t = \frac{\bar{x}_1 - \bar{x}_2}{\sqrt{\frac{s_1^2 + s_2^2}{n}}}. \tag{12.4}$$

The degrees of freedom in this case is $\mathrm{df} = 2 \cdot (n-1)$, with n being the sample size in each group.

12.7.5.3 Guidelines

The basic assumptions behind the t-test require the following guidelines to be taken into account.

- *Unequal sample sizes.* The following equation shows the correction to the t-statistic for unequal sample sizes (of course, valid only for the independent measures t-test) n_1 and n_2 (note, that for $n_1 = n_2$, this reduces to Equation (12.4))

$$t = \sqrt{\frac{n_1 n_2}{n_1 + n_2}} \frac{\bar{x}_1 - \bar{x}_2}{\sqrt{\frac{(n_1-1)s_1^2 + (n_2-1)s_2^2}{n_1+n_2-2}}}.$$

- *Unequal variances.* If the variances of the two groups are not equal (as determined by an F-test, or Levene's test for equality of variances), the t-test also needs to be corrected. In this case, the correction does not concern the t-statistic, but the degrees of freedom, which are calculated based on a more complex equation. As the correction *decreases* df, the power of the t-test will be reduced.

- *Non-normally distributed data.* Although the t-test is robust to violations of normality, very skewed data can be a problem in running the t-test. In these cases, the Mann-Whitney-U test (see Section 12.7.6) offers a non-parametric solution.

- *More measurement categories.* If more measurements are available per observation—such as when multiple post-tests were run on a group, or when a participant did multiple repetitions of one stimulus presentation—one needs to resort to analysis of variance methods that explicitly deal with these designs (see Section 12.7.7 on ANOVA).

- *Multiple t-tests.* Finally, it is worth noting that multiple testing on a given dataset leads to a greater chance of a Type I error, that is, there is a greater chance of finding a significant difference between two groups when in reality there is none. To illustrate this, see the examples that follow.

12.7.5.4 Example I

Let us return to the example given in Section 12.5.1, which we can now do properly by using the t-test. Eighteen participants were asked to recognize facial expressions of a low-realism avatar and a high-realism avatar in an experiment. The data is shown again in Table 12.8. With $n = 9$ in each group, we find $\mu_1 = 76.6\%$, $\sigma_1 = 8.7\%$ for group 1 in the low-realism condition, and $\mu_2 = 85.6\%$,

Participant	Low realism	Participant	High realism
1	80	10	90
2	70	11	80
3	80	12	80
4	80	13	90
5	90	14	90
6	70	15	80
7	60	16	80
8	80	17	90
9	80	18	90
μ	76.7	μ	85.6
σ	8.7	σ	5.3

Table 12.8. Table showing hypothetical data from an experiment in which two groups of nine participants had to recognize ten facial expressions, which were displayed by a low-realism avatar for group 1, and a high-realism avatar for group 2. Data represents percent correct.

$\sigma_2 = 5.3\%$ for group 2 in the high realism condition. Using df $= 2(9-1) = 16$, and therefore $t = -2.63$, this results in a two-tailed p value of $p = 0.0182$. Therefore we can reject the null hypothesis that the two group means are equal and conclude that they the average recognition performance in the two conditions is *different.* For a one-tailed t-test, in which we test whether the average recognition performance of group 2 is *higher* than that of group 1, the values stay exactly the same, but we simply half the p value to $p = 0.009$, reflecting the fact that a directionality in our hypothesis increases the power of the test.

12.7.5.5 Example 2

We can also test whether the second group's performance was larger than chance performance (=10%) by running a one-sample, one-tailed t-test. Using $\nu = 100$, $\mu_2 = 85.6\%$, $\sigma_2 = 5.3\%$, we get $t = 43.333$. With df $= (9-1) = 8$, we get a p value of $p < 0.001$, which is highly significant, telling us that it is highly unlikely that the second group performed at chance level.

12.7.5.6 Example 3

Looking at the variances in Table 12.8, we might need to worry that they are actually not equal in the two samples. Since we only have nine samples, however, the chances are slim that the observed difference in variances become significant—indeed, when computing both an F-test or Levene's test for equality of variances, both tests cannot reject the hypothesis that the two samples come from normal distributions with similar variances.

If we assume—just for the purpose of demonstrating the difference between the two types of tests—that the groups have unequal variance, we calculate the t-test using a corrected number of degrees of freedom. The t-value stays the same with $t = -2.63$, however, the degrees of freedom change to df = 13.21, resulting in a slightly higher p value of p = 0.010 for the one-tailed t-test. Note that this correction of df in the case of unequal variances will reduce the power of the t-test.

12.7.6 The Mann-Whitney-U Test

12.7.6.1 Summary

The Mann-Whitney-U (MWU) test compares the central tendencies of two independent groups using nonparametric statistics.

12.7.6.2 Computation

Although the t-test is fairly robust to non-normality—especially with large sample sizes, as in this case the central limit theorem states that the sample average will follow a normal distribution—for large deviations from normality one can use the nonparametric Mann-Whitney-U test for independent samples. This is in essence very similar to the t-test, but rather than checking for differences in the mean value, one compares the *rank order* of the different samples. In addition, the MWU test needs to be used for ordinal data (such as ratings without clear anchoring, ordinal categories such as easy, medium, and hard, etc.), as the data will not follow a normal distribution in this case and the concepts of mean and standard deviation are not defined.

Given two groups with n_1 and n_2 data points each, the data points of the two groups *together* first are sorted in decreasing (or increasing) order (this is possible with both ordinal and interval data). In the next step, the ranks of all data points of group 1 are summed up to yield

$$R_1 = \sum_{i=1}^{n_1} r_{1,i},$$

and, similarly, all ranks for group 2 are summed up to yield

$$R_2 = \sum_{i=1}^{n_2} r_{2,i}.$$

As a next step, we determine the two U-statistics $U_{1,2}$ to be

$$U_{1,2} = R_{1,2} - \frac{n_{1,2}(n_{1,2}+1)}{2}.$$

The smaller of the two values, $\min(U_1, U_2)$, is then used as the test statistic. Note that this test can be done with equal and unequal sample sizes.

The MWU test is a good choice in pretty much all situations because it is less sensitive to outliers than the t-test (the reason behind this follows the same logic as why the median is a less outlier-sensitive measure of the central tendency than the mean) and it can be used both for interval and ordinal data.

12.7.6.3 Guidelines

There are only two reasons why one would need to be cautious for the MWU test.

- *Less power.* As with any nonparametric test, when the underlying distributions would actually be compatible with a parametric test (in our case, a t-test), selecting the MWU test instead will result in less power, that is, the ability to detect an effect will be reduced. In most cases, however, the benefits of the test outweigh this issue.

- *Different distributions.* When the underlying distributions of the two groups are highly dissimilar, the assumptions of the MWU test do not hold anymore. If one suspects this to be the case, then one should resort to a sign test (see Section 12.7.11).

12.7.6.4 Example

Given that there are only nine samples in Table 12.8 for the avatar experiment and that the recognition accuracy is actually not a full interval variable (note that accuracy is bounded from above and below, and hence is not a continuous measurement scale), we might wish to be on the safe side and to conduct a nonparametric statistical test for the two groups instead.

Participant	Low realism	$r_{1,i}$	Participant	High realism	$r_{2,i}$
1	80	8.00	10	90	15.50
2	70	2.50	11	80	8.00
3	80	8.00	12	80	8.00
4	80	8.00	13	90	15.50
5	90	15.50	14	90	15.50
6	70	2.50	15	80	8.00
7	60	1.00	16	80	8.00
8	80	8.00	17	90	15.50
9	80	8.00	18	90	15.50
R_1		61.5	R_2		109.5
U_1		16.5	U_2		64.5

Table 12.9. Table showing hypothetical data from an experiment in which two groups of nine participants had to recognize ten facial expressions, which were displayed by a low-realism avatar for group 1, and a high-realism avatar for group 2. Data represents percent correct. The table also lists the tied ranks, as well as the rank sums and the U-statistics.

Table 12.9 shows the data together with the rank values (note that because there are many tied ranks in both groups—that is, there are many identical recognition values—the ranks need to be corrected accordingly). The table also lists the resulting rank sums $R_1 = 61.5$ and $R_2 = 109.5$ and the U-statistics $U_1 = 16.5$ and $U_2 = 64.5$. Our final test statistic is now $U = \min(U_1, U_2) = 16.5$. In order to determine the resulting p value, one could in principle list all possible combinations of rank values for the two groups and determine whether our test statistic is different from this random value. For larger sample sizes, this is not feasible and a normal approximation is used. Here, we get a p value of $p = 0.039$ for our data, which leads us to reject the null hypothesis that the two groups would come from a distribution with the same median value (note the decrease in power compared to the t-test as witnessed by the larger p value).

12.7.7 One-Way Analysis of Variance (ANOVA)

12.7.7.1 Summary

The one-way analysis of variance explains the variance of observed data as a function of one factor having different levels. Equivalently, it tests the null-hypothesis that all means across all levels of the factor are equal and thus no level has an influence on the observed data.

12.7.7.2 Computation

The ANOVA was popularized by R. Fisher in the 1920s as part of his immensely influential work on biometry in which he laid the foundations of experimental design and statistical analysis methods.[5]

A one-way ANOVA will determine whether the group means of c independent groups are different. As the name implies, an ANOVA deals with the variance of the different factors. The basic idea behind this method is that if a factor is important for the data, then the variance of the data *within* this factor across all levels should be smaller than the variance *between* factors.

In performing an ANOVA, one first has to decide whether one is interested in a so-called *fixed-effects* or *random-effects* model. The fixed-effects model (also called Model 1) is used whenever the specific factors themselves are of interest, or when the researcher is investigating the effects of specifically manipulated factors on a measurement variable. The random-effects model (also called Model 2) is used whenever the researcher is only interested in the overall contribution of the factors on the measurement variable—in the context of perceptual and cognitive studies, this model actually is relatively rarely used. Of course, one might also choose a mixed-effects model (Model 3) for analysis, which includes both random and fixed factors.

[5]It should be mentioned that t-tests, ANOVAs, and regression can all be viewed in the same framework, namely that of generalized linear models. The interested reader is referred to Cohen et al. (2003); Field (2009) for more information.

To make this distinction clear, consider an experiment in which the believability of an avatar is tested using several different facial expressions. The dependent variable is then the rating of believability by the participants and there is one factor of "expression" in the experiment, which has as different levels the different facial expressions. If the researcher is interested in specific effects of an expression on the ratings (for example, because it could be assumed that the avatar is particularly well-designed to work with negative expressions), then the analysis becomes a "Model 1" analysis. If the ANOVA results turn out to be significant, then one usually follows up with comparisons of means between expressions to learn more.

One can also consider a different experimental question, however: if the researcher sees the different expressions only as a way of introducing variance into the data and is not interested in whether the avatar performs better for happy than for sad expressions, a random-effects model (or Model 2) is used. Such an analysis could be used to determine the degree to which the variation of believability is different within expressions from that across expressions. Assuming, for example, that the ANOVA yields the result that the variance within expressions is much larger than across expressions, one can conclude that there is no single best expression that one can choose from. Alternatively, one might also use the Model 2 ANOVA in a pilot experiment. Given a minimum number of participants and trials, one subjects the results to a Model 2 ANOVA, which looks at the within-group and between-group variance. If the within group variance is much larger than the between-group variance, the follow-up experiment might include fewer participants but more trials and vice versa.

In practice, experiment designs that start from a specific hypothesis and explicitly manipulate experimental factors are Model 1 designs—if the researcher explicitly wants to acknowledge that the specific factor levels are only one particular choice among many equally good choices, a Model 2 ANOVA should be chosen. Model 2 analyses have applications in fMRI analysis or genetics, where the specific items under investigation (voxels, genes) are just one of many possible items.

If one is interested in a Model 1 analysis, it is important to determine *before* the experiment a set of planned comparisons, that is, specific combinations of factors that one is interested in comparing later on. Again, let us assume that the ANOVA found a significant influence of the factor "expression" on the ratings of believability in the Model 1 design. Here, the researcher is interested in determining which of the expressions might be the most/least believable to perhaps be able to improve the animation for these expressions in particular. As mentioned above, one of the factors known before the experiment could be that the animator put a lot of effort into the negative expressions (fear and disgust, for example) while having spent less time with the non-negative expressions (happy and surprised, for example). A *planned com-*

parison could then be conducted for these two groups with appropriate tests (see below). It might be, however, that the researcher—seeing the results of the experiment—suddenly has the insight to check also for differences between intense and weak expressions. Since such a comparison is *unplanned*, however, it calls for an *adjusted significance (or alpha) level*. This is because the number of potential groupings of all factors is rather large, and conducting any number of tests between arbitrary splits in the data will increase the chance of a Type I error, that is, the chance of getting a significant difference between two groups will increase with the number of tests run. This distinction might seem strange at first, but is really at the heart of proper experimental design—if there are clear, a priori hypotheses about how certain factors might influence the result, then planned comparisons are to be conducted; otherwise, one needs to correct for the increase in false positives. Note however, that the difference between planned and unplanned comparisons will only matter if the level of significance is barely reached with the observed data.

The following explanation presents the basic intuition behind the one-way ANOVA (see also the example below).

As mentioned above, the basic idea of the ANOVA is to determine the relation between the variance within factor levels and the variance across factor levels. The calculation of an ANOVA therefore starts with the calculation of the overall mean and the observed, overall variance of the data. If one factor were responsible for the whole variance, then obviously all samples within each factor level would need to have the same value, with differences occurring only across groups. This so-called *treatment variance* (or between-group variance) is now also calculated by replacing all samples within a factor level by their mean value and recalculating the variance. Finally, the so-called *error variance* (or within-group variance) is determined, which is given by the variance of the deviation of each sample from the group mean. Note that all of the error in our experimental design is concentrated in the within-group term—if we had a repeated-measures design (also called "within-participant design"), we would be able to split this within-group variance further.

The ANOVA proceeds by calculating various sums of squares (SS) from the data. In order to determine these, we first calculate various average sums across splits of the data. First of all, we determine the sum over all data points in each group

$$X_i = \sum_{j=1}^{n} x_{i,j}.$$

From this, we calculate the grand total sum over all data points

$$X = \sum_{i=1}^{c} \sum_{j=1}^{n} x_{i,j}.$$

Similarly, we need the grand total sum of squares over all data points

$$XS = \sum_{i=1}^{c} \sum_{j=1}^{n} x_{i,j}^2.$$

With these quantities, we can now calculate the between-group sum of squares

$$SS_{\text{between}} = \frac{1}{n} \sum_{i=1}^{c} X_i^2 - \frac{1}{cn} X^2.$$

This contrasts the squares of the sums of each group with the overall square-sum.

The within-group sum of squares is defined as

$$SS_{\text{within}} = XS - \frac{1}{n} \sum_{i=1}^{c} X_i^2.$$

This measure contrasts the overall sum of squares with the squares of the sums of each group.

Finally, the total sum of squares is defined as

$$SS_{\text{total}} = XS - \frac{1}{cn} X^2.$$

In general it holds that $SS_{\text{total}} = SS_{\text{within}} + SS_{\text{between}}$; hence we have split the sum of squares into two additive parts: one containing the within-group variation and one containing the between-group variation.

The degrees of freedom for the between- and within-group variation are $\text{df}_{\text{between}} = n - 1$ and $\text{df}_{\text{within}} = c \cdot (n - 1) = cn - n$, respectively. The degrees of freedom associated with the total variation is then $\text{df}_{\text{total}} = \text{df}_{\text{between}} + \text{df}_{\text{within}} = cn - 1$.

From the various sums of squares we can determine the two so-called *mean squares* values

$$MS_{\text{between}} = \frac{SS_{\text{between}}}{\text{df}_{\text{between}}},$$

and, accordingly

$$MS_{\text{within}} = \frac{SS_{\text{within}}}{\text{df}_{\text{within}}}.$$

These two quantities can be shown to measure the treatment variance and the error variance, respectively. If the null hypothesis of the ANOVA is true, that is, that all means are equal, it can also be shown that this must mean that the treatment variance is equal to the error variance. This test amounts to an F-test of the quotient between treatment and error variance, or equivalently

$$F = \frac{MS_{\text{between}}}{MS_{\text{within}}}.$$

If $F = 1$, the variances in both measures are equal, and hence the factor does not have any effect on the dependent variable. If $F > 1$, then the between-group variation is *larger* than the within-group variation, and the factor does have an effect on the dependent variable.

Whenever a quotient with the nominator and denominator following χ^2 statistics is used in hypothesis testing, the resulting test statistic follows the F-distribution. This distribution is characterized by two parameters which correspond to the degrees of freedom for the nominator and denominator, respectively. Therefore, for a given significance value (usually set to the standard 0.05) and the degrees of freedom df_{between} and df_{within}, this F-test then decides whether the null hypothesis can be rejected or not.

Technically, the ANOVA ends here. In cases in which one has more than two factor levels in a fixed-effects design, however, one is also interested in *which* factor level is different from the others. As the output of the ANOVA only confirms or rejects the hypothesis that one of the factors has an influence on the dependent variable, a potential comparison of the factor levels needs to be determined by these so-called *post hoc tests*. Given that there can be many factor levels in an experimental design, one again needs to be aware of multiple comparisons which—when not corrected for—increase the chance of Type I errors dramatically. There are multiple types of post hoc tests available that depend on specific ways of correcting for the chance of such false positives. The following list describes the three most prominent tests.

- *The Tukey Honestly Significant Difference (HSD) test.* A first type of corrected post-hoc test is a modified t-test. The Tukey Honestly Significant Difference (HSD) test compares multiple means in a specified order with the null hypothesis that all means are equal. For the ANOVA example, the largest and smallest means across the factor levels are first found. One then determines the difference $q = \frac{m_{\text{largest}} - m_{\text{smallest}}}{SE}$ between the largest and smallest means divided by their standard error (SE). This value is then compared against a t-distribution, with the comparison depending on the alpha level, the degrees of freedom of the first sample, and the degrees of freedom for the second sample. One then continues with comparing the largest mean against the remaining means and then repeats with the next smaller mean. It has been shown that the HSD test provides an optimal correction for Type I errors in case of balanced ANOVAs (that is, groups with equal sample sizes). Furthermore, it provides conservative estimates of effects when sample sizes are unequal in a one-way ANOVA. These properties make the HSD test the method of choice for post-hoc tests.

- *The Bonferroni test.* A more conservative, but conceptually simpler post-hoc test is the Bonferroni-test, which is based on the standard t-test and follows a simple logic. As discussed in the introduction, the chance of a false positive is 0.05 (the alpha level) for one comparison in the standard t-test, that is, it is acceptable to have a 5% chance that the observed effect occurred by chance rather than due to an experimental factor. For multiple comparisons, this alpha level is simply divided by the number of comparisons. Therefore, if a factor has ten levels and one compares all ten levels against each other, a given t-test has to be significant at least at $p = \frac{0.05}{10} = 0.005$ in order to show a statistical difference between two levels. Especially for fewer comparisons, however, this correction turns out to be too conservative and one of the other tests might be preferable.

- *The Scheffe criterion.* Finally, another type of post hoc test is the Scheffe criterion. When compared to the previous two corrections, this criterion actually provides the most conservative results. It has the advantage, however, that it directly determines a confidence interval with which to test differences between all group means simultaneously, whereas both the HSD and the Bonferroni tests are used to test between two group means at a time. Therefore, when there are many comparisons to make, the Scheffe criterion actually provides narrower confidence limits than the other two methods. The test is directly derived from the F-statistic and calculates a confidence interval according to $CI = \sqrt{(}d_f \cdot F_{\text{crit}} \cdot MS_{\text{error}} \cdot 2/n)$, where d_f is degrees of freedom of the within-factor variance, F_{crit} is the critical value of the F-distribution, for a given significance level and the two degrees of freedom of the test (e.g., $\alpha = 0.05, d_t, d_e$), MS_{error} is the between-group mean square, and n is the sample size of each group.

The effect size of the one-way ANOVA is determined directly from the sums of squares. The most popular measure is

$$\eta^2 = \frac{SS_{\text{between}}}{SS_{\text{total}}},$$

η^2	Effect size
0.01	small
0.06	medium
0.14	large

Table 12.10. Cohen's recommendations for interpreting ANOVA effect sizes.

which can be interpreted as measuring the amount of variance due to the experimental manipulation. There are different guidelines as to what value constitute how big of an effect, but the ones given by Cohen (1992) are shown in Table 12.10.

12.7.7.3 Guidelines

- *Independence of factor levels.* This assumption should not be violated, as it forms the basis of the ANOVA derivation. It simply means that if two samples are in the same group, the deviation of the first sample from the group mean should not be indicative of the deviation of the second sample.

- *Normality.* As with the t-test, the ANOVA relies on calculating variances of groups, which assumes that the data follows a normal distribution. Theoretical works show that ANOVA is usually rather robust to violations of normality given ample sample sizes per factor level. Nevertheless, one can test for normality of the data using, for example, the χ^2 goodness-of-fit test (see Section 12.7.2), or the Kolmogorov-Smirnov test (see Section 12.7.3). Another option is to plot the data in a so-called Q-Q plot in which the quantiles of the observed data are compared against quantiles from theoretical data sampled from a normal distribution with the observed mean and variance. To create the plot, one first samples N times from the normal distribution, orders this theoretical data and the observed dataset in increasing order, and then plots the two sets against each other. If the plot looks like the diagonal $y = x$, the dataset is likely to be normally distributed. If they do not fall onto the diagonal, or if any of the aforementioned tests fail, one can resort to nonparametric ANOVA methods such as the Kruskal-Wallis test (see Section 12.7.8), which do not require normally distributed data.

- *Homogeneity of variances.* The ANOVA assumes that variances for all factor levels are equal. One can test this assumption beforehand, using *Levene's test* for homogeneity of variance. It basically calculates a weighted variance as its test statistic, which is then tested against the F-distribution. If the Levene's test turns out to be significant, then there are methods which correct for this circumstance—most prominently, *Welch's ANOVA*, which weights each group by its variance thus effectively "whitening" each group. Note however, that when the sample sizes in each group are *equal*, the ANOVA is robust against violations of homogenous variances. In general, concerning unequal variances and unequal group sizes, the regular F-test underlying the ANOVA will produce false positives if the larger variances occur in groups with the smaller sample sizes.

12.7.7.4 Example

To illustrate a typical one-way ANOVA, Figure 12.16 shows two datasets of a hypothetical experiment with four groups. The data is drawn from a simulated recognition experiment with four conditions (groups). The ten "ideal" participants yield normally distributed values with a mean value of $\mu = 75\%$ and a variance of $\sigma = 10\%$. The data is plotted in a box-plot with the median at the center and the 25% and 75% quartiles marking the lower and upper end of the box, respectively. In addition, the whiskers extend out to 1.5 times the length of the box in both directions. As shown in Figure 12.16(a), all boxes include the true value of the mean. In Figure 12.16(b), however, the first group sample was shifted by 10% in the negative direction, causing the box to lie just below the true value. Note that the box-plot already indicates the possibility of an effect: the first box lies well outside the second, and the third box barely overlaps the fourth.

Before going into the calculation of the ANOVA one might want to check for the two underlying assumptions, that is, normality of the data and equal variances across the groups.

As mentioned above, a visual check for normality of the data can be conducted using a Q-Q plot, which compares the quartiles of the data against the quartiles of a normal distribution with the observed mean and observed variance. The plot shown in Figure 12.17 also includes a line connecting the first and third quartile of the data, that is, the interval that encloses the data values between 25% and 75% of the data range. As the data fits this line very well, one can conclude that it most likely conforms to the assumption of normality (which it should, of course, since it was drawn from a normal distribution in the first place). Figure 12.17 also shows that the regions outside of the first and third quantile are more prone to noise and therefore are not included in the

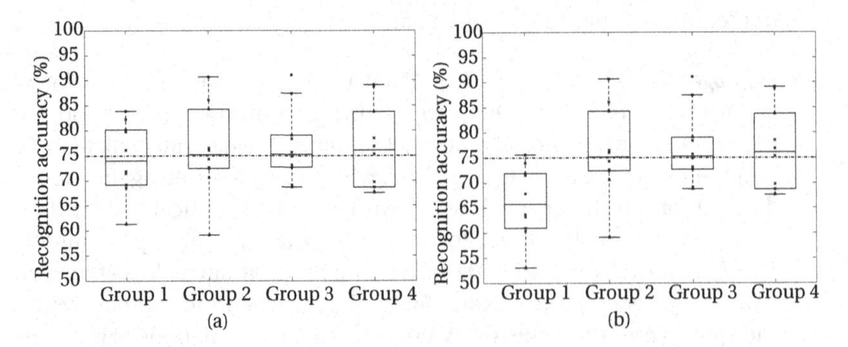

Figure 12.16. (a) Four samples from a normal distribution with mean $\mu = 75\%$ and variance $\sigma = 10\%$ shown in a box-plot with median values. The crosses show all individual measurements. Note that the upper and lower quartiles in the plots all include the true value of the mean. (b) The same data with one sample shifted by -10%.

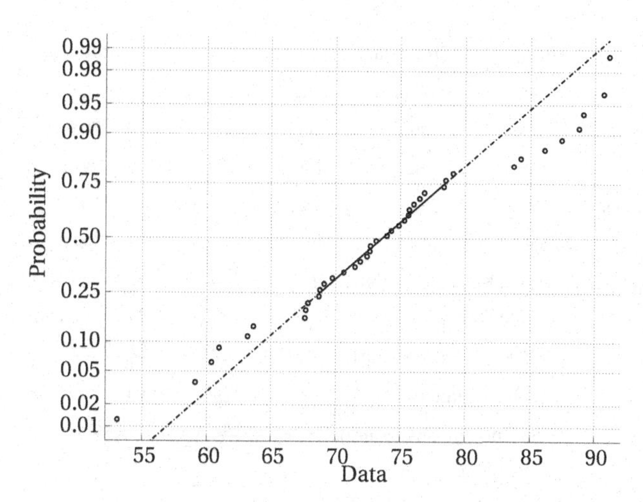

Figure 12.17. Plot showing the distribution of the quantiles from Figure 12.16(b) against probabilities of a normal distribution. The full line connects the first and third quartiles of the data and is extrapolated out as a dashed line. Note that the data fits the line very well between the first and third quartiles, indicating that the data is normally distributed.

fit of the data. If the visual analysis is not enough, the corresponding statistical test to check for normality of the data is the Kolmogorov-Smirnov test, for example. In order to do this test, we need both the mean and the variance of the assumed normal distribution—we can provide an estimate from the data itself, by taking the sample mean and the sample variance. For the data in Figure 12.16(a), we get a sample mean of $\mu = 76.112$ and a sample variance of $\sigma = 7.764$, which results in a test statistic $D = 0.088$ and a p value of $p = 0.890$. This makes it highly unlikely that the data comes from a different distribution than the normal distribution. For the data in Figure 12.16(b), we get a sample mean of $\mu = 74.035$ and a sample variance of $\sigma = 8.944$, yielding $D = 0.101$ and $p = 0.771$. Again, it is highly likely that the data comes from a normal distribution—note, however, that we evaluated *all* data together in this case, which of course is not appropriate for larger inter-group differences. In principle, one would need to repeat the test for all four groups separately.

In practice, explicit tests for normality are rarely done to this degree, as the ANOVA is rather robust against violations of normality for adequate sample sizes. In case of doubts about the normality of the data, one can always resort to a nonparametric test, such as the Kruskal-Wallis test.

The next assumption to check is whether the variances in the groups are equal. For the dataset of Figure 12.16(b), the Levene test yields a value of $L(3,36) = 0.31$ and a corresponding p value of $p = 0.82$, which clearly speaks

	Group 1a	Group 2a	Group 3a	Group 4a
Means	74.5	76.3	77.0	76.6
Variances	51.4	80.1	53.7	70.0
Sum	744.6	762.7	770.2	766.1
Sum of squares	5590.6	5889.6	5981.1	5931.3
Squared sum	55443	58170	59327	58683
SS_{within}	$= (5590.6+5889.6+5981.1+5931.3) - \frac{55443+58170+59327+58683}{10} = 2301.95$			
SS_{between}	$= \frac{55443+58170+59327+58683}{10} - \frac{(744.6+762.7+770.2+766.1)^2}{40} = 38.26$			
SS_{total}	$= (5590.6+5889.6+5981.1+5931.3) - \frac{(744.6+762.7+770.2+766.1)^2}{40} = 2340.21$			
MS_{within}	$= \text{mean(variances)} = \frac{SS_{\text{within}}}{\text{df}_{\text{within}}=N-c} = \frac{2301.95}{36} = 63.94$			
MS_{between}	$= \text{variance(means)} \cdot \text{sample size} = \frac{SS_{\text{between}}}{\text{df}_{\text{between}}=c-1} = \frac{38.26}{3} = 12.76$			
$F(3,36)$	$= \frac{MS_{\text{between}}}{MS_{\text{within}}} = 0.20$			
p	$= 0.69$			
η^2	$= \frac{SS_{\text{between}}}{SS_{\text{total}}} = \frac{38.26}{2340.21} = 0.016$			

Table 12.11. ANOVA for data from Figure 12.16(a).

in favor of equal variances among the groups. Artificially increasing the variance within Group 3 by a factor of two, for example, already brings the Levene statistic into the significant range with $p = 0.04$.

The first thing to do for the ANOVA is to calculate the average of these sample variances; this is called the within-group mean square MS_{within} and represents an estimate of the within-group variance. There is another way of estimating the within-group variance using the sample means: assuming that all sample means are equal, the four repeated measurements are distributed with a standard deviation that is equivalent to the within-group standard deviation divided by the square-root of the sample size n. This simply follows the definition of the standard error of the mean, $SEM = |SD_{\text{means}}| = \frac{SD_{\text{within}}}{\sqrt{n}}$. Squaring this equation and re-arranging then yields $Var_{\text{within}} = |Var_{\text{means}}| \cdot n$. Therefore, the second step in the ANOVA is to simply calculate the variance of the means and multiply it by the sample size n. This quantity is called the between-group mean square MS_{between}. If all samples are drawn from the same population, all means and variances should be equal, therefore requiring that $MS_{\text{within}} = MS_{\text{between}}$, or, equivalently, $F = \frac{MS_{\text{between}}}{MS_{\text{within}}} = 1$.

To see what this means for the data, consider the first case: as is shown in Table 12.11, the four sample means are rather similar with values around the true mean. The sample variances also show a similar pattern. The within-group mean squares is $MS_{\text{within}} = 63.9$, whereas the between-group mean squares is $MS_{\text{between}} = 12.8$. This yields an F-value of $F(3,36) = 0.20$, which

	Group 1b	Group 2b	Group 3b	Group 4b
Means	64.5	76.3	77.0	76.6
Variances	51.4	80.1	53.7	70.0
Sum	662.1	762.7	770.2	766.1
Sum of squares	4430.0	5889.6	5981.1	5931.3
Squared sum	43838	58170	59327	58683
SS_{within}	$= 2301.95$			
$SS_{between}$	$= 817.60$			
SS_{total}	$= 3119.55$			
MS_{within}	$=$ mean(variances) $= 63.9$			
$MS_{between}$	$=$ variance(means) $\cdot$ sample size $= 272.5$			
$F(3,36)$	$= \frac{MS_{between}}{MS_{within}} = 4.26$			
p	$= 0.01$			
η^2	$= \frac{SS_{between}}{SS_{total}} = \frac{817.60}{3119.55} = 0.27$			

Table 12.12. ANOVA for data from Figure 12.16(b).

for the given degrees of freedom is equivalent to saying that the null hypothesis is confirmed with a p value of 0.90 and, therefore, that all means most likely are from the same population.

The situation is different for the data in Figure 12.16(b), however, as Table 12.12 shows: MS_{within} is still the same, but $MS_{between}$ now is an order of magnitude larger, resulting in an F-value of $F(3,36) = 4.26$. This corresponds to a p value of 0.01, which when compared with the default level for significance $\alpha = 0.05$ is an indication to reject the null hypothesis of equal group means.

Assuming that the ANOVA followed a Model 1 design, the next step would be to conduct specific comparisons of the means. As mentioned above, there are two types of comparisons: planned and unplanned.

For a planned comparison, one would need to have a specific *a priori* hypothesis about the behavior of the groups. Returning to the avatar experiment, for example, group 1 could be an experimental condition in which a different animation style based on a lower resolution of the animation was used. Therefore, there are good reasons to expect a difference in this condition. To conduct the planned comparison of whether group 1 is really different from the remaining conditions, another one-way ANOVA is run with two groups (of unequal sample sizes): group 1 and groups 2–4 combined. This ANOVA yields a highly significant result with $F(1,38) = 13.43$, $p < 0.001$.

For an example of an unplanned comparison, the avatar experiment could again have four conditions in which four different animation styles were compared for recognition accuracy. This time, however, there is no particular reason to assume that one animation style was different from the others. The experimental result therefore seems quite surprising to the researcher and to really check whether the suspicious group 1 condition is different from the others, an unplanned comparison needs to be run. As mentioned above, there are three main types of comparisons one can do as post hoc tests. The Tukey HSD tests show group 1 to be different from all three remaining groups, with no significant differences among groups 2–4. Similar results are obtained for the Bonferroni-corrected tests, the difference being a smaller overall effect as the Bonferroni test is more conservative than the HSD test. Finally, the Scheffe criterion—the most conservative of all criteria—only lists group 1 as different from group 3. In the present case, as stated above, the Tukey HSD test is actually the recommended one, as a balanced ANOVA was conducted prior to testing. One can therefore conclude that, indeed, the first animation style proved to be different from the remaining animation styles in the experiment. Also note that the results of the Bonferroni and Scheffe tests show clearly that the planned comparison conducted above has a vastly higher statistical power.

Finally, we note that the effect size found in this ANOVA is $\eta^2 = 0.27$, showing that 27% of the variance between *all* groups is explained by the change in groups. As per Cohen's recommendation (always to be taken with caution), this would be a large effect.

12.7.8 The Kruskal-Wallis Test

12.7.8.1 Summary

The Kruskal-Wallis test is a nonparametric extension of the one-way ANOVA for testing whether two or more (at least ordinal) samples come from the same distribution.

12.7.8.2 Computation

The Kruskal-Wallis (KW) test is an extension of the Mann-Whitney-U test for comparing the central tendencies (that is, medians) of more than two groups of data. The main advantage of this test is that it can be used for both ordinal and interval data. Accordingly, this test also does not rely on assumptions of normality.

Similarly to the Mann-Whitney-U test, the KW test relies on the statistics computed on the ranks of the observations, and, similarly to the one-way ANOVA, the test computes the ratio of between-group rank differences to within group rank differences.

Given one experimental factor with c levels or groups, where each level has n_i datapoints, we first gather all the data and determine the rank of each sample data point $r_{i,j}$, where i runs over all levels $1 < i < c$ and j runs over all data points per level $1 < j < n_i$.

From these $r_{i,j}$, we now need to calculate the average rank sums r_i for each level

$$r_i = \frac{1}{n_i} \sum_{j=1}^{n_i} r_{i,j},$$

and the overall rank-sum r for all data points

$$r = \sum_{i=1}^{c} r_i = \frac{c+1}{2}.$$

The test statistic is now calculated as the ratio of between group differences versus within group differences, that is

$$K = \frac{c-1}{t} \frac{\sum_{i=1}^{c} n_i (r_i - r)^2}{\sum_{i=1}^{c} \sum_{j=1}^{n_i} (r_{i,j} - r)^2}. \tag{12.5}$$

The denominator of Equation 12.5 can be simplified as the summation runs over all groups and all elements. Given that $r = \frac{c+1}{2}$, the equation can be simplified to contain only average group rank sums r_i as follows

$$K = \frac{1}{t} \frac{12}{c(c+1)} \sum_{i=1}^{c} \frac{r_i^2}{n_i} - 3(c+1). \tag{12.6}$$

This equation is, of course, easier to calculate, but obscures the fact that the original formulation of K is given by the ratio of between- to within-rank differences.

The distribution of K can be approximated with a χ^2 distribution if the sample sizes in each level are sufficiently large (usually, $n_i > 5$). If sample sizes are smaller (that is, $n_i \leq 5$), the distribution needs to be calculated exactly. The degrees of freedom for the χ^2 distribution test is $\mathrm{df} = c - 1$.

The factor t in Equations 12.5 and 12.6 is a correction factor for tied ranks. Whenever there are a lot of tied ranks—as will happen often with ordinal data as obtained from rating experiments—this factor will tend to make K smaller, hence reducing the power of the test.

12.7.8.3 Guidelines

- *Difference in distributions* Similarly to the Mann-Whitney-U test, the Kruskal-Wallis test assumes that the only difference in the groups lies in their median value, that is, that all groups have identically shaped distributions. If this assumption is violated, the test might result in false positives.

12.7.8.4 Example

In order to gauge the differences between a standard one-way ANOVA and the KW test, we can repeat the example with the random four-group data shown in Figure 12.16.

	Group 1a	Group 2a	Group 3a	Group 4a
Medians	74.0	75.2	75.3	76.1
Lower quartile	69.2	72.4	72.7	68.7
Upper quartile	80.2	84.3	79.1	83.7
r_i	18.7	21.3	21.8	20.2
$\chi^2(3)$	= 0.41			
p	= 0.94			

Table 12.13. Kruskal-Wallis test for data from Figure 12.16(a).

Table 12.13 lists the relevant statistics for the data from Figure 12.16(a) with four groups of random variables drawn from the same normal distribution.

As the test shows very clearly, we cannot reject the null hypothesis that the four groups come from the same distributions with p = 0.94.

Table 12.14 lists the relevant statistics for the data from Figure 12.16(b) with four groups of random variables drawn from the same normal distribution—the data for group 1, however, was lowered by 10%.

Again, similarly to the one-way ANOVA, the test leads us to reject the null hypothesis that the four groups come from distributions with similar medians. Hence, in this case, both the one-way ANOVA and the Kruskal-Wallis test lead to the same results.

Similarly to the one-way ANOVA, the next step after having found significant differences is to determine *which* groups differ. As we are dealing with potentially non-normally distributed data, we cannot simply take t-tests for our post hoc analyses. However, we can take the Kruskal-Wallis test and compare the different group medians using a post hoc correction such as the Bonferroni correction. Another option available in most statistics programs is compares differences in rank sums using a nonparametric test. If we choose the first option, we would need to correct the α-level to $\alpha' = \frac{\alpha}{6} = 0.0083$ for each of the six possible comparisons. Doing so yields only one significant group difference for the data in Table 12.14: the difference between group 1 and group 3. All other

	Group 1b	Group 2b	Group 3b	Group 4b
Medians	65.8	75.2	75.3	76.1
Lower quartile	60.1	72.4	72.7	68.7
Upper quartile	71.9	84.3	79.1	83.7
r_i	10.5	23.6	24.9	23
$\chi^2(3)$	= 9.89			
p	= 0.02			

Table 12.14. Kruskal-Wallis test for data from Figure 12.16(b).

differences fail to reach significance. Note the obvious difference in power between the nonparametric post hoc tests and the parametric post hoc tests.

12.7.8.5 Example 2

One of the advantageous properties of nonparametric tests is that they tend to be less sensitive to outliers. In order to illustrate this, let us modify the data in Figure 12.16(b) by modifying one data value in group *2* to 0. This means that one participant in group 2 had a recognition accuracy of 0%—a rather extreme outlier, and one that usually would warrant closer inspection of the data and possible removal of that participant's data from the experiment. However, if we just run the one-way ANOVA on that data, we will find the following statistics: $F(3,36) = 1.40$, $p = 0.26$, and hence we would conclude that all four groups come from distributions with similar mean values. Running the Kruskal-Wallis test yields $\chi^2(3) = 8.57$ with $p = 0.04$, which is still a significant result.

12.7.9 The Paired t-Test

12.7.9.1 Summary

A paired t-test is commonly used to test whether two means of datasets consisting of dependent measures are equal or not.

12.7.9.2 Computation

There is another important concept for t-test applications when comparing two groups, namely whether the samples are paired or not. This refers to the experimental design, as a paired t-test is used for dependent measures, that is, for comparing two groups in which the sampling was not done independently. The most common application is in a repeated-measures design, when the same participants are tested twice; for example, face-recognition performance before and after explicit training (in this case, this is also a within-participant design with a pre-test and post-test phase).

A paired t-test also needs to be applied when members of the two groups can be paired according to additional criteria; that is, there are obvious subgroups that might have an influence on the result. An example would be when testing students who worked together in pairs on a problem, or data from left-handed versus right-handed participants. If the two groups are sampled independently—such as when there is a between-participants design with a "before" group and a different "after" group of participants, the independent measures, or unpaired t-test needs to be used. Note that the within-participant design also then implies that the dependent measures t-test has more statistical power compared to the independent measures t-test. It can be applied to reliably detect differences between groups that are small in comparison to the variance within the groups.

Instead of calculating the means and the variances for each group separately, in the case of paired samples the relevant t-statistic is determined on the *differences between the paired values* for the two samples $\{x_1\},\{x_2\}$. That is, we first calculate the differences

$$\Delta_i = x_{i,2} - x_{i,1}.$$

From these differences, we determine the mean $\bar{\Delta}$ and variance s_Δ

$$t = \frac{\bar{\Delta}}{\frac{s_\Delta}{\sqrt{n}}}. \tag{12.7}$$

The degrees of freedom value for a paired t-test is simply df $= n-1$, with n being the sample size of either group.

As in Equation (12.3), we can extend the t-statistic to test for a specific value of the differences, by inserting this value v as follows

$$t = \frac{\bar{\Delta} - v}{\frac{s_\Delta}{\sqrt{n}}}.$$

12.7.9.3 Guidelines

- *Normality.* The paired sample t-test assumes normally distributed samples. If this assumption cannot be met, other nonparametric alternatives such as the Wilcoxon test (see Section 12.7.10) or the sign test (see Section 12.7.11) should be used.

12.7.9.4 Example

As in most paired experiments, the typical use case is when there is a before–after scenario in which the same population responds twice for a certain question. In order to illustrate the difference in power between an independent-sample t-test and a dependent (paired) sample t-test, we can change the data for the avatar experiment from a between-group to a within-group design. Let us assume that a total of nine participants were asked to rate the effectiveness of an avatar before and after exposure to the computer animations. That is, we have nine participants do the recognition task before they know the avatar, and then after they have played around with it, have listened to stories read by it, etc.

We take the same data values as in Table 12.8 for our test. However, now we know that the first data point in the first session belongs to the first data point in the second session—that is, our test becomes more constrained.

Using Equation (12.7) and the data from Table 12.15, we get $t = \sqrt{9}\frac{8.89}{6.09} = 4.43$. With the associated degrees of freedom of df $= n - 1 = 8$, the two-tailed p value becomes p $= 0.002$, which is almost an order of magnitude lower than the p value of the independent t-test for the same data values.

Participant	Pre-test	Participant	Post-test	Δ_i=Post-test-Pre-test
1	80	1	90	10
2	70	2	80	10
3	80	3	80	0
4	80	4	90	10
5	90	5	90	0
6	70	6	80	10
7	60	7	80	20
8	80	8	90	10
9	80	9	90	10
Mean of differences $\bar{\Delta} = 8.89$				
Variance of differences $s_\Delta = 6.09$				

Table 12.15. Table showing hypothetical data from an experiment in which a group of nine participants had to recognize ten computer-animated facial expressions before and after exposure to the avatar. Data represents percent correct.

12.7.10 The Wilcoxon Test

12.7.10.1 Summary

The Wilcoxon test is a nonparametric test to determine whether the difference between two paired, dependent measurement samples of at least ordinal type comes from a distribution with median 0.

12.7.10.2 Computation

Whenever paired measurement variables are ordinal or one suspects that their underlying distribution deviates from the normal distribution, the Wilcoxon test should be used. The Wilcoxon test uses the ranks of the absolute differences between the paired values to determine whether the two samples come from the same distribution.

The first step for two paired data samples $\{x_1\}, \{x_2\}$ is to determine their absolute differences

$$\Delta_i = |x_{i,2} - x_{i,1}|.$$

These absolute differences are then ranked with ties receiving the weighted rank as usual. Next, two signed rank statistics $W_{+,-}$ are formed by summing over all positive differences and all negative differences:

$$W_+ = \sum_i \Delta_i \quad \forall i : x_{i,2} - x_{i,1} > 0,$$
$$W_- = \sum_i \Delta_i \quad \forall i : x_{i,2} - x_{i,1} < 0.$$

Note that both sums exclude terms for which $\Delta_i = 0$.

The final test statistic W is then $W = \min(W_+, W_-)$. The p value for small sample sizes $n < 15-20$ is determined from exact tables, whereas larger sample sizes use a normal approximation.

12.7.10.3 Guidelines

- *Symmetry of distributions.* The Wilcoxon test assumes two equal, symmetric distributions around the median of the samples. If this is not valid, a less constrained test such as the sign test needs to be run.

- *Same median in the two groups?* Strictly speaking, the Wilcoxon test evaluates whether the sample differences $x_1 - x_2$ come from a distribution with median 0. It should be noted that this is *not* equivalent to testing whether the two samples $\{x_1\}, \{x_2\}$ have the *same median.*

12.7.10.4 Example

As in most paired experiments, the typical use case is when there is a before/after scenario in which the same population responds twice for a certain question. As an example, consider the responses of students to a general feedback questionnaire at the middle and at the end of a class. Let us assume that the instructor is only interested in the response to one particular question (How well do you feel that you have mastered the topic of the course?), which then gives us two paired samples of feedback ratings for the student population.

Response data is shown in Table 12.16, which gives the values of Δ_i, the sign of Δ_i, and the rank of Δ_i. From this, $W_+ = 6$ and $W_- = 72$, indicating that the majority of ratings went up. The final test statistic is $W = \min(W_+, W_-) = 6$, which yields a p value of p = 0.0078, a highly significant value. Hence, we can reject the null hypothesis that the difference between the two samples has a median different from 0 and conclude that the ratings, indeed, are different between the two sessions.

Student	Rating 1	Rating 2	Δ_i	Sign(Δ_i)	Rank(Δ_i)
1	2	4	2	-	8.5
2	3	5	2	-	8.5
3	3	4	1	-	3.0
4	3	5	2	-	8.5
5	4	5	1	-	3.0
6	4	3	1	+	3.0
7	3	5	2	-	8.5
8	1	4	3	-	12.0
9	2	4	2	-	8.5
10	5	4	1	+	3.0
11	4	5	1	-	3.0
12	3	5	2	-	8.5

Table 12.16. Response data showing two, repeated ratings from 12 students for the same feedback question (How well do you feel that you have mastered the topic of the course?). The rating scale is 1=not at all, 2=with difficulties, 3=somewhat, 4=well, 5=very well.

Note that if the instructor had wanted to test for whether the answers for the second round of questions had *higher* ratings than for the first round, a one-tailed test would have yielded a p value of $p = \frac{0.0078}{2}$.

12.7.11 The Sign Test

12.7.11.1 Summary

The sign test is a nonparametric test to determine whether two paired, dependent measurement samples come from the same distribution. The measurements need to be at least of ordinal type.

12.7.11.2 Computation

The sign test was developed by F. Wilcoxon as a nonparametric alternative to the paired sample t-test. The test only looks for the sign of the difference between the data pairs and determines whether there is a significant effect for the number of positive versus negative differences. It can be used for asymmetric distributions in the samples, unlike the t-test or the Wilcoxon test, giving this test wide applicability.

In order to calculate the sign test for two paired data samples $\{x_1\}, \{x_2\}$, each with n items, we first create an indicator function that singles out all positive differences

$$f+_i = \begin{cases} 1 & : x_{i,2} - x_{i,1} > 0 \\ 0 & : x_{i,2} - x_{i,1} \leq 0. \end{cases}$$

The negative indicator function $f-$ is defined accordingly as

$$f-_i = \begin{cases} 1 & : x_{i,2} - x_{i,1} < 0 \\ 0 & : x_{i,2} - x_{i,1} \geq 0. \end{cases}$$

We then count the number of positive and differences

$$W_+ = \sum_{i=1}^{n} f+_i,$$
$$W_- = \sum_{i=1}^{n} f-_i.$$

Assuming that all data pairs were drawn independently and that both positive and negative differences are equally likely (in other words, that there is no effect), $W_{+,-}$ will follow a *binomial distribution* with parameters n' and probability $p = 0.5$. It is important to note in this context that n' is the number of data pairs for which the difference is $x_{i,2} - x_{i,1} \neq 0$, or equivalently $n' = \sum_{i=1}^{n} f+_i + \sum_{i=1}^{n} f-_i$. From the binomial distribution we can determine the p value for our test for both one-tailed and two-tailed testing by taking the minimum of the two counts $W = \min(W_+, W_-)$ as our test statistic.

12.7.11.3 Guidelines

- *Power.* The sign test only assumes that the each paired data sample (for each person, for example) was drawn independently, thus giving it the largest applicability—albeit at reduced power. The Wilcoxon test is more powerful than the sign test and is applicable in many cases.

- *Size of the differences.* Note that by using only the sign of the differences, the sign test will effectively discard any notion of the absolute values of the differences, such that it will not matter *how much* the results of the two samples differ, but only *that they differ.* If the experimenter is only interested in changes in measures the sign test should be used; however, if the size of the change also needs to be taken into account, the Wilcoxon test, or the paired sample t-test (if applicable) should be used instead.

12.7.11.4 Example

As an example, let us return to the questionnaire responses of the student feedback given in Section 12.7.10. Students were given a feedback questionnaire at the middle and at the end of a class, and we want to evaluate the response to one particular question (How well do you feel that you have mastered the topic of the course?), which results in two paired samples of feedback ratings.

Example data is shown in Table 12.17, which gives the two ratings and the values of $f+_i$, $f-_i$ for each data pair. From this, we find that $W_+ = 10$ and $W_- = 2$, indicating that the majority of the 12 ratings went up. Taking $W = \min(W_+, W_-) = 2$ as our test statistic, we find a p value of $p = 0.039$ from the binomial distribution. Hence, using the sign test, we can also reject the null hypothesis that the difference between the two samples has a median different from 0 and conclude that the ratings, indeed, are different between the two sessions.

Student	Rating 1	Rating 2	$f+_i$	$f-_i$
1	2	4	1	0
2	3	5	1	0
3	3	4	1	0
4	3	5	1	0
5	4	5	1	0
6	4	3	0	1
7	3	5	1	0
8	1	4	1	0
9	2	4	1	0
10	5	4	0	1
11	4	5	1	0
12	3	5	1	0

Table 12.17. Response data showing two, repeated ratings from 12 students for the same feedback question (How well do you feel that you have mastered the topic of the course?). The rating scale is 1=not at all, 2=with difficulties, 3=somewhat, 4=well, 5=very well.

This test result would not be changed as long as the sign of the rating differences stayed the same. Hence, the fact that Student 8 actually seemed to have made a very large boost in terms of understanding is lost in the test results.

Also note that if we flip one more rating from a positive change to a negative change (such as changing student 5's result from (4, 5) to (5, 4)), the test would not be significant anymore with $p = 0.15$ for a two-tailed test. However, the Wilcoxon test would still be significant ($p = 0.016$, two-tailed), which illustrates the reduced power of the sign test as compared to the Wilcoxon test.

12.7.12 Repeated Measures One-Way ANOVA

12.7.12.1 Summary

The repeated-measures one-way ANOVA is used to evaluate whether a group of more than two paired, dependent, interval-measurement samples come from the same distribution.

12.7.12.2 Computation

The repeated measures one-way ANOVA is a straightforward extension of the one-way ANOVA to dependent samples. It should be used whenever the samples were repeated, when the same participants were tested in several different conditions with the same measure (that is, in a within-participant design), or when measures are highly related to each other (such as questions concerning the same topic on a questionnaire). In all of these cases, the measurements were not taken independently of each other and hence need to be evaluated as dependent samples.

Like the independent one-way ANOVA, the repeated measures one-way ANOVA partitions the data in several sums-of-squares values that measure different variations in the data. As before, the variation in the experiment can be partitioned into the treatment-, or between-group, variation SS_{between}, and into the within-group variation SS_{within}. However, this time the within-group variation can be split further into the *participant* variation SS_{part} and the remaining error variation SS_{error}. The additional participant variation term can be estimated because we know which data points in which group belong to which source of variation. In the analysis, the between-group variation stays the same; however, the unspecific within-group variance for the independent one-way ANOVA is now split further into variance that is associated with participants and the remainder that is the within-group, or error variation.

In order to calculate the sums of squares, we now also sum over the "rows" of the data, that is, we also determine the sum for each participant:

$$X_j = \sum_{i=1}^{c} x_{i,j}.$$

As before, we get the group sum

$$X_i = \sum_{j=1}^{n} x_{i,j}.$$

From this, we determine the grand total sum over all data points

$$X = \sum_{i=1}^{c} \sum_{j=1}^{n} x_{i,j}.$$

Similarly, we need the grand total sum of squares over all data points

$$XS = \sum_{i=1}^{c} \sum_{j=1}^{n} x_{i,j}^2.$$

With these quantities, we can now calculate the between-group sum of squares as before

$$SS_{\text{between}} = \frac{1}{n} \sum_{i=1}^{c} X_i^2 - \frac{1}{cn} X^2.$$

The within-group sum of squares is still defined as

$$SS_{\text{within}} = XS - \frac{1}{n} \sum_{i=1}^{c} X_i^2.$$

One part of this variation, however, is the participant sum of squares, which we determined as

$$SS_{\text{part}} = \frac{1}{c} \sum_{j=1}^{n} X_j^2 - \frac{1}{cn} X^2.$$

If we know the SS_{within} and SS_{part}, by definition the difference between the two types of variation is the error, or residual term we want

$$SS_{\text{error}} = SS_{\text{within}} - SS_{\text{part}}.$$

And as before, the total sum of squares can be determined as

$$SS_{\text{total}} = XS - \frac{1}{cn} X^2 = SS_{\text{part}} + SS_{\text{between}} + SS_{\text{error}}.$$

Each variation is associated with a corresponding degrees of freedom value. The degrees of freedom value for the participant variation is, of course, $\text{df}_{\text{part}} = n-1$, which means that the between-group value reduces to $\text{df}_{\text{between}} = c-1$ and the within-group value to $\text{df}_{\text{within}} = (c-1)\cdot(n-1)$. The degrees of freedom associated with the total variation is still $\text{df}_{\text{total}} = \text{df}_{\text{part}} + \text{df}_{\text{between}} + \text{df}_{\text{within}} = cn-1$.

As before, we calculate the corresponding means of square values

$$MS_{\text{between}} = \frac{SS_{\text{between}}}{\text{df}_{\text{between}}}.$$

What was the mean square value for the within-group variation before, now is split into two parts

$$MS_{\text{part}} = \frac{SS_{\text{part}}}{\text{df}_{\text{part}}}$$

and the actual error part

$$MS_{\text{error}} = \frac{SS_{\text{error}}}{\text{df}_{\text{error}}}.$$

With these values, the F-ratio becomes:

$$F = \frac{MS_{\text{between}}}{MS_{\text{error}}}.$$

In effect, the additional variable MS_{part} removes the effect of the within-participant design from the equation, as it does not form part of the final F-value. This is why the within-participant design has more power. In addition, note that the resulting degrees of freedom also are reduced for the calculation.

Since the data is paired, we cannot talk about within-group variances anymore, and the concept of homogeneity of variances turns instead into the requirement of *sphericity*. Sphericity means that the variance for any set of two difference scores should be the same across the dataset. The correct test for this is the Mauchly sphericity test (which is based on the covariance matrix of the dataset). If the test is significant, the assumption of sphericity across the dataset is most likely violated and the repeated measures ANOVA cannot be applied.

The pairwise effect size for comparing two groups i and j is given by the standard effect size measure

$$ES_{i,j} = \frac{\bar{x}_i - \bar{x}_j}{\sqrt{MS_{\text{within}}}}.$$

The overall effect size is, as before

$$\eta^2 = \frac{SS_{\text{between}}}{SS_{\text{total}}},$$

which specifies the amount of variance due to the experimental manipulation.

12.7.12.3 Guidelines

- *Overall guidelines.* Since the repeated measures one-way ANOVA is an extension to the regular ANOVA, the guidelines from Section 12.7.7 also apply here. Most importantly, this concerns the normality of the samples, as calculation of variances assumes that the data comes from normally distributed sources. If this assumption cannot be satisfied, a non-parametric alternative to the ANOVA needs to be selected, such as the Friedman test (see Section 12.7.13).

- *Sphericity.* Enforcing the strict sphericity criterion often leads to problems, as the assumption of equal variance across *all* possible sets is frequently violated. Rather than not using the ANOVA at all, one can correct for the violation of sphericity by adjusting the degrees of freedom to a lower value (similar to the case of unequal variance in the independent sample t-test), thereby changing the cumulative distribution function for determining the critical F-value. There are several corrections available, of which the most drastic correction changes $\mathrm{df}_{\mathrm{between}}$ from $\mathrm{df}_{\mathrm{between}} = c-1$ to $\mathrm{df}_{\mathrm{between}} = 1$ and $\mathrm{df}_{\mathrm{within}}$ from $\mathrm{df}_{\mathrm{within}} = (c-1)(n-1)$ to $\mathrm{df}_{\mathrm{within}} = n-1$. This provides a very conservative test of equality of means across the conditions. A less conservative and hence more often used test is the *Greenhouse-Geiser correction,* which is included in most statistical software packages—again, as with all corrections, this alteration will reduce the power of the test. The Greenhouse-Geiser measures the violation of sphericity and ranges from the lower bound of $gg = \frac{1}{c-1}$ to the upper bound $gg = 1$. If $gg = 1$, the data is fully spherical and there is no problem. Otherwise the assumption of sphericity is violated, and one needs to correct all degrees of freedom to $\mathrm{df}' = \mathrm{df} \cdot gg$. The lower bound also suggests that for $c = 2$ levels, there is no problem with the repeated measures ANOVA, as two levels are by default spherical ($gg = 1$, when $c = 2$).

12.7.12.4 Example

Let us return to the avatar experiment and change the same data into a repeated-measures design. Ten participants were invited to interact with an avatar, and the recognition performance was measured at four points in time with the data values being exactly the same as in Figure 12.16(b). We now wish to know whether there was a significant effect of time of testing on the recognition results.

Table 12.18 shows the results of the calculation. We can see that a third of the variation is taken out of the equation with the within-participant design. The calculation results in an F-value of $F(3,27) = 4.73$ and a p value of $p = 0.009$, which is slightly lower than the one obtained with the independent one-way ANOVA.

12.7.13 Friedman test

12.7.13.1 Summary

The Friedman test is a nonparametric extension of the repeated-measures ANOVA. It determines whether a group of two or more paired, dependent ordinal measurement samples come from the same distribution.

	Time 1b	Time 2b	Time 3b	Time 4b
Means	64.5	76.3	77.0	76.6
Variances	51.4	80.1	53.7	70.0
Sum	662.1	762.7	770.2	766.1
Sum of squares	4430.0	5889.6	5981.1	5931.3
Squared sum	43838	58170	59327	58683
$SS_{between}$	$= 817.60$			
SS_{within}	$= 2301.95$			
SS_{part}	$= 744.87$			
SS_{error}	$= SS_{within} - SS_{part} = 1557.08$			
SS_{total}	$= 3119.55$			
$MS_{between}$	$=$ variance(means) $\cdot$ sample size $= 272.5$			
MS_{part}	$=$ mean(variances)$= 82.76$			
MS_{error}	$=$ mean(variances)$= 57.67$			
$F(3,27)$	$= \frac{MS_{between}}{MS_{error}} = 4.73$			
p	$= 0.009$			
η^2	$= \frac{SS_{between}}{SS_{total}} = \frac{817.60}{3119.55} = 0.27$			

Table 12.18. Repeated measures ANOVA for data from Figure 12.16(b) re-interpreted as four different points in time.

12.7.13.2 Computation

Whenever the assumption of normality cannot be guaranteed for groups of dependent measure data, a Friedman test needs to be applied. Similarly to the Kruskal-Wallis test, the Friedman test uses rank sums to test for significant differences between groups. And, similarly to the repeated measures ANOVA, the Friedman test pulls out the variations within participants.

More specifically, given c groups of data, each with n samples from n participants, we rank all elements *within each group.* For data points $x_{i,j}$ with i indexing groups and j indexing participants, we get for each $x_{i,j}$ an associated rank $r_{i,j}$. From these $r_{i,j}$, we now need to calculate the average rank sums r_i for each group as

$$r_i = \frac{1}{n} \sum_{j=1}^{n} r_{i,j}$$

and the overall rank sum r for all data points

$$r = \frac{1}{cn} \sum_{i=1}^{c} \sum_{j=1}^{n} r_{i,j}.$$

The test statistic now is calculated as the ratio of between-group differences versus within-group differences, that is

$$Fr = n^2(c-1)\frac{\sum_{i=1}^{c}(r_i - r)^2}{\sum_{i=1}^{c}\sum_{j=1}^{n_i}(r_{i,j} - r)^2}.$$

There are no corrections necessary for tied ranks in this case.

Similarly to the Kruskal-Wallis test, the distribution for the Friedman test statistic follows a χ^2 distribution with degrees of freedom $\text{df} = c - 1$ for suitably large sample sizes.

Specialized post hoc tests for the Friedman test are available, but not all software packages include them. Of course, it is possible to follow up a significant result with a Wilcoxon test and a suitable Bonferroni correction.

12.7.13.3 Guidelines

- *Power.* The Friedman test has considerably less power for normally distributed data than the corresponding repeated measures ANOVA.

12.7.13.4 Example

Let us assume that we have five participants, each of which has rated five different computer animations on a scale of 1 (not realistic) to 5 (fully realistic). Of course, given this experimental design, we want to know whether the factor "animation style" has an effect on the ratings. As the same participants rated each animation, this is clearly a repeated-measures, or within-participant design. Because we have very few participants, we might want to avoid the assumption of normality and resort to a nonparametric test instead.

The data is shown in Table 12.19 and plotted as a box-plot in Figure 12.18. If we run the Friedman test, the result is $\chi^2(4) = 15.26$ with $p = 0.004$. Hence, despite the considerable variation in the data, we have a highly significant result and can conclude that the factor "animation style" does have an effect on the ratings. Post hoc tests reveal that animations 2 and 5 have significantly higher ratings than group animation 3.

	Animation 1	Animation 2	Animation 3	Animation 4	Animation 5
Participant 1	3	3	2	2	3
Participant 2	4	4	2	2	4
Participant 3	3	2	1	2	3
Participant 4	3	4	3	3	4
Participant 5	4	3	1	1	2

Table 12.19. Data for a rating experiment in which five participants rate the realism of five animations.

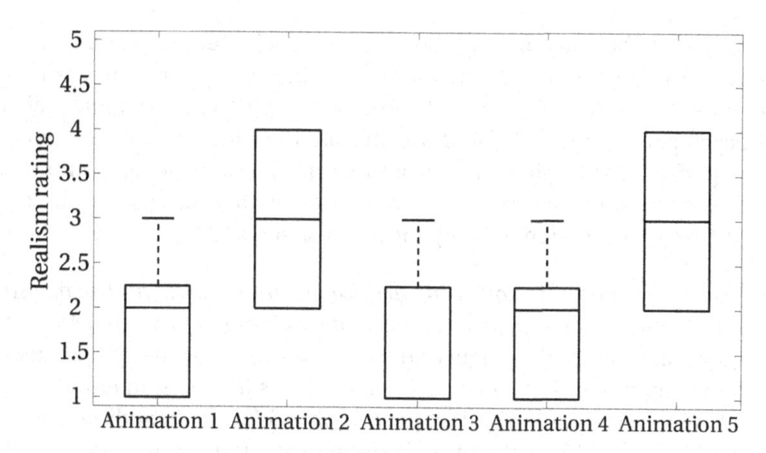

Figure 12.18. Box-plot of the rating data from Table 12.19.

Note that because the variation across participants is very large, the Kruskal-Wallis test will not result in a significant value for the same data: $\chi^2(4) = 8.53$ with $p = 0.07$. Since each participant, however, has a reliable variation along the animations the paired-sample test will highlight the correct pattern.

12.7.14 Two-Way ANOVA

12.7.14.1 Summary

The two-way ANOVA is a straightforward extension of the ANOVA that includes two factors, each having multiple factor levels. Most importantly, the two-way ANOVA can be used to investigate *interactions* between factor levels of the two factors.

12.7.14.2 Computation

One of the most common paradigms in experimental design is the two-by-two setup in which the experimenter wants to investigate the influence of two factors that each have two levels on the dependent variable. These factors are almost always fixed; that is, the experimenter controls the two factors and hence the resulting ANOVA is a *Model 1, two-by-two ANOVA.*

What is important about this type of ANOVA is that not only can the so-called *main* effects be investigated—that is, whether factor 1 and/or factor 2 influence the dependent measure—but also whether the result depends on specific combinations of factor 1 *and* factor 2. In other words, we can investigate the interactions of the two factors. This represents an important step forward in terms of the complexity of the analysis, as it treats the factors not as independent but as potentially co-dependent.

In addition, both factors can be within- or between-participant (in any combination). Depending on the experimental design, this will modify the power for either or both factors as it allows the ANOVA to also estimate effects within each participant and split those off from the within-factor variations.

Going into all the calculation of all relevant sums of squares and mean squares is beyond the scope of this book. Hence, here we just list the components that are estimated from a typical two-by-two ANOVA.

- *Two independent factors with two independent levels and n measurements per level of each factor.* Note that this design means that $4n$ participants are randomly assigned to each group. For this design, we get the following terms. As before, each factor enters its variation into the overall, observed variation and we get SS_{factor1}, SS_{factor2}. In addition, we also get $SS_{\text{factor1,factor2}}$ as the interaction between the two factors. Together with the usual total variation SS_{total}, we get the remaining term as the error term $SS_{\text{error}} = SS_{\text{total}} - (SS_{\text{factor1}} + SS_{\text{factor2}} + SS_{\text{factor1,factor2}})$. The associated degrees of freedom for these terms are: $\text{df}_{\text{factor1}} = 1$, $\text{df}_{\text{factor2}} = 1$, $\text{df}_{\text{factor1,factor2}} = 1$, and $\text{df}_{\text{error}} = 4n - 1$.

- *Two-way repeated-measures design in which two factors with two levels and n measurements per level of each factor were measured from n participants.* In this context, we get the following terms. Again, each factor enters its variation into the overall, observed variation and we get SS_{factor1}, SS_{factor2}, and we also get $SS_{\text{factor1,factor2}}$ as the interaction between the two factors. Because all elements include a within-participant component, we can estimate an additional error term for all main factors *and* for the interaction. That is, we also have $SS_{\text{error;factor1}}$, $SS_{\text{error;factor2}}$, and $SS_{\text{error;factor1,factor2}}$. The associated degrees of freedom for these terms are $\text{df}_{\text{factor1}} = \text{df}_{\text{factor2}} = \text{df}_{\text{factor1,factor2}} = 1$ and $\text{df}_{\text{error;factor1}} = \text{df}_{\text{error;factor1}} = \text{df}_{\text{error;factor1,factor2}} = n - 1$.

- *Mixed-participant design with one between-participant factor and one within-participant factor.* In such a design, the experimenter is usually interested in evaluating the between-participant factor as a function of the within-participant factor. Typical designs include pre- and post-test studies in which two measurements are taken with one experimental group and one control group. In this case, the experimenter will be looking for an interaction, as the post-test results will be different due to the experimental manipulation for the experimental group, but *not* for the control group. Another design is a generalization of the repeated-measures, one-factor design (in which one variable is measured repeatedly for many participants) to two factors that are measured at two different points in time. Since there are neither specific experimental nor control groups in this case, any type of result could be interesting, depending on the research question. A mixed-design analysis with both within- and between-participant factors is also known as a split-plot ANOVA. For the mixed two-by-two ANOVA in which factor 1 is a between-participant factor and factor 2 is a within-participant factor, we get the following terms and associated degrees of freedom values. As before, each factor enters its variation into the overall, observed variation and we get SS_{factor1}, SS_{factor2}. In addition, we also get $SS_{\text{factor1,factor2}}$ as the interaction between the two factors. With a within-participant factor, we can calculate two more error terms, one for each factor: $SS_{\text{error;factor1}}$, $SS_{\text{error;factor2}}$. The associated degrees of freedom for these terms are $\text{df}_{\text{factor1}} = \text{df}_{\text{factor2}} = \text{df}_{\text{factor1,factor2}} = 1$ and $\text{df}_{\text{error;factor1}} = \text{df}_{\text{error;factor2}} = 1$.

Effect size measures for fully independent designs can be estimated using the usual η^2

$$\eta^2 = \frac{SS_{\text{effect}}}{SS_{\text{total}}},$$

where SS_{factor1}, SS_{factor2}, or $SS_{\text{factor1;factor2}}$ are substituted for the wanted effect sum-of-squares measure.

Whenever there is a within-participant design, that is, whenever some of the cells in the ANOVA are dependent, it is better to report partial η^2, which is defined as:

$$\eta_p^2 = \frac{SS}{SS_{\text{effect}} + SS_{\text{error}}}.$$

Again, we can substitute SS_{factor1}, SS_{factor2}, or $SS_{\text{factor1;factor2}}$ for the effect measure. SS_{error} relates to the relevant error term for each factor in the calculation of the ANOVA. The interpretation of η_p^2 is similar to η^2, however, the amount of variance is not related to the total error term, but rather as a proportion related to the sum of the effect and the associated error term.

12.7.14.3 Guidelines

- *Interpretation.* Interpretation of the ANOVA results should always start at the highest interaction. As an example, take an Avatar experiment in which two fixed factors were tested with recognition accuracy as the dependent variable: animation style (with two different types of animation styles A_1 and A_2) and expression (with five expressions). The ANOVA found significant main effects of animation style and expression, as well as an interaction of animation style and expression. Since there are only two animation styles, one of the animation styles (say A_1) at first glance seems to perform worse than the other one. This statement, however, is true only on average: the interaction might, for example, reveal that for the surprised and happy expressions, animation style A_1 actually did equally well or even better than A_2, but for the remaining three expressions A_1, indeed, fared far worse. It is therefore important to carefully analyze the interaction and to summarize the experiment at that level rather than at the levels of the main effect.

- *Unequal sample sizes/missing data values.* For unequal sample sizes the ANOVA needs to be corrected. Similarly, for missing data values in some cells the problem becomes tricky, as the cells might be either in an independent or a dependent factor. Most statistics packages can deal with missing data and unequal sample sizes by adjusting the degrees of freedom using a more sophisticated regression method.

- *Homogeneity of variances.* Levene's test for homogeneity of variances should be run to test for this assumption. In a mixed design, the Levene test needs to be run for the between-group factor.

- *Sphericity.* For a repeated measures ANOVA, we had mentioned the additional requirement of sphericity across the levels. However, note that in the case of a two-by-two ANOVA, we have only two levels for each factor and hence the Greenhouse-Geiser correction need not be applied, as $gg = 1$ automatically.

12.7.14.4 Example 1

As a first example, let us consider a between-participants design in which we collect attractiveness ratings of face pictures with two factors, gender (male and female) and eye size (large and small). Because it is a full between-participants design, we need different participants for each combination of factors. Given our usual rule of thumb for measuring large effect sizes, we chose ten participants for each cell, yielding a total of 40 participants.

In doing the two-by-two ANOVA, we are interested in measuring three types of variability in the data:

- variability between male and female (effect of factor 1: gender)

- variability between large and small (effect of factor 2: eye size)

- variability between male and female for large versus small, or equivalently, in the variability between large and small for male versus female (interaction effect)

We want to measure not only what the effect of each factor is on ratings, but also, if there is an effect, whether it is the same for all levels of the other factor.

Table 12.20 shows the average data for each cell, and Figure 12.19 shows bar plots of the data for each factor. Note that the plots do not give the full picture, as they need to average over one factor and hence cannot display all data.

As we can see from the table and the figure, on average both male and female faces are judged as equally attractive (see column means of Table 12.20,

Eye Size	Gender: Male	Gender: Female	Row Means
Small	3.30	2.20	2.75
Large	4.70	5.80	5.25
Column Means	4	4	4

Table 12.20. Average data for a two-by-two experiment on attractiveness of faces with the factors gender (male and female) and eye size (large and small).

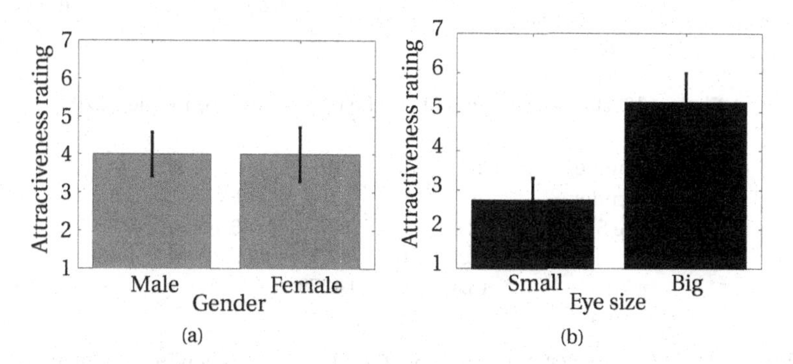

Figure 12.19. Bar plots of the rating data from Table 12.20.

and Figure 12.19(a)); in addition, small eyes seem to be judged as less attractive than large eyes (see row means of Table 12.20, and Figure 12.19(b)). So far, the results are not particularly surprising. However, the full data also shows that the difference between small and large eyes for female faces seems to be larger than that for male faces.

This interaction of the factor levels can be plotted as shown in Figure 12.20, either as a function of eye size (Figure 12.20(a)), or as a function of gender (Figure 12.20(b)). In Figure 12.20(a) the lines diverge, and in Figure 12.20(b) the lines cross each other, indicating that the effect depends on both factors. If there were no interaction, the lines would be parallel.

First of all, we want to make sure that the variances in all cells are equal. We perform a Levene test for checking this for all four groups and find $L(3,36) = 1.05, p = 0.38$, with which we can conclude that the sample variances likely are equal. Hence, we can continue with the ANOVA.

Table 12.21 shows the typical output of our two-by-two ANOVA with all relevant statistics. With the data that we have gathered, we find no main effect of gender $F(1,36) = 1$, $\text{p} = 1$, and $\eta^2 = 0$, a significant main effect of eye size $F(1,36) = 164.1$, $\text{p} < 0.001$, and $\eta^2 = 0.69$. The main effects are qualified by a significant interaction $F(1,36) = 28.29$, $\text{p} < 0.001$, and $\eta^2 = 0.13$. In analyzing the results, we cannot say that gender has no influence on attractiveness, but

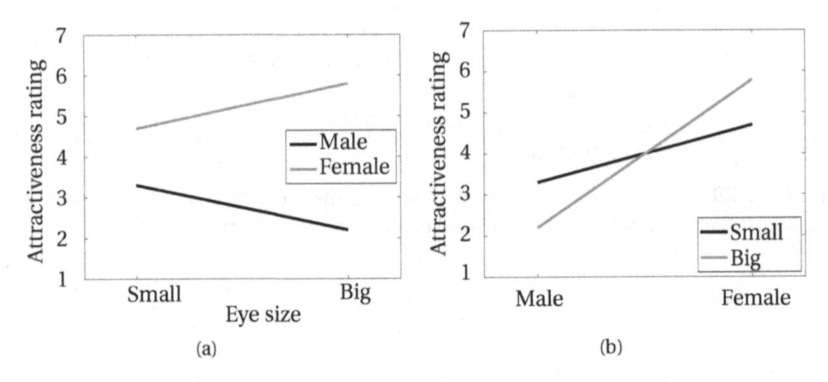

Figure 12.20. Interaction plots of the rating data from Table 12.20.

Source	SS	df	MS	F	p	η^2
Gender	0	1	0	1	1	0
Eye Size	62.5	1	62.5	146.1	0	0.69
Gender * Eye Size	12.1	1	12.1	28.29	0	0.13
Error	15.4	36	0.4278			
Total	90	39				

Table 12.21. Typical ANOVA output table for a two-by-two between-participants design.

we must conclude that attractiveness depends on both factors: gender *and* eye size.

Hence 69% of the variance in the experiment is related to one main effect (eye size), and 13% is related to the interaction. The remaining 18% is then "hidden" in the participant variance in each cell.

12.7.14.5 Example 2

Example 1 dealt with a full between-participants design in which each cell contained data from a different set of participants. What would happen if we repeated the experiment as a full within-participant design, that is, if one group of participants saw all four combinations of factor levels? Of course, in such an experimental design, we would need to worry about learning and carry-over effects, as responses to one condition might be dependent on having already experienced another condition. To address this problem, we could present trials of all four conditions randomly, or we could present them blockwise with random ordering of blocks for different participants.

Table 12.22 shows the output for the same data as Table 12.21, but now used in a within-participant design in which we have ten participants who saw all four conditions. Because of the within-participant design, it becomes possible to split the overall, unspecific error term in Table 12.21 (Error = 15.4) into three

Source	SS	df	MS	F	p	η_p^2
Participants	8.5	9	0.944	(3.7)	(0.004)	
Gender	0	1	0	1	1	0
Gender error	1.5	9	0.167			
Eye size	62.5	1	62.5	112.5	0	0.93
Eye size error	5	9	0.556			
Gender * Eye size	12.1	1	12.1	272.25	0	0.97
Gender * Eye size error	0.4	9	0.044			
Total	90	39				

Table 12.22. Typical ANOVA output table for a within-participant design.

error terms related to the factors (gender error = 1.5, eye size error = 5) and the interaction (gender * eye size error = 0.5), as well as the error term associated with the variation in each of the ten participants (participant variation = 8.5). The table also reports the partial η^2 for all individual terms. Note that because the participant variation accounts for a considerable amount, hence reducing the associated error terms for the factors and interactions, all η_p^2-values are close to 1, indicating that whatever variation there is, is due almost exclusively to the variation in the factor.

In addition, because for a within-experiment the variation of participants can be estimated from the data, it is in principle possible to determine an F-value for the effect of participants as well. Unless one is explicitly interested in this variation, the value is usually not reported, however. Note that here we would have an effect indicating that some participants responded differently than others. Full, multiple comparison-corrected, post hoc tests would need to be run to determine exactly which participants those are.

12.7.14.6 Example 3

The final example concerns the mix of a between- and within-participant design, in which all levels of one factor were seen by all participants, but separate groups of participants saw the two levels of the other factor. Of course, it is possible to do two different versions of this mixed design, depending on which factor is chosen as the within-participant factor. Since within-participant designs increase the power of the experiment, one would choose the factor that would need additional power as the within-participant factor.

If we run the Levene test on the two between-participant groups, we actually do get a significant violation with $F(1,38) = 21.1$, $p < 0.001$. We therefore could either reduce the degrees of freedom in the analysis that follows, or—as we do here—proceed with caution.

Table 12.23 shows the output for the same data as Table 12.20, but now for a mixed design in which we have 20 participants split across the two factors. So, instead of 40 participants in the first experimental design, and ten partici-

Source	SS	df	MS	F	p	η_p^2
		Between-participant				
Gender	0	1	0	1	1	0
Gender error	10	18	0.556			
		Within-participant				
Eye Size	62.5	1	62.5	208.3	0	0.92
Gender * Eye size	12.1	1	12.1	40.3	0	0.69
Eye size error	5.4	18	0.556			
Total	90	39				

Table 12.23. Typical ANOVA output table for a mixed design.

pants in the second one, we now have 20 participants. We chose gender as the between-participant factor (that is, ten participants saw male faces, and ten different participants saw female faces), and eye size as the within-participant factor. Because of the one within-participant factor, it becomes possible to split the overall, unspecific error term in Table 12.20 (error = 15.4) into two error terms related to each single factor (gender error = 10, eye size error = 5.4). The latter error now includes both the main effect and the interaction, as we cannot split the interaction effect further since it contains both within- and between-participant variation. Again, the table also reports the partial η^2 for all individual terms. Because of the mixed design, the effect size for the interaction is reduced compared to the previous design.

Finally, all terms are highly significant, hence, even reducing the degrees of freedom due to the potential violation of the homogeneity of variance assumption will not change the results in this case. Note however, that if the results were less clear, caution would need to be applied to the interpretation. In most cases, such a violation will be due to outliers, and hence could be corrected by pruning the data accordingly.

12.7.15 The Pearson Correlation

12.7.15.1 Summary

The Pearson correlation evaluates the strength of the (linear) association between two interval variables.

12.7.15.2 Computation

The Pearson correlation is used whenever two variables x, y are measured and one wishes to assess the degree with which they are associated with each other; or, in other words, the degree with which they are correlated. The Pearson correlation coefficient r is the most common and also most simple measure of this correlation. The coefficient r evaluates the degree with which variations in the variable y can be explained *linearly* by variations in the variable x. More specifically, when there is a high degree of positive correlation between x and y, we expect high values for y to occur for high values of x, and low values of y for low values of x. For a high degree of negative correlation, we expect the reverse, namely high values of y for low values of x and vice versa. Note that the correlation does not assess the *absolute* scale of the variables x and y—just the relative covariation.

Whereas the variability of one variable is measured by its variance, the co-variability of two variables x and y is measured by their *covariance*. In addition, it is easy to see that the maximum covariance of the two variables is actually provided by the two separate variances of each variable—if one variable's variance, for example, was 0, we would also not expect the two variables to co-vary at all. The maximum covariance is actually given by the geometric mean of the two variances, given that we are talking about averaging squared measures. Hence, we can define the Pearson correlation coefficient as

$$r = \frac{\text{cov}(x, y)}{\sqrt{\text{var(x)var(y)}}}.$$

For the purposes of calculating r, we can relate it directly to the sum of squared deviations from the definition of the variances, and by extending the same concept to the covariance, that is,

$$r = \frac{\sum_{i=1}^{n}(x - \bar{x})(y - \bar{y})}{\sqrt{\sum_{i=1}^{n}(x - \bar{x})^2 \sum_{i=1}^{n}(y - \bar{y})^2}},$$

with the usual definition of $\bar{x}, \bar{y}$ being the sample means.

The Pearson correlation coefficient follows $-1 \leq r \leq 1$, and its sign, of course, gives the directionality of the correlation.

From the correlation coefficient, one usually also determines the so-called *coefficient of determination*, which is simply r^2. This coefficient follows $0 \leq r^2 \leq 1$, and its interpretation is that it yields the strength of the linear association between x and y, or in other words, it provides the amount of variance in y that is explained by x. Hence, the directionality of the correlation is provided by the sign of r and the strength given by r^2, with both measures important in the analysis to report.

The significance of the correlation is equivalent to asking whether the slope of the line is significantly different from 0. The significance value is obtained from a t-statistic with

$$t = \frac{r}{\sqrt{\frac{1-r^2}{n}}} \tag{12.8}$$

and an associated degrees of freedom value of df $= n - 2$.

12.7.15.3 Guidelines

- *Sample size.* Small sample sizes can result in very large correlation values (see Example 1), conversely, very large sample sizes can result in very small, but highly significant correlation values (the t-value is dependent on $\sqrt{n}$; see Equation (12.8)). The experimenter needs to look at the effect size in either case to interpret the results correctly.

- *Outliers.* The correlation coefficient is highly sensitive to outliers, in that one misplaced data value can have a large influence on the result (see Example 2). The nonparametric Spearman correlation is less sensitive to outliers, but measures only monotonic relationships between the two variables x and y rather than a linear relationship.

- *Non-linear relationship.* In the case of a curvilinear relationship, the correlation might result in a serious misinterpretation of the data. Again, plotting the data and looking at the relationship in detail can prevent this in most cases (see Example 3).

- *Causation.* Everyone knows the saying that correlation does not equal causation. However, it is worth noting that in cases in which one does not have experimental control over the variables (that is, it is not possible to form a control group, for example), correlating the variables is the only statistical option left for model building. Hence, special care should be taken to limit implications of causality in presenting data from correlational analysis. At the very least, the final interpretation should include a statement with regards to the limits of correlational analysis; ideally, it should also contain an in-depth analysis of the possible confounding factors that might influence the results.

- *Confirmation bias.* In any correlational study it is very easy to pay attention only to the factors that confirm one's suppositions—for example, the introspective correlation between arthritis and bad weather is supposed to be very high. The actual correlation in a controlled study, however, was found to be rather low. This is because we tend to overlook the data points that do not confirm the correlation, that is, when pain occurs in good weather, or no pain occurs in bad weather.

- *Regression versus correlation.* There is an intimate relation between correlation and linear regression—in effect, the correlation coefficient r is related to the slope of the best-fitting line that runs through the x, y data points. Hence, a correlation analysis is a necessary step of a linear regression, and mathematically, a part of the linear regression is implied when doing a correlation analysis. There are some slight, perhaps philosophical differences between the two: a regression analysis will create a functional relationship between y and x and thus make the causation assumption much more explicit, whereas the correlation analysis talks only about an association between the two variables.

12.7.15.4 Example 1

Figure 12.21 illustrates the effect of sample sizes on correlation values. Shown are three histograms that were created by correlating 100 paired samples of uniformly distributed, random numbers with different sample sizes of 5, 25, and 100 samples. For smaller sample sizes, the (random) correlation values actually span the whole range from -1 to 1, whereas the spread of the histogram becomes much smaller for larger sample sizes. Hence, for small sample sizes large correlation values are to be expected, even by chance.

12.7.15.5 Example 2

Figure 12.22(a) shows the dangers of outliers. A perfectly linear relationship between the two variables x and y is disturbed by one outlier, which results in a large change in the correlation value.

12.7.15.6 Example 3

Figures 12.22(b), and (c) illustrate what happens if the linear correlation analysis is applied to nonlinear data. In Figure 12.22(b), the data is monotonically increasing due to an exponential relationship $y = e^x$; however, due to the highly nonlinear nature of the function, the correlation value yields $r = 0.72$. Similarly, although the relationship between x and y is given by a perfect sinusoidal function in Figure 12.22(c), the correlation still yields a relatively large value for r. None of these functions can actually be fit properly with a linear trend, making these correlations invalid to interpret.

12.7.16 Spearman's Rank Correlation

12.7.16.1 Summary

Spearman's rank correlation evaluates the strength of the monotonic association between two variables that are at least of ordinal type.

12.7.16.2 Computation

Given the general idea of nonparametric evaluations, Spearman's rank correlation determines the how well the ranks of two sets of variables $\{x\}, \{y\}$ of n items correlate with each other. In order to calculate the coefficient, we first need to convert the data points into ranks. For this, the ratings are sorted either in decreasing or increasing order, and the position of each rating in this table is the rank of that rating. Note that when ranks are tied, the rank for each tied item is the average of its position in the table; this will happen frequently, especially for ordinal data as commonly encountered during rating tasks, for

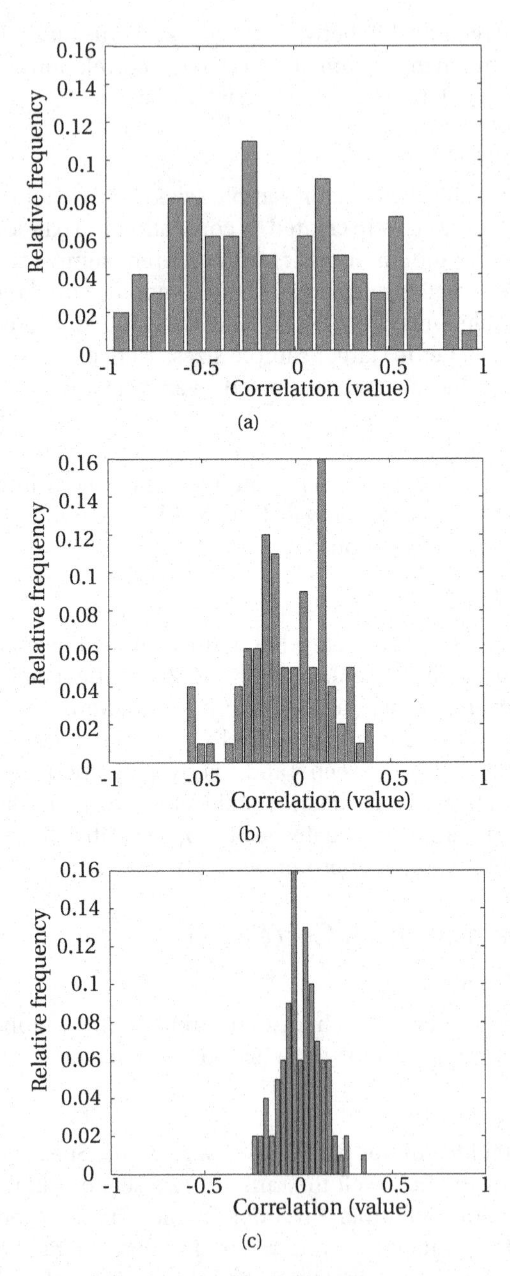

Figure 12.21. Histograms of the correlation values obtained for 100 samples of random numbers with a sample size of (a) 5, (b) 25, and (c) 100. Note that the larger the sample size, the less spread we get in terms of correlation values.

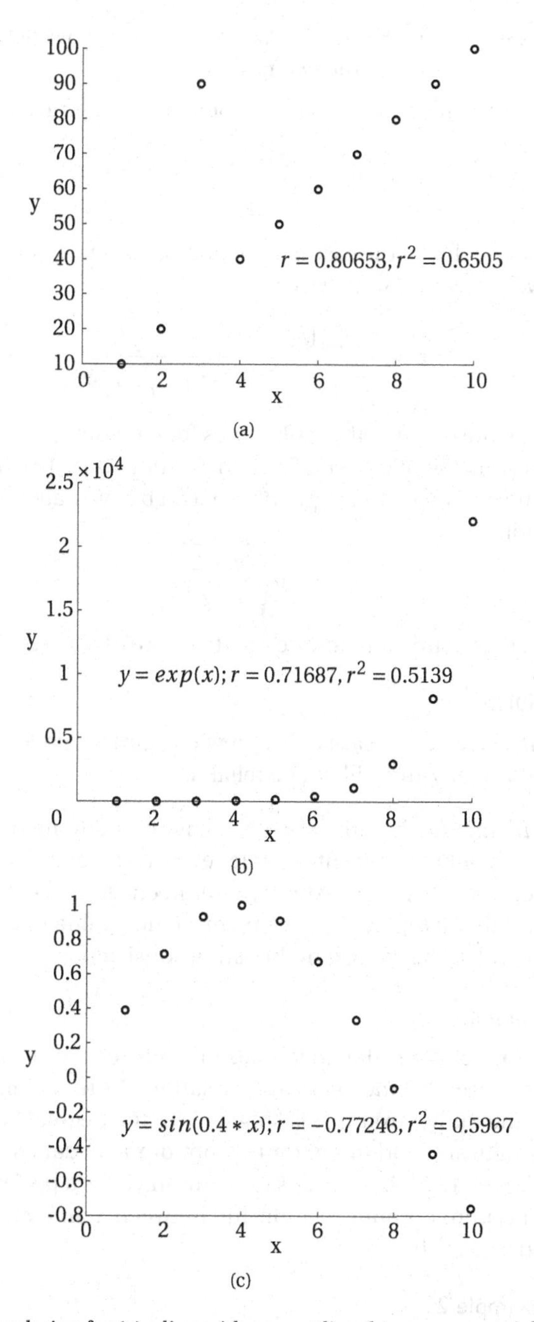

Figure 12.22. Correlation for (a) a line with one outlier, (b) an exponential relationship, and (c) a sinusoidal relationship.

example. This will result in a set of ranks $\{r_x\}, \{r_y\}$. Next, we determine the rank differences $d_i = r_{x,i} - r_{y,i}$ for the two rank sets.

Finally, Spearman's rank correlation coefficient is defined as

$$\rho = 1 - \frac{6\sum_i d_i^2}{n(n^2-1)}.$$

Another possibility in the case of tied ranks is to determine the Pearson correlation between the *ranks*, such that

$$\rho = \frac{\sum_i (r_{x,i} - \bar{r}_x)(r_{y,i} - \bar{r}_y)}{\sqrt{\sum_i (r_{x,i} - \bar{r}_x)^2 \sum_i (r_{y,i} - \bar{r}_y)^2}},$$

where $\bar{r}_x, \bar{r}_y$ are the mean of the rank values for each set.

In all cases, the significance of the correlation (whether the two datasets show a statistically significant dependence) can be evaluated via an associated t-statistic as follows

$$t = \rho\sqrt{\frac{n-2}{1-\rho^2}},$$

where the degrees of freedom for the t-statistic are $\mathrm{df} = n-2$.

12.7.16.3 Guidelines

- *Few categories* In the case of very few categories and few items, the rank correlation measure will not be reliable.

- *Many items.* At the opposite end, having many items will always result in a highly significant p value, even if the actual correlation value is rather low. Again, the experimenter needs to make the judgment call as to whether a highly significant correlation value of $\rho = 0.1$ is actually meaningful in the context of the rating consistency.

12.7.16.4 Example 1

Figures 12.23(a)–(c) show the Spearman rank correlation using the same examples as were used for the Pearson correlation. Note that in Figure 12.23(a) the value for the outlier example is "higher" (to the degree that one can compare the two values). In addition, for the monotonous data of the exponential function in Figure 12.23(b), the rank correlation yields a perfect score of $\rho = 1$, whereas the nonmonotonous relationship between x and y in Figure 12.23(b) also results in a "bad" fit.

12.7.16.5 Example 2

For another example illustrating the comparison of ordinal ratings, see Section 13.2.3.

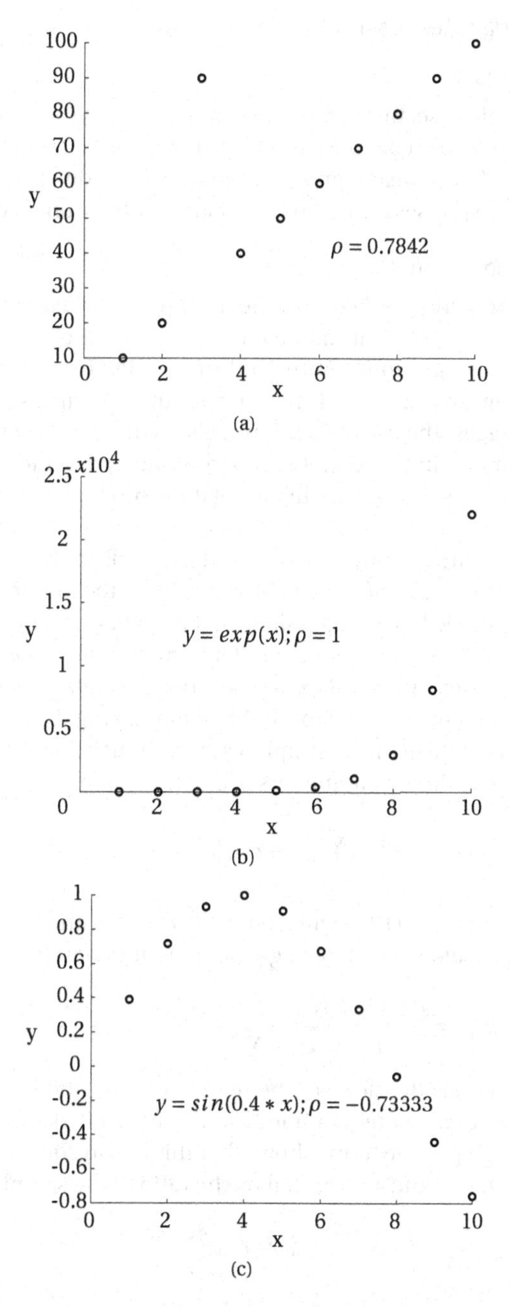

Figure 12.23. Spearman rank correlation for (a) a line with one outlier, (b) an exponential relationship, and (c) a sinusoidal relationship.

12.7.17 Linear Regression

12.7.17.1 Summary

In the most simple case, for a set of measurements of a dependent variable y_i, and given a set of known datapoints x_i, linear regression will fit a linear model, such that $y = a + b \cdot x$. Linear regression can be used to determine a linear trend in participants' responses or to make predictions about unobserved data.

12.7.17.2 Computation

In general, linear regression is one of the most well-studied and widely applied methods in data analysis. This mainly has to do with the fact that linear models are particularly easy to interpret and to handle. In cases where the relationship between two variables is not known, the principle known as *Occam's razor*[6] would suggest the use of linear models, as these represent the most simple, nontrivial modeling choice. Indeed, the underlying model of analysis of variance methods also assumes a linear relationship between the modeled parameters.

Whereas there are a number of different methods for linear regression, depending on the properties of the modeled variables, the most common method is to perform a simple linear regression (that is, to fit only one dependent variable) using an ordinary least squares (OLS) approach. The resulting linear model is a simple line $y = a + b \cdot x$ whose slope b is related to the correlation coefficient between the two variables. When employing the OLS approach, the line is fitted to the data in the least squares sense; that is, the following equation is minimized to obtain the parameters a, b of the model

$$\sum_{i=1}^{n} (y_i - a - b \cdot x_i)^2. \tag{12.9}$$

Differentiating Equation (12.9) with respect to a and b, and setting the resulting equations to zero yields the following equation for the optimal parameters

$$b_{\text{opt}} = \frac{n \sum_{i=1}^{n} x_i y_i - \sum_{i=1}^{n} x_i \sum_{i=1}^{n} y_i}{n \sum_{i=1}^{n} x_i^2 - (\sum_{i=1}^{n} x_i)^2} = \frac{\bar{xy} - \bar{x}\bar{y}}{\bar{x^2} - (\bar{x})^2}.$$

This equation defines the slope of the regression line and is equivalent to the sample covariance calculated for both variables x, y divided by the variance in x. Some rearranging of the terms shows that this is also equal to the correlation coefficient between x and y weighted by the ratio of the sample variances

$$b_{\text{opt}} = r_{x,y} \frac{s_x}{s_y}.$$

[6] Occam's razor is the maxim that when faced with two equally likely explanations for a phenomenon, one should choose the simpler one, or the one with fewer parameters, all other things being equal.

Having obtained b_{opt}, a_{opt} can be determined as

$$a_{\text{opt}} = \frac{1}{n}\sum_{i=1}^{n} y_i - b_{\text{opt}}\frac{1}{n}\sum_{i=1}^{n} x_i = \bar{y} - b_{\text{opt}}\bar{x}.$$

Hence, the resulting line passes through the center of mass $(\bar{x}, \bar{y})$ of the points.

The OLS method can be shown to be an optimal estimator, provided that the errors in the observed data have a mean expected value of zero, are uncorrelated, and have equal variance. In addition, the OLS also requires the observed variables to be uncorrelated, which in turn means that experimental data coming from (for example) repeated measurements needs to be analyzed with a different method.

Note that it is important to realize that the regression is linear in the parameters a, b and *not* in the variable x. Therefore, as long as the observed variable y is modeled as a linear combination of functions of x, the linear regression can still be applied. For example, a quadratic trend could be modeled as: $y = a + b \cdot x + c \cdot x^2$. Of course the optimal parameters a, b, c need to be calculated differently in this case.

We can calculate the significance of the fit using the t-statistic. Given our linear fit estimates of the intercept a and slope b of the line $y = a + b \cdot x$, which passes optimally through all y_i for a set of known points x_i with associated standard errors of the fit SE_a and SE_b, the related t-distributed variable as compared against a specific value of slope b_{given} is then

$$t = \frac{b - b_{\text{given}}}{SE_b}.$$

12.7.17.3 Guidelines

As linear regression and correlation are intimately related, many of the same guidelines apply. See Section 12.7.15 for more details.

12.7.17.4 Example 1

Figures 12.24 returns to the examples from the Pearson and Spearman correlation sections, but plot the actual regression lines together with the resulting line equation. For Figure 12.24(a), the line clearly bends away from the nine non-noisy data points due to the influence of the one outlier. The regression lines for Figures 12.24(b) and (c) clearly have little to do with actually describing the functions—except for the global upwards and downwards trends perhaps. Again, none of these functions can actually be fit properly with a linear trend.

12.7.17.5 Example 2

In experiments involving both reaction times and accuracy measurements, it is often the case that the two variables correlate. This phenomenon is known as

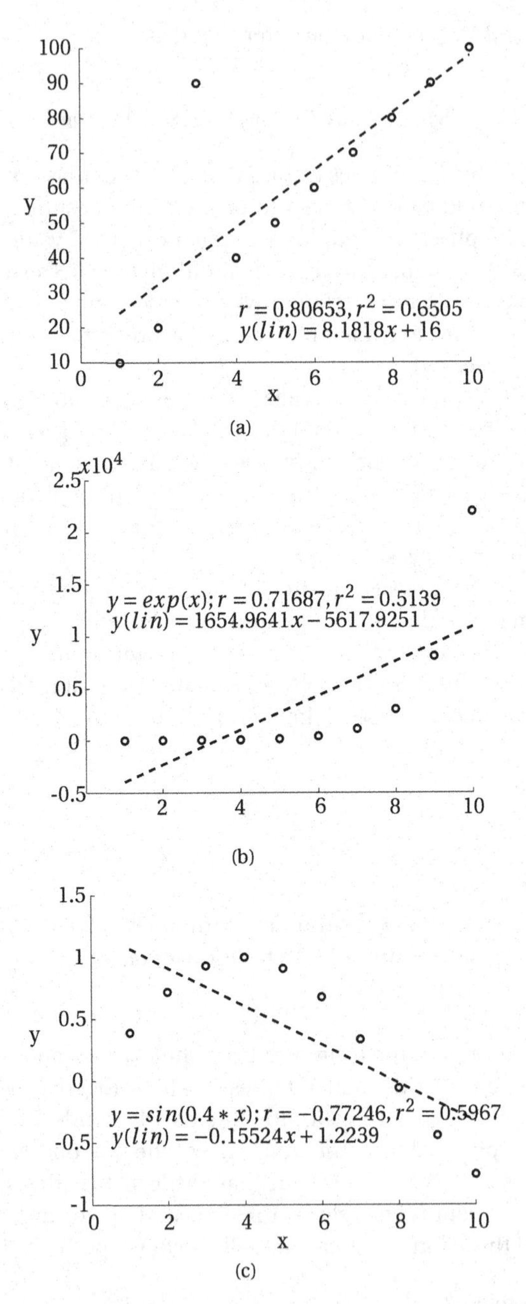

Figure 12.24. Linear regression lines for (a) a line with one outlier, (b) an exponential relationship, and (c) a sinusoidal relationship.

the speed-accuracy tradeoff and usually manifests itself in a positive correlation between accuracy and reaction time; that is, the longer participants take, the better they are. Obviously, this relationship is bounded from above by the maximum accuracy achievable and from below by the chance level and therefore will not be linear over the whole range.

In the following example, the accuracy data taken from the avatar experiment discussed in the ANOVA sections was complemented by simulated reaction time data. The reaction time data was created by multiplying the accuracy data by a factor of 10 and adding zero-mean, uniformly distributed noise of ±250 ms, resulting in a strictly linear relationship between the two variables: $rt = 10 \cdot \text{accuracy} + \epsilon$.

The plot of the data is shown in Figure 12.25 together with its best fit line obtained through the use of ordinary least squares as described above. The resulting equation is $rt_{fit} = 11.634 \cdot \text{accuracy} - 133.37$, with the slope of 11.6 being very close to the true value of 10. Figure 12.25 also shows the residual of the fit (that is, the difference between the observed and predicted y-value). This plot is good for cross-checking that the linear model provides a good, overall explanation of the data. In this case, the residuals are distributed evenly around the zero line with no obvious trends confirming a good fit.

With the resulting linear equation, we would be able to predict the reaction time for arbitrary values of accuracy; it would be $rt(100) = 1030$ ms for 100% accuracy in this case. Note also that the equation would be equally valid for non-sensical accuracy values larger than 100% and also for negative values, confirming the need to restrict the valid range for this (as well as any) linear model.

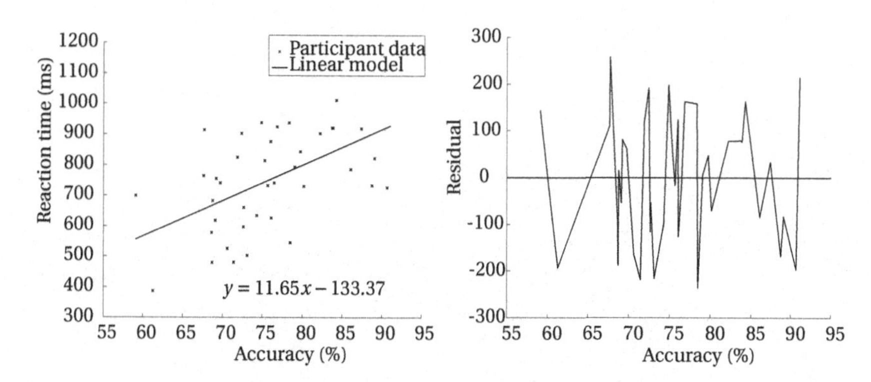

Figure 12.25. Plot showing simulated reaction time data as a function of accuracy. The reaction time data are created with a linear relationship from the accuracy data. The linear fit is shown with its corresponding equation. The second plot shows the residual of the fit.

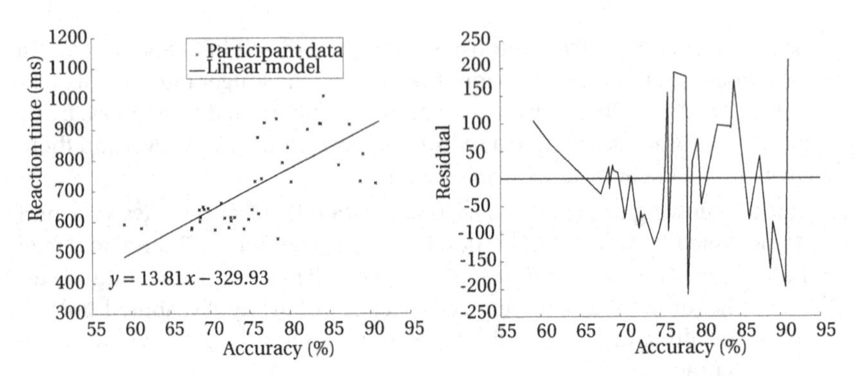

Figure 12.26. Plot showing simulated reaction time data as a function of accuracy. The reaction time data are created from the same function as in Figure 12.25 for the first 20 data points, but saturate at higher accuracies. The linear fit is shown with its corresponding equation. The second plot shows the residual of the fit.

In contrast, Figure 12.26 shows a different situation: here, the reaction time data was generated by the same method as above for the first 20 accuracy values. For the lower accuracies, however, a value of 600 ms with a small standard deviation was chosen to introduce an artificial lower bound on reaction times. Such a lower bound might reflect the fact that participants have to spend at least 600 ms to solve the task at all, with higher reaction times affording only more accuracy if they reach a critical threshold.

Note that here, the residual plots show a clear systematic deviation of the line from the data, indicating a bad fit for the initial portion of the data—as would be expected from the data generation process.

Chapter 13

Free Description, Questionnaires, and Rating Scales

This chapter gives an overview of some analysis techniques for the open-ended tasks of free descriptions and questionnaires, as well as for the more constrained rating scales.

13.1 Analysis of Free-Description Data

On many occasions, the experimenter wishes to get a less constrained answer from participants—perhaps because the exact response categories will be determined from the free descriptions themselves, or because the experimental question necessitates this approach. Whereas free-description experiments are comparatively easy to design (a typical instruction could be to write down, what you think about this painting in no more than three sentences?) and to run, analysis of the results is difficult, as the responses need to be coded into specific categories first. Obviously, in most cases, this recoding cannot be done automatically and requires human interpretation skills, thus introducing a subjective element into the process.

13.1.1 Text Analytics

In some cases—for example, if the free description contains a story, a script, or even a series of associative words—the experiment will generate a lot of text, and the experimenter might wish to use automatic tools to simplify some tasks. Typical tasks for such results would be

- parse the text to generate lists of tokens (that is, words),

- process these tokens to get rid of common English stop-words (the, and, an, etc.) to focus only on semantically important word classes,
- extract word stems to reduce variation further,
- classify words into word classes, such as nouns, verbs, adjectives, etc., and
- use the word occurrences as a feature-vector to automatically classify the document content (positive versus negative review, spam mail, etc.)

These tasks can all be done automatically and help greatly with analysis of large text corpuses. The field in (computational) linguistics that is associated with tasks like this is called *text analytics* and has become a very important research area due to the vast amounts of data that have become easily accessible online and in electronic libraries.

13.1.2 RapidMiner

One of the most successful free tools for text analytics (and, indeed, for data mining in general) is RapidMiner, a Java-based data integration, data analysis, and reporting suite. Apart from being available for free,[1] the nice thing about this suite is its user interface, which allows graphical, drag-and-drop design of data-mining processes. In addition, it offers plugins for importing and exporting a large range of different file and database formats, as well as extensive visualization tools. Data can be transformed, processed, modeled, and classified in a number of ways by interfacing to the WEKA machine-learning library.

As an example, Figure 13.1 show screenshots from the setup of a typical RapidMiner process that is used for analyzing and processing a document. The RapidMiner window consists of a list of processing categories on the left (including import and export plugins, web mining, data series calculations and transformations, as well as text processing operators), with the main window showing the current process flow chart. Operators are dragged from the left window into the middle, and corresponding input and output slots are connected up via the mouse. Operators, when selected, show their parameter choices in the tabs on the right, with documentation about the operator appearing on the bottom right. Once data has been processed by clicking the Play button, RapidMiner switches to ResultView, in which the results of the process can be viewed either as text, or as different types of plots if numerical data is available.

In our case, the document is a variant of the well-known GNU Public License (GPL), which is included with RapidMiner. This document has 618

[1] In addition to the Community Edition of RapidMiner, there is also an Enterprise Edition available which includes support and certified processing tools.

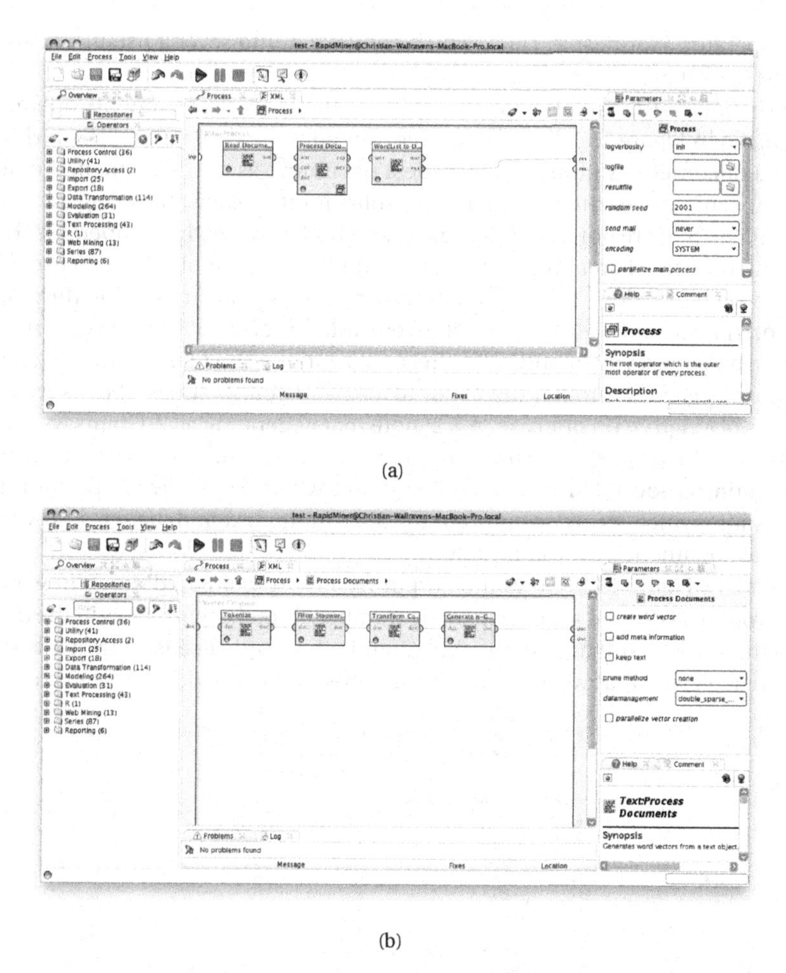

Figure 13.1. (a) and (b) RapidMiner windows showing the process flow chart for a simple document analysis.

lines, 5,178 words, and 32,387 characters, as counted with the `wc` command available on all UNIX operating systems. If we put this document into RapidMiner, we need to define a process flow that consists of linking up several operators. The overall process flow chart is shown in Figure 13.1(a) and consists of only three operators: Read Document, which reads in the license as a text file from disk, Process Documents, which is actually a container for the many processing steps shown in Figure 13.1(b) and takes as input the read-in document and converts it into a RapidMiner WordList. This WordList is taken by the last operator, WordList to Data, which converts it into several useful statistics about the list.

The interesting bit all happens inside of the Process Documents operator as shown in Figure 13.1(b). Here we see four additional operators, all of which take a document as input and provide a document as output. The first operator Tokenize takes the document and splits it into tokens, throwing away all characters that are not letters—basically, we find words with this operator. The second operator Filter Stopwords (English) filters the list of tokens to exclude words such as ("the," "a," "in," etc.) according to a built-in list. The third operator Transform Case converts all tokens into lower-case tokens to get rid of duplicate words, which only differ in capitalization. The final operator Generate n-Grams goes through the list of tokens and counts series of consecutive tokens of length n (in our case, $n = 2$; note that generating all 2-Grams will also generate all 1-Grams as the process is recursive). With this operator, we are investigating co-occurring word combinations, which might give important insights into which terms occur often together in the text and, hence, into what kind of document we are analyzing.

Finally, Figure 13.2 shows the output results view of RapidMiner, in which the WordList is sorted according to the number of total occurrences in the document. The token that occurs most often is the word "license," which is not surprising given that we are looking at a software-license document. That it is

Result Overview | ExampleSet (WordList to Data)

Meta Data View | Data View | Plot View | Annotations

ExampleSet (2758 examples, 0 special attributes, 3 regular attributes)

Row No.	word	in documents	total
1221	license	1	100
2659	work	1	97
2305	source	1	48
584	covered	1	42
304	code	1	37
1858	program	1	37
587	covered_work	1	36
2575	version	1	33
2446	terms	1	28
562	corresponding	1	27
563	corresponding_source	1	27
460	convey	1	26
541	copyright	1	26
516	copy	1	24
1696	patent	1	23
1561	object	1	21
1562	object_code	1	21
1840	product	1	21
921	general	1	20
2280	software	1	20
2523	use	1	20
1418	means	1	19
1482	modified	1	17
481	conveying	1	15
2150	rights	1	15
2309	source_code	1	15
2542	user	1	15
63	additional	1	14

Figure 13.2. Output results of the analysis in Figure 13.1.

a software license could be inferred from the fact that words such as "source," "program," "code," and "software" are relatively high-ranking in this list. In addition, the list also gives information about the most frequently occurring 2-Grams as series of words joined by an underscore. The most high-ranking 2-Gram is "covered work" followed by "corresponding source." Note also, that the number of occurrences for "corresponding" and "corresponding source" are exactly the same, which means that the adjective always precedes the noun in this document.

13.2 Rater Agreement and Rater Reliability

In experiments in which items are judged or rated by one or several raters the results will need to be evaluated for consistency. In this context, one can talk about two concepts that are closely related but nevertheless important to distinguish: the first concept is *reliability*, which refers to the degree with which the raters provide the same ordering of item properties. The second concept is *agreement*, which is how well raters agree in the values of their ratings of the item properties. Hence, reliability is of interest whenever only the *relative* consistency of ratings is important, whereas agreement compares the *absolute* values.

As an example, we will discuss data collected by Wallraven et al. (2009) in a study on measuring the aesthetic appeal of paintings from different art periods. The paintings used in the study spanned 11 movements in art history, ranging from Gothic to Postmodern. For each of these art periods five major artists were identified, and for each artist five representative paintings were chosen. For the experiment, all 275 paintings were shown to a total of 20 participants split into two groups of ten participants each. Participants viewed the paintings while eye-tracking data was collected. Each of the 275 (randomized) trials consisted of a fixation cross that was shown for two seconds, followed by an image that was shown for six seconds. After the image, participants in one group (the complexity task) answered how complex the image was visually, whereas for the other group (the aesthetics task), the question was, how aesthetically appealing the image was. Both values were rated on a scale of 1 (not complex or not aesthetic) to 7 (highly complex or highly aesthetic) with 4 being the neutral middle point on the scale. In the following we will discuss behavioral data (consisting of two different ratings).

Figure 13.3 shows all ratings for the ten participants in each group for all paintings; from this broad overview plot one can see that aesthetics ratings seem to be a bit more variable than complexity ratings. However, in both tasks there is considerable variation among raters. To make matters a little easier, Table 13.1 shows ratings for 20 selected images for only two participants in the aesthetics task.

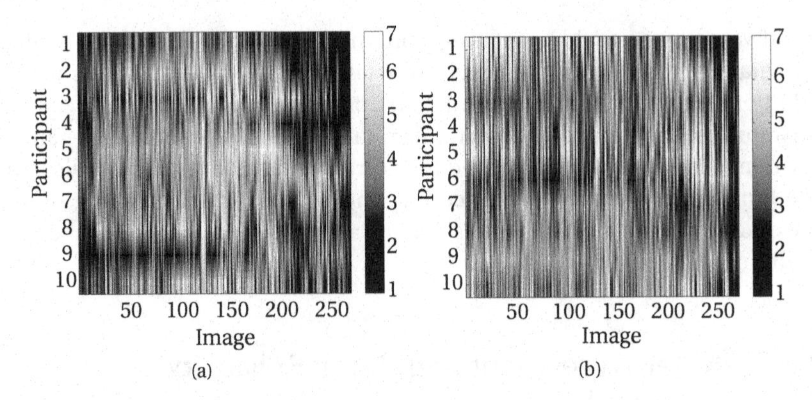

Figure 13.3. Ratings of ten participants for (a) the aesthetics task and (b) the complexity task. Data from Wallraven et al. (2009).

As the data show, there is considerable variation between the two raters. However, it also seems as if participant 2 rated most images consistently *higher* than participant 1. Therefore, although the agreement of the absolute values might be low, the overall consistency in terms of the relative ranking of images might actually be higher (note that image number 18 is rated as the most aesthetically pleasing and image number 8 as the least aesthetically pleasing image).

There is a simple change, however, which illustrates another important concept in assessing reliability: imagine if we had asked the same participant *twice* to rate the *same* images. The (hypothetical) data now would look as shown in Table 13.2 and we could start to look for that participant's *internal* agreement or reliability. Again, we would be able to say that participant 1's agreement was low but that the data does show a certain reliability. Nevertheless, in this particular case the results warrant caution, as the data suggests that the two scales that participant 1 applied over the two sessions seem to differ considerably—a value of 2 in one session becomes a value of 5 in another, and so on. Hence, *intra*-rater consistency might be a trickier issue in some cases,

Image	1	2	3	4	5	6	7	8	9	10
Participant 1	2	2	2	4	3	3	3	1	3	3
Participant 2	3	3	3	4	4	3	5	2	4	2

Image	11	12	13	14	15	16	17	18	19	20
Participant 1	4	3	5	3	3	2	4	3	5	3
Participant 2	5	5	7	6	5	5	4	6	4	5

Table 13.1. Aesthetic ratings for two participants for 20 paintings.

Image	1	2	3	4	5	6	7	8	9	10
Participant 1	2	2	2	4	3	3	3	1	3	3
Participant 1	3	3	3	4	4	3	5	2	4	2

Image	11	12	13	14	15	16	17	18	19	20
Participant 1	4	3	5	3	3	2	4	3	5	3
Participant 1	5	5	7	6	5	5	4	6	4	5

Table 13.2. Repeated, hypothetical aesthetic ratings for one participant for 20 paintings.

as one would like to claim repeatability of the results at perhaps a tighter level than for *inter*-rater consistency, that is, one would look for higher intra-rater agreement than for inter-rater agreement.

In general, one also needs to pay attention to the types of responses in rating studies; in principle, raters can answer on a nominal, ordinal, or even continuous scale. For our previously introduced example of judging the quality of a computer-animated avatar, we might ask observers to simply choose adjectives from a list that relate to the perceived quality of the animations. This will result in a nominal scale with observers dividing the stimuli/questions into categories from a list including adjectives such as "jerky," "believable," "annoying," "great," etc. A priori, there is no intrinsic way of sorting or ordering these adjectives, and as such one can analyze the degree of agreement in the categorical responses among observers. In the most simple case, we can simply count how many people agree for each category, which then relates directly to a measure of inter-rater agreement. Note, however, that agreement in a particular category might arise simply by chance, such as when an observer has chosen the wrong category, or the label is not clearly understood, etc. Such chance agreement will happen more often if there are only a few categories to choose from, and will need to be taken into account for a valid agreement measure.

Another possibility would be to have observers rate the animations on a ranked scale—either from 1 to 5, or with ranked categories from low quality to medium quality to high quality, for example. For these categories, there exists a clear ordering that results in an ordinal scale of responses. If one observer rates the animation as "high quality," and another one rates the same animation as "medium quality," the degree of disagreement between the two raters is somehow measurable. Whereas for nominal scales raters are either in agreement or in disagreement, for ordinal scales, rating ceases to be a binary decision. Again, for measuring agreement, chance assignment needs to be taken into account, but also the fact that there is an ordered notion of agreement/disagreement.

Finally, for continuous scales, we might choose to assign arbitrary quality ratings (for example, from 0% to 100%) to the perceived quality of the avatar.

However, as previous research on rating scales has shown, this will not result in a truly continuous scale (see Chapter 5) as raters will not use the full scale. It might be possible to assign a continuous scale to the quality of the avatar by, for example, using the reaction time or recognizability as a rating. Of course, agreement here ceases to be exact, as the probability of two reaction times being exactly the same is rather low—in this case, one would rather resort to a standard analysis of variance (see Chapter 12) to compare the variation of the raters among each other. As this is a somewhat unusual approach, in the following sections, we will focus only on techniques suitable for nominal and ordinal ratings.

The treatment of the various κ statistics follows the treatment proposed in the excellent book by Gwet (2010), which we recommend for further reading, as it contains a wealth of information on inter-rater agreement statistics.

13.2.1 Percentage Agreement

13.2.1.1 Summary

The simplest agreement measure for ratings.

13.2.1.2 Computation

Given a set of categories, and a set of ratings from more than one rater for each item in question, this measure simply counts how many times raters agree for each item. The overall agreement is averaged over all single-item agreements.

13.2.1.3 Guidelines

As stated earlier, this measure tends to overestimate the actual degree of agreement, especially when there are only a few categories, as it does not take into account chance agreements. Nevertheless, for an easy first analysis of rater agreement, this will give a good feeling for the degree of consistency in the experimental data. Its actual use in publications with inter-rater agreement statistics, however, is rather low as one might assume.

When there is a larger number of raters and an ordinal rating scale, percentage agreement might not result in any useful data, as the chances of only one rater choosing a different rating are rather high. In the strictest sense, this one rating will then result in there being no agreement at all for the item in question, even if all other raters agree exactly.

13.2.1.4 Example

For the data in Table 13.2, the percentage agreement measure is rather low—only 3 out of 20 ratings of the two raters were actually the same, resulting in an agreement score of 15%.

Following up on the comment from above about the brittleness of this measure, the percentage agreement for both the aesthetic ratings and the complexity ratings shown in Figure 13.3 is actually 0% for all ten raters and all 275 paintings. If we relax the assumptions about exact agreement and allow two different ratings to be present for each painting, we still find only 2 out of 275 paintings for the aesthetic ratings and 7 out of 275 for the complexity ratings.

13.2.2 Cohen's Kappa for Nominal Ratings

13.2.2.1 Summary

Cohen's kappa (κ) measures the degree of inter-rater agreement for nominal ratings. It comes in many different forms depending on the number of raters and the number of categories in question. It is based on counting the number of category agreements, but includes a correction for chance agreement. The following lists the different types of κ separately.

13.2.2.2 Computation of κ_{22} for Two Raters with Two Categories

If we have two raters, each of whom have rated items into two nominal categories, we can summarize the result in a table such as in Table 13.3. Each cell shows the number of items that were put into the respective categories by the two raters. The totals shown in the last row and the last column obviously need to sum up to the total count of items n_{**}.

The contingency table can be used to calculate the naive percentage agreement—or equivalently the probability of both raters putting the items in the same category—by summing up the diagonal elements, such that

$$p(\text{agree}) = \frac{n_{11} + n_{22}}{n_{**}}.$$

As outlined above, however, the raters could have put the items into the same category purely by chance, and we need to correct the agreement measure for this scenario. Following a proposal from Cohen, the probability of chance agreements occurring can be estimated as the sum of the two agreement probabilities for each of the two categories. The agreement probability

	R1 C1	R1 C2	Overall
R2 C1	n_{11}	n_{12}	n_{1*}
R2 C2	n_{21}	n_{22}	n_{2*}
Overall	n_{*1}	n_{*2}	n_{**}

Table 13.3. Contingency table for two raters and two categories. Raters 1 and 2 are abbreviated R1 and R2, respectively; categories 1 and 2 are abbreviated as C1 and C2, respectively. Each cell lists how many items were put into the same categories by each rater.

for category 1, for example, is proportional to the product of the row and column sum for category $1(n_{1*} \times n_{*1})$. For both categories, we therefore have

$$p(\text{chance}) = \frac{n_{1*}}{n_{**}} \times \frac{n_{*1}}{n_{**}} + \frac{n_{2*}}{n_{**}} \times \frac{n_{*2}}{n_{**}}.$$

And with that the corrected measure, Cohen's κ for two raters and two categories becomes

$$\kappa_{22} = \frac{p(\text{agree}) - p(\text{chance})}{1 - p(\text{chance})}.$$

Here the denominator $1 - p(\text{chance})$ is related to the number of ratings for which agreement would not be expected by chance, whereas the nominator measures the non-random agreement.

13.2.2.3 Computation of κ_{2c} for Two Raters with c Multiple Categories

For multiple nominal categories, the general principle for evaluating agreement does not change, and Table 13.4 shows the contingency table for this case.

Again, the overall probability of agreement between the raters is given by the diagonal elements

$$p(\text{agree}) = \sum_{i=1}^{n} \frac{n_{ii}}{n_{**}}.$$

The chance agreement also is given by the sum of all single category agreement probabilities

$$p(\text{chance}) = \sum_{i=1}^{n} \frac{n_{i*}}{n_{**}} \times \frac{n_{*i}}{n_{**}}.$$

Consequently, the definition of κ stays the same with

$$\kappa_{2c} = \frac{p(\text{agree}) - p(\text{chance})}{1 - p(\text{chance})}.$$

	R1 C1	R1 C2	...	R1 Cc	Overall
R2 C1	n_{11}	n_{12}	...	n_{1c}	n_{1*}
R2 C2	n_{21}	n_{22}	...	n_{2c}	n_{2*}
...			...		
R2 Cc	n_{c1}	n_{c2}	...	n_{cc}	n_{n*}
Overall	n_{*1}	n_{*2}	...	n_{*c}	n_{**}

Table 13.4. Contingency table for two raters and c categories. Raters 1 and 2 are abbreviated R1 and R2, respectively; categories are abbreviated as C1 to Cc, respectively. Each cell lists how many items were put into the same categories by each rater.

13.2.2.4 Computation of κ_{rc} for r Multiple Raters with c Multiple Categories

In many cases, we have more than two raters for the items under question. With more than two raters we encounter the problem alluded to above in the context of analyzing the aesthetic ratings for ten raters, as we need to come up with a definition of what agreement means: when all raters give exactly the same rating, or when the majority of raters give the same rating.

One possibility is to consider all possible pairs of raters from the whole group. For a group of r raters, this is $(r-1)!$. For a group of three raters, either no pairs, one pair, or all three pairs can be in agreement. This results in a weighting of agreement, with full agreement being reached when all rater pairs agree, and partial or no agreement reached for the other cases. The average of these agreement values over all items then gives the overall agreement score—or, in other words, the probability of agreement $p(\text{agree})$. Again, this score needs to be adjusted by the probability of agreement occurring by chance.

With the data from Table 13.5, which now lists for each item how many raters put it into which category, one can define the agreement probability as

$$p(\text{agree}) = \frac{1}{nr(r-1)} \sum_{k=1}^{i} \sum_{l=1}^{c} r_{lk}(r_{lk}-1).$$

Similarly, the chance agreement is defined as

$$p(\text{chance}) = \frac{1}{n} \sum_{k=1}^{i} \frac{r_{lk}}{r}.$$

And with these two probabilities, we have the usual definition of κ as

$$\kappa_{rc} = \frac{p(\text{agree}) - p(\text{chance})}{1 - p(\text{chance})}.$$

	C1	C2	...	Cc	Overall
I1	r_{11}	r_{12}	...	r_{1c}	r
I2	r_{21}	r_{22}	...	r_{2c}	r
...			...		
Ii	r_{i1}	r_{i2}	...	r_{ic}	r
Overall	r_{*1}	r_{*2}	...	r_{*c}	r

Table 13.5. Contingency table for r raters, c categories, and i items. Categories are listed from C1 to Cc, and items from I1 to Ii. Each cell lists how many raters put a given item into the category.

13.2.2.5 Guidelines

- *Ordinal and interval data.* The κ statistic as presented here was developed only for nominal categories. As such, if used on ordinal categories it might underestimate the degree of agreement, as it does not take into account partial agreement between categories—on a rating scale of 1 to 5, if two raters rate one item with a 1 and a 2, respectively, they will be in as much disagreement in the κ sense, as if they had rated the same item with a 1 and a 5. There are extensions to the κ measure that can deal with ordinal and interval data; for more information, the interested reader is referred to Gwet (2010). In the next section, we will briefly review the use of correlations for judging inter-rater agreement on ordinal data.

- *Number of categories.* In general, κ is influenced by the number of categories in that the fewer categories there are, the higher κ will be, hence leading to a potential bias.

- *Interpreting the value of* κ. As with any measure in statistics, the interpretations and guidelines of what constitutes high agreement versus low agreement, given a particular value of κ, are diverging. There are several different scales in use for judging the quality of the agreement; perhaps the most common of which is the one proposed by Landis and Koch (1977), in which they recommend the breakdown of κ-values shown in Table 13.6.

- *Missing ratings.* In some experiments, not all raters will be able to rate all items—because of a missing rating due to an experimental error, or because of the sheer number of items. All measures in this section can be corrected for missing values by inserting a virtual row and column into the tables, which collects the missing ratings for each rater-category pair. Both the p(agree) and p(chance) will have to be adjusted to take into account missing values, such that only items that both raters actually have rated are taken into account for p(agree). The reader is referred to Gwet (2010) for more details on how that affects the three types of κ discussed here.

κ	Interpretation of agreement
< 0.0	Poor
0.00 - 0.20	Slight
0.21 - 0.40	Fair
0.41 - 0.60	Moderate
0.61 - 0.80	Substantial
0.81 - 1.00	Almost Perfect

Table 13.6. Recommended interpretation of κ-values according to Landis and Koch (1977).

- *The κ-paradox.* Sometimes, the value of κ will be far lower than expected with respect to the overall probability of rater agreement. In the two-rater, two-category case this happens, for example, when the agreement counts in one cell are very high with little or no disagreements in other cells. The problem goes back to the definition of chance agreement, which hinges on the observed probabilities with which items are sorted into the two categories. There are measures that circumvent this problem, some of which are discussed in Gwet (2008).

13.2.2.6 Example 1: κ_{22}

Consider an experiment in which the qualities of an avatar are judged by having observers classify the animations into two qualities: lifelike and artificial. Now, these categories do contain an ordinal quality in that one might say that lifelike would be "better" than artificial in terms of the realism of the avatar; however, from the outset one might not want that, and indeed, for some applications an artificial avatar might actually be advantageous. In addition, since there are only two categories and therefore raters can either disagree or agree, we can use the original definition of κ for this data.

Table 13.7 shows hypothetical data from two raters who viewed 100 different animations and rated each animation as lifelike or artificial. The overall agreement for the two categories is given by the diagonal in the table $p(\text{agree}) = \frac{39+44}{100} = 83\%$—a rather high level of agreement. However, this needs to be corrected for chance agreement, which is given by $p(\text{chance}) = \frac{52}{100} \times \frac{43}{100} + \frac{48}{100} \times \frac{57}{100} = 50\%$, resulting in $\kappa_{22} = \frac{83\%-50\%}{100\%-50\%} = 66\%$ and hence in a non-negligible reduction of the overall agreement. According to Table 13.6, this data has "substantial" agreement strength.

	R1 Lifelike	R1 Artificial	Overall
R2 Lifelike	39	13	52
R2 Artificial	4	44	48
Overall	43	57	100

Table 13.7. Contingency table for two raters rating 100 animations from different types of avatars as lifelike or artificial.

13.2.2.7 Example 2: κ_{22}

The data from Table 13.2 can be converted into two-category data by collecting all ratings from 1–3 into a "low-aesthetic value" and from 4–7 into a "high-aesthetic value" category. Of course, doing so will be dangerous, as one might come up with other splits of the data—and furthermore, as the original ratings were actually on an ordinal scale, the quantization will hide the partial

	R1 LowAes	R1 HighAes	Overall
R2 LowAes	6	0	6
R2 HighAes	9	5	14
Overall	15	5	20

Table 13.8. Contingency table for two raters and two aesthetic categories. Data were obtained by categorizing the ratings from Table 13.2 into low and high aesthetic value categories.

agreements. The resulting contingency table for this conversion is shown in Table 13.8. Overall agreement is $p(agree) = \frac{6+5}{20} = 55\%$. However, chance agreement is $p(chance) = \frac{6}{20} \times \frac{15}{20} + \frac{14}{20} \times \frac{5}{20} = 40\%$, resulting in $\kappa_{22} = \frac{55\%-40\%}{100\%-40\%} = 25\%$ and hence in a rather dramatic reduction of the overall agreement value. According to Table 13.6, this data has only "slight" agreement strength.

13.2.3 Spearman's Rank Correlation Coefficient

13.2.3.1 Summary

Whereas the original formulation of Cohen's κ measure only allows for nominal rating scales; for ordered data, one can use rank correlation measures. The Spearman rank correlation coefficient ρ is a non-parametric measure that determines the amount of dependence (and hence, agreement) between two ordinal variables.

13.2.3.2 Computation

Given two sets of ratings $\{r_1\}, \{r_2\}$ of n items, we first need to convert the ratings into ranks. For this, the ratings are sorted either in decreasing or increasing order and the position of each rating in the table is the rank of that rating. Note that when ranks are tied, the rank for each tied item is the average of their position in the table—for most ordinal-rating scales, this will happen frequently and thus is the normal case. This will result in a set of ranks $\{x_1\}, \{x_2\}$. Next, we determine the rank differences, $d_i = x_{1,i} - x_{2,i}$, for the two rank sets. Finally, Spearman's rank correlation coefficient is defined as

$$\rho = 1 - \frac{6\sum_i d_i^2}{n(n^2-1)}.$$

Another possibility in the case of tied ranks is to determine the Pearson correlation between the ranks, such that

$$\rho_2 = \frac{\sum_i (x_{1,i} - \bar{x}_1)(x_{2,i} - \bar{x}_2)}{\sqrt{\sum_i (x_{1,i} - \bar{x}_1)^2 \sum_i (x_{2,i} - \bar{x}_2)^2}},$$

where $\bar{x}_{\{1,2\}} = \frac{\sum_i x_{\{1,2\},i}}{n}$ is the mean of the rank values for each set.

In all cases, the significance of the correlation, that is, whether the two datasets show a statistically significant dependence, can be evaluated via an associated t-statistic as follows:

$$t = \rho\sqrt{\frac{n-2}{1-\rho^2}},$$

where the degrees of freedom for the t-statistic are $df = n - 2$.

13.2.3.3 Guidelines

- *Few categories.* In the case of very few categories and of few items, this measure (as with any other statistical measure) will not be reliable.
- *Many items.* At the opposite end, having many items will always result in a highly significant p value, even if the actual correlation value is rather low. Again, the experimenter needs to make the judgment call, whether a highly significant correlation value of $\rho = 0.1$ is actually meaningful in the context of the rating consistency.
- *Relative agreement.* The Spearman rank correlation coefficient only quantifies a monotonic relationship between two variables; hence, the absolute differences between the variables are not important. Again, however, in many cases in which raters use slightly offset scales, this is actually a beneficial property of the correlation.

13.2.3.4 Example

The data from Table 13.2 has two raters and seven categories for 20 paintings. Table 13.9 shows the rank values $\{x_{1,2}\}$ and the associated rank differences that are used to calculate the Spearman correlation coefficient.

Given the rank differences and $n = 20$, we can evaluate the equation; the resulting $\rho = 0.49$ shows that despite absolute disagreements, there is a correlation between the two raters. The associated p value for the correlation is $p = 0.01$, which is significant. The second method of calculating ρ_2 (by determining the correlation on the rank values themselves) yields a similar value of $\rho_2 = 0.5$ with an associated p value of 0.02.

If one calculates the Spearman correlation coefficient using statistical software, the software usually corrects the above value to take into account the tied ranks in the data. The actual ρ for this data then will be lower, with $\rho_{tied} = 0.435$ and an associated p value of 0.05. The result is just significant.

Overall, we can conclude that the two raters' judgments do agree on a relative basis, but that their absolute agreement is low (as per the κ analysis in Section 13.2.2, for example). This hints at the possibility that the two raters might have used different scales, such that one participant always judged the paintings a little higher than the other.

Rating r_1	Rating r_2	Rank x_1	Rank x_2	$d = x_1 - x_2$
2	3	3.50	4.50	-1.00
2	3	3.50	4.50	-1.00
2	3	3.50	4.50	-1.00
4	4	17.00	9.00	8.00
3	4	10.50	9.00	1.50
3	3	10.50	4.50	6.00
3	5	10.50	14.50	-4.00
1	2	1.00	1.50	-0.50
3	4	10.50	9.00	1.50
3	2	10.50	1.50	9.00
4	5	17.00	14.50	2.50
3	5	10.50	14.50	-4.00
5	7	19.50	20.00	-0.50
3	6	10.50	18.50	-8.00
3	5	10.50	14.50	-4.00
2	5	3.50	14.50	-11.00
4	4	17.00	9.00	8.00
3	6	10.50	18.50	-8.00
5	4	19.50	9.00	10.50
3	5	10.50	14.50	-4.00

Table 13.9. Aesthetic ratings from Table 13.2 with associated ranks and rank differences.

13.2.4 Kendall's Coefficient of Concordance

13.2.4.1 Summary

Kendall's coefficient of concordance, or Kendall's W, is a generalization of the Spearman correlation coefficient for evaluating the inter-rater agreement between r raters judging n items on an ordinal scale.

13.2.4.2 Computation

If we have more than two raters, one possibility would be to evaluate ρ for all possible pairs and then average the resulting values. However, there are methods that deal with this scenario more explicitly, as relying too much on averaging might inflate the correlation values. One of those methods is Kendall's W, which is based on r raters rank-ordering n items.

The underlying idea is the same as for the Spearman correlation in that for each item i, its rank $x_{i,j}$ given by each rater j is determined first. The sum over all rater's ranks,

$$x_i = \sum_j^r x_{i,j},$$

then yields the overall rank for that item.

On average, there are

$$\bar{x} = \frac{1}{2} r(n+1)$$

ranks in the set, which value can be used to determine the average deviations from the mean rank as

$$d = \sum_{i}^{n} (x_i - \bar{x})^2.$$

Finally, with these values, Kendall's W is defined as

$$W = \frac{12d}{r^2(n(n^2-1))}.$$

For tied ranks, W needs to be corrected accordingly, as W will be affected by the large number of similar difference-values d calculated over tied ranks. The formula for the corrected W will not be given here, but the tie-corrected Kendall's W is accessible through all major statistical software.

The associated test-statistic for W is the F-statistic, with

$$F = \frac{W(r-1)}{1-W}$$

and the two necessary degrees of freedom values of $df1 = n - 1 - \frac{2}{r}$ and $df2 = df1(r-1)$.

13.2.4.3 Example

Table 13.10 shows aesthetic ratings from four raters for the 20 paintings. Evaluating Kendall's W *without* tied rankings from the equations above yields $W = 0.69$, with a highly significant associated p value of $p < 0.001$. Using the proper Kendall's W *with* corrections for tied rankings yields a lower value of $W = 0.55$, which is still highly significant at $p < 0.001$. Hence, we can conclude that the four raters do indeed rate the 20 paintings similarly on a *relative scale.*

As a side note, the full Kendall's W for all ten raters and all 275 paintings is $W = 0.33$, $p < 0.001$ for the aesthetic ratings and $W = 0.42$, $p < 0.001$ for the

Participant 1	2	2	2	4	3	3	3	1	3	3
Participant 2	3	3	3	4	4	3	5	2	4	2
Participant 3	2	2	6	2	1	2	4	1	2	2
Participant 4	2	2	2	5	4	4	4	3	4	3

Participant 1	4	3	5	3	3	2	4	3	5	3
Participant 2	5	5	7	6	5	5	4	6	4	5
Participant 3	6	3	6	3	2	3	3	3	5	6
Participant 4	4	4	4	4	3	4	4	4	5	3

Table 13.10. Aesthetic ratings from four raters for 20 paintings.

complexity ratings. First of all, the high p values illustrate the influence of the large $n = 275$ on the test statistic. Secondly, the overall results show that raters are still fairly consistent in both tasks, with a slight advantage of complexity over aesthetic ratings—a finding which is in accordance with the general assumption that aesthetic judgments will have a higher degree of subjectivity than complexity judgments.

13.2.4.4 Guidelines

The same guidelines apply as for the Spearman correlation.

13.2.5 Intraclass Correlation Coefficient (ICC)

13.2.5.1 Summary

Whereas the rank correlation coefficients work on paired data values, the various ICCs assess rater agreement based on group membership and are a measure of the within-group variance. In addition, the ICCs presented here take into account absolute rater agreements rather than relative ones. The ICCs are closely related to ANOVAs.

13.2.5.2 Computation

The original formulation of ICCs comes from the seminal paper by Shrout and Fleiss (1979) that discusses several ways to assess rater reliability. The paper discusses a total of six coefficients, which are described briefly next.

Case 1 deals with a set of items to be rated that are seen as a random sample of all items of interest. In other words, each item is rated by r raters, but since these are drawn randomly from a large population of raters, we cannot determine the identity of the rating for each item. In this case what we are interested in assessing, is to determine for each item how each rater's judgment differs from the average ratings for that item. In effect, this is very similar to a one-way ANOVA, as it determines the within-group variance. Hence, in Case 1 we focus on the *items* more than on the judges. In most inter-rater agreement studies this will be a fairly rare setup, but it might be appropriate for a large online study in which control of the raters or the raters' identity is not possible.

In Case 2, both the items and the rates are viewed as a random sample. Here, there are r raters, all of whom rate all n items. Hence, the raters also introduce a random effect into the model as the source of each rating can be uniquely identified. This model is similar to a two-way ANOVA and is probably the most appropriate model for most rating studies.

In Case 3, the effects of the raters are fixed since we are only interested in the random effects introduced by the ratings. This model does not see the raters themselves as a random sample, but instead postulates that these are the only raters we are interested in. This kind of model might be suitable for expert ratings, in which raters bring a special kind of expertise to the study, which makes them unique.

Finally, all three cases come in two versions: one in which individual ratings are analyzed and one in which the mean of the ratings is analyzed. All coefficients are much higher when the mean ratings are used, as one important source of variation is averaged out before assessing the agreement of the ratings. These coefficients are therefore of interest when there is too much uncertainty associated with each individual rating, such as when the items under question are very hard to judge.

It is usual practice to label the coefficients according to the model (the first number in the bracket is the case number from above) and whether they are calculated based on individual ratings (the second number in the bracket is 1) or on ratings averaged over r judges (the second number in the bracket is r), resulting in the six ICCs: ICC(1,1), ICC(1,r), ICC(2,1), ICC(2,r), ICC(3,1), ICC(3,r).

As with any ANOVA design, for the calculation of the different coefficients, we can determine four sources of variation in our data: variance between targets, within targets, between raters, and a residual error term. In an ANOVA calculation, each of these variances gives rise to a mean square (MS) value, resulting in the four values: *BMS*, *WMS*, *RMS*, and *EMS*.

With these four MS values, the three different cases are determined as follows below.

- *Case 1*

$$\mathrm{ICC}(1,1) = \frac{BMS - WMS}{BMS + (r-1) * WMS}.$$

- *Case 2*

$$\mathrm{ICC}(2,1) = \frac{BMS - EMS}{BMS + (r-1) * EMS + k(JMS - EMS)/n}.$$

- *Case 3*

$$\mathrm{ICC}(3,1) = \frac{BMS - EMS}{BMS + (r-1) * EMS}.$$

The test-statistics for each ICC measure are based on the F-statistic and are as shown next.

- *Case 1*

$$F = \frac{BMS}{WMS}, df1 = n-1, df2 = n(r-1).$$

- *Case 2*

$$F = \frac{BMS}{EMS}, df1 = n-1, df2 = (n-1)(r-1).$$

- *Case 3*

$$F = \frac{BMS}{EMS}, df1 = n-1, df2 = (n-1)(r-1).$$

In each case, F is the value of the F-statistic, and $df1$ and $df2$ are the two degrees of freedom.

13.2.5.3 Guidelines

- *Incorrect use of ICC*$(1,1)$. As the model for Case 1 assumes that the identities of the r raters are unknown, there is no way of grouping the ratings. Hence the value of ICC$(1,1)$ will always be lower than the values for ICC$(2,1)$ or those for ICC$(3,1)$, potentially underestimating the degree of agreement in the ratings.
- *Misuse of ICC*$(*,r)$. Since the ICCs based on mean ratings will always be higher, it might be tempting to use these measures in all cases. However, it is probably appropriate to do so only if there are doubts about the reliability of each single rating.

13.2.5.4 Example

We are going to list the ICCs for the data from Table 13.2 (two raters and seven categories for 20 paintings for the aesthetic task), the data in Table 13.10 (four raters and seven categories for 20 paintings for the aesthetic task), and the full data for both tasks shown in Figure 13.3 (ten raters, 275 paintings). All six different ICCs are listed in Table 13.11 for each dataset (p values not shown).

As the data shows, the single item correlations are rather low in all cases, and with ten raters, the mean coefficients do give a very large boost to the agreement scores. Interestingly, the scores for the complexity rating do *not* look more consistent—again, this is because the ICC does care about absolute agreement for ratings. Hence, a pre-normalization might be a good idea for these ratings if the experimenter suspects that different scales might have been used by the raters.

Dataset	ICC $(1,1)$	ICC $(1,r)$	ICC $(2,1)$	ICC $(2,r)$	ICC $(3,1)$	ICC $(3,r)$
Table 13.2	0.32	0.90	0.33	0.91	0.49	0.95
Table 13.10	0.11	0.72	0.13	0.75	0.19	0.83
Figure 13.3(a)	0.10	0.97	0.11	0.97	0.14	0.98
Figure 13.3(b)	0.04	0.91	0.04	0.92	0.06	0.95

Table 13.11. Different intraclass correlation coefficients for four datasets.

13.3 Analysis of Semantic Differentials, Similarity Ratings, and Multiple Scales

In this section, we present two types of analysis that are designed to find underlying structure in a large set of ratings. The first is factor analysis (first mentioned in Chapter 5), and it requires a dataset where each item is rated along multiple scales (as is done in semantic differentials and intelligence tests). The goal is to find out whether there are covariations that can be used to summarize many of the scales with one or more common factors. The scores that specific stimuli receive on these common factors can be used as coordinates in a semantic space. The second kind of analysis is multidimensional scaling (also introduced in Chapter 5) and can be a great deal more unspecific. Here, the experimenter evaluates the pair-wise (dis)similarities between all stimuli in a dataset and uses this matrix to reconstruct a lower-dimensional space in which each stimulus will be given a coordinate—that is, given similarity ratings, we can reconstruct a kind of "perceptual" or "cognitive" space. Both techniques require the experimenter to interpret the results wisely, as the number of factors in factor analysis and the number of dimensions in multidimensional scaling needs to be divined from the data—a task that is often not easy. Both techniques, however, are extremely valuable tools for pilot experiments as they allow the experimenter to hone in on interesting dimensions of the task.

13.3.1 Factor Analysis

13.3.1.1 Summary

Factor analysis (FA) refers to family of techniques that are used for either dimensionality reduction or for discovering the structure of a series of measurements. The primary goal of FA is to examine the structure between the measured variables. FA makes the assumption that all of the observable variables can be explained by linear combinations of unknown, unobservable latent variables called *factors*. It takes as input a series of N d-dimensional observations and a specification as to how many factors there are. Its outputs are the weightings (eigenvalues or *loadings*) of the original observations onto the factors.

13.3.1.2 Use in Perceptual Research

In the early part of the twentieth century, Charles Spearman noticed that the students' scores on very different, seemingly unrelated topics tended to be correlated. He suspected that much of the variation in the scores could be explained by the variance in some underlying variable, which he called general mental ability or general intelligence (Spearman, 1904). Based on the correlation analysis invented by Karl Pearson and Francis Galton (Galton, 1888), he designed a new technique to try to extract the hidden, underlying source of variance. Subsequent work by Spearman and a number of other researchers have shown that there does not seem to be one general intelligence factor, but many—anywhere from three to eight have been proposed (see, e.g., Gardner, 1983; Sternberg, 1985; Thurstone, 1938). In the 1920s and 1930s, Thurstone systematically organized and improved factor-analysis techniques, including coining the term "factor analysis" (Thurstone, 1931). Factor analysis is used heavily today in the study of intelligence, in the design and analysis of scholastic tests, and in marketing. In fact, factor analysis is used in the design and validation of just about any questionnaire-like instrument.

Another common use is in the study of the formation of meaning, or how people represent concepts and objects. In the 1950s Charles Osgood suggested that people are very good at comparing very disparate and unconnected things because people represent everything in the same set of conceptual dimensions (Osgood, 1957). He developed a new experimental task (called the semantic differential) based on Likert scales (see Chapter 5). In the semantic differential task, a series of N ratings from bipolar Likert scales are obtained for S stimuli from P participants. The scores are averaged across the participants to ensure a more accurate estimate of each scale's value for the stimuli, resulting in a $S \times N$ matrix. Factor analysis on such matrices tries to uncover the underlying structure. Just as Spearman felt that the variation in his students' test scores were really being caused by a hidden factor (general intelligence), Osgood thought that the variations in the semantic-differential ratings might be reflecting variation in some other, hidden variables. After having countless participants use a large number of scales to rate many different words, he found that three factors can explain about 67% of the variance (Osgood, 1964). These three factors are evaluation (good versus bad), potency (strong versus weak), and activity (fast versus slow). Sometimes a fourth common factor (predictability) is also found. These factors have been found in many different cultures (Osgood, 1964). Semantic differentials and factor analysis have been used in an overwhelmingly large range of topics including marketing (see, e.g., Gao and Lee, 1995), emotions (see, e.g., Fontaine et al., 2007), meaning of words (see, e.g., Osgood, 1964), and the quality of specific baroque music performances (see, e.g., Schubert and Fabian, 2006). Naturally, these four basic factors are not the only ones involved in representing objects and concepts. Each specific domain has its own factors as well.

One of the assumptions of FA is that the underlying factors are orthogonal. Thus, FA can be used to take a series of N measurements and convert them into F factors that are guaranteed to be independent. Some fields take advantage of this by using FA as a preprocessing technique.

Regardless of the field of study, there are two basic forms of factor analysis. The most common form is *exploratory factor analysis* (EFA). The goal of EFA is to examine the pattern of relationships between a number of variables (usually the dependent measures) to look for sources of common variance. That is, it tries to detect structure among the variables or to classify the variables. EFA assumes that any variable may be associated with any factor. EFA returns an indication of how many factors are needed to explain the data as well as how strongly related each variable is with each factor. Another way of looking at this last point is to say that EFA groups the variables and provides a measure of how well each variable fits into its group. This information can be invaluable in designing a questionnaire, as it will allow you to select the best questions to cover the domain of interest.

The second form of FA is *confirmatory factor analysis* (CFA). The goal of CFA is to verify whether a specific, hypothesized set of factors and weights can explain the variance in the variables. To perform a CFA, you need to at least specify how many factors there are supposed to be. You generally should also specify which variables are supposed to match which factors.

13.3.1.3 Computation

FA tries to explain the variation in a set of variables using fewer factors. In other words, factor analysis is a dimension-reduction technique. FA assumes that variations in a variable reflect variations in some hidden, or latent, variable. In general, it is assumed that these latent variables, which are called factors, are not in principle observable. For example, Spearman felt that the fact that some students performed well on tests for a number of topics while others performed badly on all the tests is due to the fact that some were smarter than others. That is, the correlations variations in the test scores were really caused by a single unobservable factor (general intelligence).

EFA decomposes variance of the measured variables into two parts: *common variance* and *unique components*. The common variance is the portion of variance shared between variables and is called a *commonality*. The commonalities are often calculated based on the squared multiple correlation of a variable with all other variables. The common variance is used to determine the weights for the factors. In other words, EFA models the variance in the measured variables as a linear combination of the factors (and the weights are determined by the commonalities) and some error terms. The error terms are components unique to a specific variable. They are called the *specific variances* and reflect the variance in a given variable that cannot be explained by the common variance.

Interestingly, the fact that FA models the variables as a linear combination of underlying factors makes it—to some degree—the opposite of a principal component analysis (PCA), which tries to make up new components based on linear combinations of the measured variables. The direction of combination is reversed for the two techniques. If the specific variances for each of the measured variables are all equal, then FA and PCA are essentially the same technique.

There are six steps in a factor analysis.

1. *Collect the data.* To perform a factor analysis, you need to first collect different measurements of the same thing. For example, you might present a number of participants P with a questionnaire consisting of N rating scales. Each scale measures something different (e.g., how often do you engage in your hobby, or how happy are you). This will yield a $P \times N$ matrix. You might also show P people S pictures of expressions and have participants rate the images on N different Likert scales. The scores are averaged across the participants to ensure a more accurate estimate of each scale's value for the stimuli, resulting in a $S \times N$ matrix.

2. *Decide on the type of analysis.* Normally in a EFA you will examine the relationship between the N ratings. This is commonly referred to as a R-factor analysis. In the questionnaire example, you would look at the common variances of the different questions across participants to see how the questions might be grouped. You could, of course, choose to examine the inter-relations among the participants. Are there participants who are particularly happy or who engage in their hobbies often? This is a Q-factor analysis.

3. *Prepare the data.* Factor analysis decomposes the variance in the variables into common and unique components. To do this we need to examine how well the variables co-vary. Thus, the next step is to calculate a variance covariance matrix. The diagonal of this matrix contains the communalities. Although it is possible to perform the FA on the variance covariance matrix, it is often not a good idea. It is possible, for example, that the different scales have greatly different ranges. Some questions may be on 7-point Likert scales (such as ratings of happiness) while others may be on much larger scales (such as the age of the participants). Thus, it is common to standardize the variables. This is done by ensuring that the mean of the variables is 0 and the variance is 1. In practice, this means using the correlation matrix.

4. *Deciding on the number of factors.* The next step is one of the more subjective aspects of FA, and often a source of considerable complaint. Before you determine how well the variables group together, you need to decide how many groups you want. That is, you need to specify how many factors you think there are. There are four common ways of doing this, which appear below.

    - *Existing theory.* You can decide on the number of factors based on existing theory. For example, decades of work with semantic differentials has shown that there are three major factors (evaluation, potency, and action) that usually jointly explain about 67% of the variance. So, if you are performing a FA on some semantic-differential data, you would be reasonably justified in extracting three factors.

- *Variance explained.* You might decide that you wish to have enough factors to explain a fixed percentage of the variance. Usually, enough factors are chosen to explain $\frac{2}{3}$ (67%) of the variance. How do you know how much of the variance the factors explain if you do not have them yet? The answer is simple: you can use the eigenvalues from the correlation matrix calculated in the previous step.
- *Kaiser criterion.* The most common way of choosing the number of factors was introduced by Kaiser (1960), and states simply that you use only those factors whose eigenvalues are greater than 1.
- *Scree plots.* Cattell (1966) noted that if you plot the eigenvalues in descending order, the resulting curve starts with a sharp decline and quickly levels off. In the scree-plot method, you examine the curve to find the point where the initial, smooth decrease levels off. You then choose all of the factors to the left of this break point.

Obviously the maximum number of factors that you can have is the same as the number of observed variables that you have. That would, of course, not result in any dimension reduction. If you wish to use fewer factors, any of the above methods are appropriate. A number of studies have compared the two objective methods (the Kaiser criterion and the scree plots) (Browne, 1968; Cattell and Jaspers, 1967; Hakstian et al., 1982; Linn, 1968). In general, the Kaiser criterion sometimes suggests more factors than the scree method. In most common cases, however, both are very reasonable, especially if there are few factors and a lot of variables. It is important to remember that the resulting solution will have to be interpreted, so it can help to use the solution that makes the most sense.

5. *Calculate the weights.* The next step is to calculate which set of weights for the prespecified number of factors best explains the variables. These factor weights are often called the *factor loadings* of the different variables onto the factors. There are a number of different methods for calculating the weights. These include PCA (which seeks to maximize the total variance explained), and the principal axis method (which tries to explain as much of the common variance as possible, and maximum likelihood.

6. *Rotation.* The extraction of weights is an under-determined problem. For a given set of factors, there are an infinite set of weights that can explain the data. Each of the different sets, however, is a rotation of the original set. Since rotating the solution alters the weights, it changes which variables are grouped together. Each rotation is just as valid as the others mathematically, but some "make more sense" than others. In this step,

the data are rotated until a solution is found that is appropriate. Fortunately, one does not have to search through the infinite possibilities. Several standard rotations exist that follow different criteria. These include varimax (which maximizes the variance explained by the first few factors; this is the most common rotation), quartimax, equimax, direct oblimin, and promax.

The steps just described are more accurately described as exploratory factor analysis. For a confirmatory factor analysis, additional steps are necessary to measure the goodness of fit of the prespecified model.

13.3.1.4 Guidelines

- FA is only as reliable as the data. As mentioned in Chapter 4, self-reports are notoriously unreliable.
- The interpretation of a given factor is entirely subjective. What precisely is the common characteristic of a group of items? Since the factor itself cannot be observed, there is no objective way of saying which interpretation of the factor is correct.
- It is best to have at least 20 participants for every variable.
- You should try to have at least three variables for each factor.
- You need to have more participants (for questionnaires) or stimuli (for semantic differentials) than variables.

13.3.1.5 Example

As described in Chapter 5, we examined the semantic clustering of facial expressions using a semantic-differential task. We selected four scales for each of the first two theorized semantic dimensions (evaluation and potency) and three scales for the remaining two factors (activity and predictability) based on Fontaine et al. (2007)'s analysis of 144 scales, yielding 14 bipolar Likert scales. We presented 54 stimuli (nine expressions from six actors) to 20 participants. We collapsed the data across participants to obtain a 54×12 matrix. After calculating the correlation matrix, it was clear that the first three factors had Eigenvalues over 1 and these three factors jointly explain 91% of the variance. This suggests that three factors are sufficient to explain the results. Extracting a three-factor solution using varimax rotation yielded factor loadings consistent with existing theory. The first factor, which accounts for about 51% of the variance, appears to be a fusion of the activity and predictability factors. There was insufficient variation in the stimuli along the activity and predictability factors for them to require separate factors. The second dimension, accounting for about 30% of the variance, is clearly potency (the scales chosen to load onto

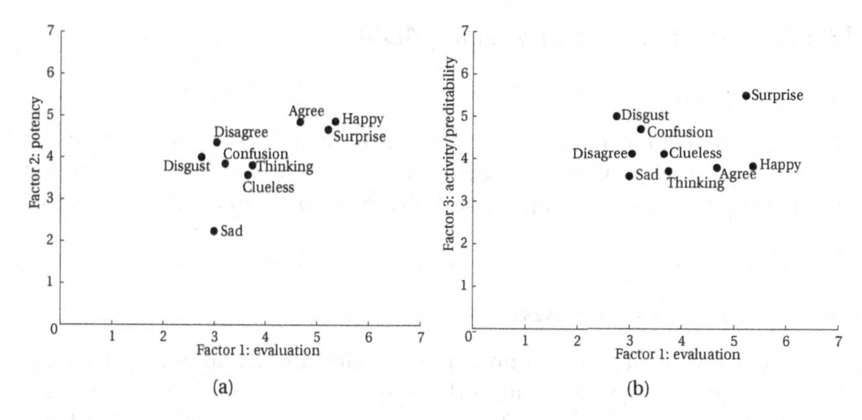

Figure 13.4. Location in semantic space of the facial expressions. The average score of each expression along each of the three semantic dimensions is plotted. (a) Factor 1 versus factor 2, full scale. (b) Factor 1 versus factor 3, full scale.

potency did in fact do so). The final factor contains the scales representing Evaluation, and accounts for 12% of the variance.

The factor loadings for each expression were averaged across actors, resulting in the coordinates of those expressions in semantic space. Following tradition, these average scores are plotted pairwise in Figure 13.4 using the full scale (1 to 7). A quick glance at the figure shows that the scores all tend towards the middle; that is, participants did not use the extremes all that often. This is one of the most common findings with any form of Likert scale. The semantic structure of the expressions is more clear in the closeup plots seen in Figure 13.5.

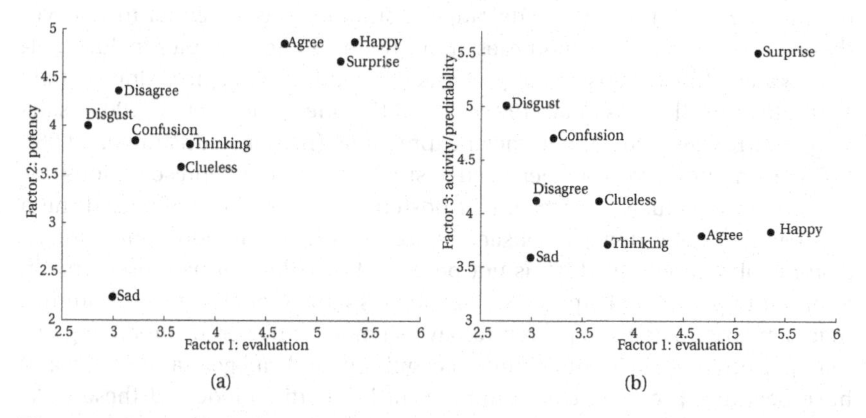

Figure 13.5. Location in semantic space of the facial expressions. The average score of each expression along each of the three semantic dimensions is plotted. (a) Factor 1 versus factor 2, closeup. (b) Factor 1 versus Factor 3, closeup.

13.3.2 Multidimensional Scaling (MDS)

13.3.2.1 Summary

MDS refers to a family of algorithms that for a set of N data points take as an input a $N \times N$ matrix of (dis)similarity data, and return—for a given number of dimensions d—the coordinates of the N data points in the d-dimensional space.

13.3.2.2 Use in Perceptual Research

In a recent review of research on similarity, Goldstone and Son (2005) provide a long list of cognitive abilities that depend on the ability to judge similarity; including reasoning, problem solving, perceptual recognition, and categorization. They divide similarity research into two classes: research on conceptual stimuli, such as theories or stories, and research on perceptual stimuli, such as colors, textures, sounds, odors, and tastes. In studying how humans judge similarity, it is particularly important to have a well-defined input space. This is due to a long tradition of visualizing similarity data using spatial models, in which the (perceptual) distance between two items is related to the (perceptual) similarity between them. This approach was pioneered by Richardson's characterization of a perceptual color space based on comparisons made amongst Munsell color samples (Richardson, 1938). Research on the structure of such psychological spaces has been closely connected to the development of MDS techniques (seminal work was done by Torgerson, 1952; Shepard, 1962). MDS techniques operate on pairwise similarity ratings taken over a set of stimuli and return a map of objects in which distances have been fit to similarity data. The techniques have been applied to study mental representations of a wide range of stimulus classes, either to discover them or to test specific hypotheses about them. Some examples include colors (Ekman, 1954), Morse code patterns (Shepard, 1962), spices (Jones et al., 1978), textures (Hollins et al., 1993), synthetic tones (Caclin et al., 2005), salts (Lim and Lawless, 2005), and facial expressions (Bimler and Paramei, 2006). MDS techniques have also been used to study how mental representations differ across individuals. For example, Bosten et al. (2005) recently used them to show that color-deficient observers are sensitive to an additional dimension of color variation which is not perceived by color-normal observers. In addition to providing helpful visualizations, spatial models derived from human similarity data have proven to have significant power for predicting human performance in identification, recognition, and categorization tasks and have been used to formulate a number of influential models of these tasks (Edelman, 1999; Gillund and Shiffrin, 1984; Nosofsky, 1991). Lately, these techniques have also been used to identify perceptual-processing strategies in multisensory data—for example, by having participants judge similarity by touch and by vision and then using MDS to compare the two perceptual spaces (Cooke et al., 2007).

13.3.2.3 Computation

In the case of perceptual experiments, the most common application is to gather similarity ratings from participants (for example, on a Likert scale of 1–7, where 1 means fully dissimilar and 7 fully similar) for a set of stimuli. If one has N stimuli, this will result in $N \times N$ comparisons for a full design which compares stimulus A to stimulus B and vice versa. Alternatively, one could run a timesaving version that only compares stimulus A to stimulus B, thus resulting in $N + N \times (N-1)/2$ comparisons (note that this assumes perceptual symmetry in the comparison of stimulus A to stimulus B).

The first step is to create the matrix of perceptual dissimilarites. A good sanity check is to confirm how many times participants rated the pair A-A as being a 7. If this fails for a larger number of cases, something must have gotten mixed up in the data analysis or even in the experimental design. All MDS algorithms require a symmetric matrix as input, for which the diagonal elements are all 1. This means that the experimental data has to be renormalized and modified to fit this assumption. Usually, one has a similarity matrix for each participant, so one can either do a MDS for each participant and average the data to create a joint similarity matrix for all participants, or use weighted MDS (see below).

The second step is to choose an MDS algorithm. MDS algorithms come in three flavors: classical MDS, metric MDS (both of which deal with interval, or ratio data), and non-metric MDS, which deals with ordinal data.

Classical MDS minimizes a particular loss-function called strain and returns as output the configuration of points in a d-dimensional space, which minimizes strain.

Metric MDS is a superclass of classical MDS, in which several loss functions and known dissimilarity configurations are available.

Non-metric MDS assumes only ordinal relationships between data points and therefore tries to find a non-parametric, monotonic explanation of the (dis)similarity data. It therefore allows the use of general Minkowski metrics.

The choice of the algorithm depends on the data at hand. Often the metric MDS gives good enough results in perceptual contexts. Note that the non-metric MDS has to solve a much more complex optimization problem and is therefore more susceptible to noise in the data.

All of the above flavors are also available for a weighted MDS (often also called INDSCAL) in which k similarity matrices are used to create *one* underlying (perceptual) space. This algorithm uses weighted metrics, in which each (perceptual) dimension can be weighted differently. For example, in a face-comparison task such an algorithm can be used to find out whether the first perceptual dimension (say, age) matters more than the second (say, head shape) for participants.

All algorithms require the choice of dimensions as an input parameter. Usually, however, this is an experimental unknown—that is, one would like to know how many perceptual dimensions are best suited for explaining the data. A post-hoc analysis consists of running the MDS algorithm with different number of dimensions and looking for a sharp dip in the stress output. For metric MDS and non-metric MDS, the stress value is normalized between 0 and 1, and previous simulations have shown 0.2 to be an acceptable value (Borg and Groenen, 2005).

If one wishes to compare different MDS maps (that is, to investigate whether they might be statistically different), one can resort to Monte-Carlo type strategies. By creating a large number of similarity matrices through random perturbations of the base similarity matrix (for example, by using the standard deviation of the similarity ratings across participants), one can create a large number of solutions. Each point in the d-dimensional space will therefore also be perturbed accordingly and one can gain a feeling for how stable the perceptual solutions really are.

13.3.2.4 Guidelines

- *Interpretation of higher dimensions.* Higher dimensions are often difficult to interpret perceptually and thus it is a good idea to run experiments on stimuli that differ only along few, well-defined, parametrized physical dimensions.

- *Uniqueness of solution.* Classical MDS is only defined up to a rotation in the final space, making interpretation even more difficult.

- *Averaging of matrices.* When running multiple participants, averaging of the similarity matrices only produces a good result if the correlation of the similarity matrices is high. In any case, it is a better idea to resort to INDSCAL algorithms as these provide additional information about the data.

13.3.2.5 Example

The canonical example for MDS is to reconstruct a map (that is, the location of objects in two dimensions) when all pair-wise distances are known. As an example, we take as input the pair-wise distances between four Italian cities shown in Figure 13.6(a). Given this similarity matrix, MDS will produce different topologies of the four cities in an up to four-dimensional space. Obviously, the more dimensions we give MDS the better it will be able to explain the distances, a fact that is shown in the stressplot (see Figure 13.6(b)), which charts the stress of the classical MDS solution as the dimensionality increases. Given the sharp falloff that occurs when moving from a one-dimensional solution to a two-dimensional solution, we might assume that a reconstruction in two dimensions provides a good representation of the data. Why, then, is the stress

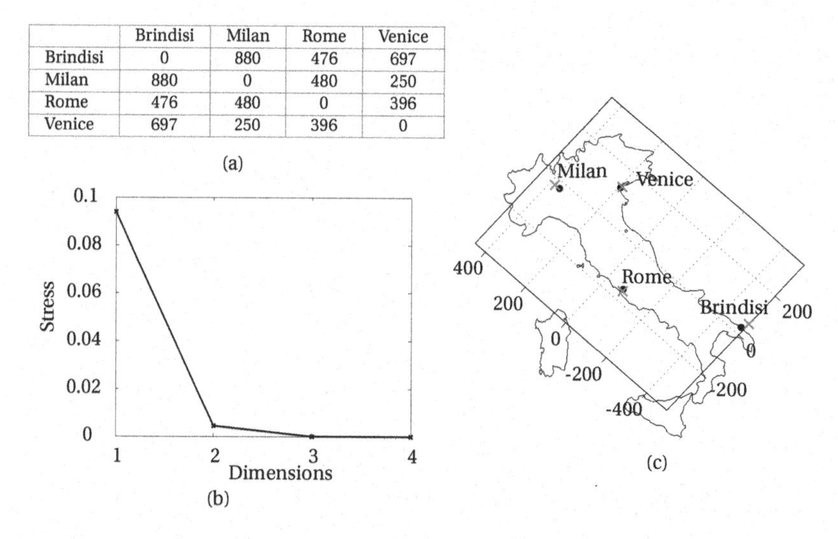

	Brindisi	Milan	Rome	Venice
Brindisi	0	880	476	697
Milan	880	0	480	250
Rome	476	480	0	396
Venice	697	250	396	0

Figure 13.6. (a) Matrix showing the distances between four Italian cities measured roughly by hand on a map, (b) stress plot showing that two dimensions explain the data fully, (c) reconstructed map (crosses) overlaid over a map showing the true locations of the four Italian cities (dots).

not exactly zero for two dimensions? There are two possible reasons. First, the distances between the cities might be subject to measurement noise, which will contaminate the reconstructed result. Second, even if we measured perfectly, the reconstruction in a two-dimensional Euclidean space will be wrong for very long distances measured on the earth's surface since the surface of the earth becomes non-Euclidean due to the non-negligible influence of the earth's curvature. Be that as it may, we can safely choose a two-dimensional solution for our MDS and assign each city a point on the two-dimensional map. Figure 13.6(c) shows the result of the MDS solution with the reconstructed map (as crosses) overlaid on top of a map of Italy with the actual locations of the cities (as dots). Note that in order to align the MDS solution with the real world, we had to rotate the coordinate frame (also shown in Figure 13.6(c)). This demonstrates the fact that a classical MDS solution is only defined up to a rotation.

Chapter 14

Forced and Multiple Choice

The expert in these tasks will need to know about signal detection theory and how to calculate d', the receive operator characteristic (ROC), and threshold sigmoids.

14.1 Signal Detection Theory

In the field of psychophysics, signal detection theory (SDT) is one of the core mathematical frameworks with which to understand and analyze certain types of experimental paradigms. Signal detection theory—as the name implies—is about detecting a signal; more specifically, it is about detecting a potentially noisy signal amid a potentially noisy background. Therefore, the classic SDT experiment involves two conditions, a *signal present condition*, in which the noisy signal is shown, and a *signal absent condition*, in which only a noisy background is shown. The noise serves as a tool to make the detection difficult, as otherwise the task would become too trivial.

Perhaps the most basic experimental setup in SDT is related to contrast detection, in which the target (the signal) is a square of a brightness level s shown on a background of brightness b. The task is to detect the presence of the square as a function of the difference in brightness levels $s - b$ between the square and the background (the contrast). If we make the experiment very simple, we can choose one relatively low-contrast level, and show 100 trials in which the square is present, and 100 trials in which the square is not present.

As another possibility, consider a standard two-alternative forced-choice (2-AFC) paradigm. In one version of this paradigm, a participant first studies a series of items (faces, objects, etc.). In the main part of the experiment, the participant's task is to view an item on the screen, and to decide whether that item had been previously seen or not. In many cases, one will refer to the old items as *targets* and to the new items as *distractors*. As an example, we can take an experiment, in which participants first have to learn the names of ten faces. After the learning phase participants see 20 faces—ten of which are the

	Participant answered "seen"	Participant answered "not seen"
Item was seen	correct (hit)	incorrect (miss)
Item was not seen	incorrect (false alarm)	correct (reject)

Table 14.1. 2×2 matrix of responses for a 2-AFC paradigm.

old faces that were presented before, and ten of which are new faces that were not presented before. Basic statistics tells us that if participants would flip a coin in each trial and answer "seen," whenever the coin showed heads, and "not seen," whenever the coin showed tails, their answers will be correct in 50% of all cases, on average.

If a participant takes part in the experiment and correctly identifies eight faces as "seen," we would consider that to be a good performance. Conversely, a participant who identified only two faces correctly would be judged as almost hopeless at face recognition. However, these numbers tell only one part of the story. To illustrate this, consider the case in which the participant *always* answers "seen": in this case, the participant would be 100% correct on all "old" faces, but also 100% wrong on all "new" faces. Clearly, when trying to characterize performance in such a task, it is vitally important to pay attention not only to the correct identifications, but also to the errors!

In general then, this experimental situation gives rise to four possibilities, which are summarized in Table 14.1.

As we have seen, when analyzing the results of such an experiment the final performance should take into account not only how often the participant was correct (the hit rate), but also how often the participant incorrectly identified an object as "seen," when it had not been part of the learning set (the false alarm rate). This combined performance measure is known as *sensitivity.* Note that since each line needs to add up to 100%, an equivalent way of describing performance would be to pair the correct rejection rate with the incorrect miss rate.

14.1.0.6 Recommended Reading

There are many classic textbooks on signal-detection theory, including Gescheider (1997) and Green and Swets (1974). A more concise introduction can be found in Macmillan (2002).

14.2 Receiver Operator Characteristic

The above example experiment examines the possible reasons for performance differences between people. Imagine that the ten target faces that participants had to learn were similar to the distractor faces, making the decision of whether a face is "seen" or "not seen" rather difficult. In this case, a participant might choose to be conservative and try to avoid making too many errors, thus rejecting faces even though they seem to bear some resemblance to the learned faces. This strategy would result in many misses but few false alarms. As an example, let us assume that this particular participant scores three hits and one false alarm. Another participant might set the decision threshold differently and score more hits but also more false alarms; let us say in this example, the participant scores seven hits and three false alarms. Finally, a participant who wants to score a lot of hits might decide that many faces look like the learned ones, resulting in many more hits, but also many more false alarms—let us assume nine hits and seven false alarms.

Note that the three participants differed in their *criterion*, that is, in the similarity threshold between two faces at which they decided that the presented face was similar to one of the faces stored in memory. Note also, that as this example shows, when considering *both hits* and *false alarms*, we get the impression that all three participants seem to do equally well—regardless of the criterion or threshold they set themselves, and very much different from the performance when considering only hit rates, for example. Indeed, this notion of a threshold-independent performance is captured in the receiver operator characteristic (ROC), which plots the hit rate against the false alarm rate. The data for the three participants is shown in Figure 14.1 together with an interpolated curve. Regardless of how the participants set their criterion, their combined performance (or sensitivity) falls on the curve shown in the plot.

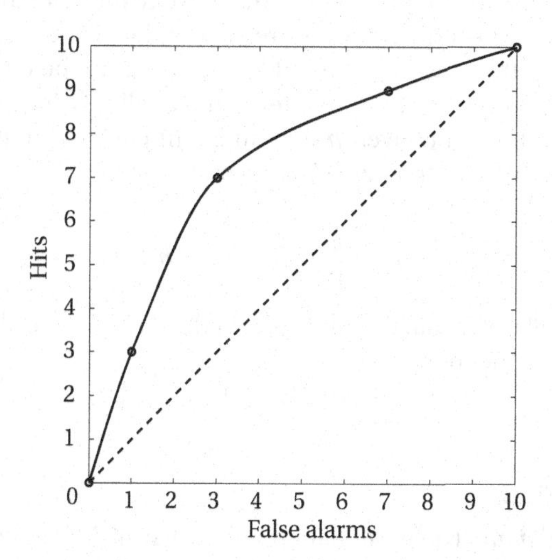

Figure 14.1. Receiver operating characteristic for a hypothetical face recognition experiment using three observers.

A good way of manipulating the criterion for participants is to give differential rewards (monetary rewards usually work best) for the four performance characteristics (*hits, misses, false alarms,* and *correct rejections*). Given the basic premise that participants want to maximize their rewards, they will modify their criterion so as to maximize the relevant performance measure. Note that while adjusting the threshold will change the relative contributions of the different performance measures, the assumption for many psychophysical experiments is that the participants' resulting *sensitivity* (the combined performance of hits and false alarm rates) will stay the same—regardless of the criterion.

One of the most straightforward measures of sensitivity is the area under the ROC curve, which we will describe briefly in the Section 14.2.1; in most cases, however, the more constrained d'-criterion is used (see Section 14.3).

14.2.1 Area under the Curve (AUC)

14.2.1.1 Summary

The AUC is a non-parametric measure of sensitivity based on a ROC curve.

14.2.1.2 Computation

Whereas the ROC curve plotted in Figure 14.1 is symmetric, in general a ROC curve is any curve that plots the connection between hits and false alarm rates. If certain assumptions (see Section 14.3) about the underlying data-generation distributions are not met, the curve might look different—indeed, any monotonically increasing relationship between hit rate and false alarm rate can be viewed as ROC data. Consequently, the most simple measure that is related to the sensitivity of the experiment is the area under the ROC curve.

In most cases, the area under the ROC curve cannot be determined analytically and therefore must be estimated numerically. Using straightforward numerical integration, and given n measurement pairs of hit and false alarm rates (h_i, f_i) on the ROC curve, the AUC becomes

$$\text{AUC} = \sum_{i=1}^{n-1} (f_{i+1} - f_i) \times h_i,$$

which sums up all the rectangles underneath the data. Obviously, the area under the curve is bounded by

$$0 \leq \text{AUC} \leq 1.$$

14.2.1.3 Guidelines

- *Amount of data.* The more data points are available, the more precisely this measure will estimate the true ROC curve. This is because of the finer quantization of the area.

14.2.1.4 Example

For the data shown in Figure 14.1, the AUC becomes

$$\mathrm{AUC} = \frac{1}{10 \times 10}((1-0) \times 3 + (3-2) \times 7 + (7-3) \times 9 + (10-7) \times 10) = \frac{83}{100} = 0.83.$$

14.3 Measuring Sensitivity Using d'

How can the sensitivity of a participant be explained theoretically? The answer to this lets us return to the basics of signal-detection theory. Again, assume that we have a detection experiment in which a target is either present or absent. Because both conditions are contaminated with noise, detectability will not be perfect and hence in some cases, the participant will wrongly answer present even though the target is absent, and vice versa. Assuming that in both conditions the noise is distributed similarly and normally, we can plot the two probability distributions around the two true detection values as shown in Figure 14.2(a).

The plot shows two different distributions that reflect the fact that based on some level of stimulus property (here, contrast), the detectability varies accordingly. The task for the participant is to decide—given a particular, observed stimulus value on the x-axis—whether the target was present or not. This can be achieved by setting a threshold t, above which the participant will reply present. However, setting this threshold needs to be done with care, as setting it too low will ensure that the full target distribution is captured, but will also result in many different values from the noise distribution being identified incorrectly (false alarms).

Figure 14.2(a) shows two distributions that are offset by a value of 4, resulting in a relatively low degree of overlap between the two. In addition, seven hypothetical thresholds are indicated with numbers, each of which gives rise to a certain number of hits (namely, the area under the "Target+Noise" curve to the right of the threshold) and a certain number of false alarms (namely, the area under the "Noise" curve to the right of the threshold). Hence, each threshold gives a pair of hit/false alarm values. If we plot these pairs, we obtain Figure 14.2(b), which is the ideal ROC curve. We see that for a threshold of 1, for example, there will be a large number of hits, but also a large number of false alarms. A threshold of 5 is exactly in the middle of the overlap region, resulting in the same number of misses as false alarms (in many applications this value is called the *equal error rate* (EER)). We can also see that the final ROC curve is extremely close to being ideal in that it almost shoots up vertically to full hit-rate at zero false-alarm rate, which would of course be the ideal situation.

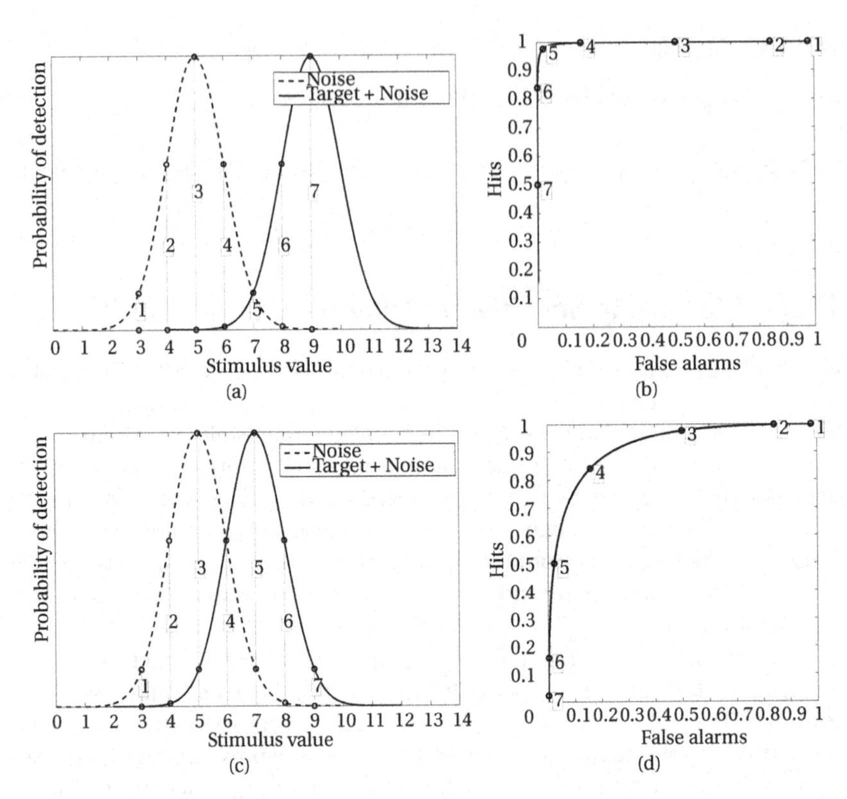

Figure 14.2. Two different examples for a detection experiment with (a) easy and (c) hard difficulty level. The lines in the plots (a) and (c) show different thresholds or criteria, (b) and (d) show the corresponding ROC curves for those value pairs.

In contrast, Figure 14.2(c) shows two distributions that are offset by a value of 2, resulting in a much larger degree of overlap between the two—this discrimination task is therefore much more difficult. Accordingly, the same seven thresholds from Figure 14.2(a) will result in much larger error rates overall, causing the resulting ROC curve to bend down much more prominently in Figure 14.2(d). For example, a threshold of 7 that guaranteed an almost perfectly zero false alarm rate at a hit rate of 0.5 in Figure 14.2(b) now yields almost no hits.

There is another measure hidden in Figure 14.2 that relates to the distance between the peaks of the noise distribution and the target+noise distribution. Obviously, the closer the two distributions are together, the harder the task will be, and accordingly the more bent the ROC curve will get. This measure is known as *d-prime*, or d' and is exactly that—the distance between the two distributions.

What makes d' such a useful measure is that if the two distributions are well-behaved, and if our participants are equally well-behaved, we can use this one number as the indicator of the whole ROC curve. In other words, each ROC curve belongs to certain d' value. The higher the d' value becomes, the more sensitive the participant; that is, the more the target can be separated from the noise distribution.

14.3.1 The Measure d'

14.3.1.1 Summary

The d' measure is one of the most common measures of sensitivity for perceptual and cognitive studies.

14.3.1.2 Computation

Given two distributions, each of which are normally distributed and share the same underlying noise process, d' is the distance between the distributions in units of standard deviation. That is

$$d' = \frac{\text{separation}}{\text{spread}}.$$

Given an experiment in which hit rates and false-alarm rates are available, d' is calculated as the difference between the two z-transformed performance scores, or

$$d' = z(\text{hits}) - z(\text{falsealarms}).$$

The z-transform $z(p)$ is the inverse of the cumulative Gaussian distribution. Basically, this goes from the ROC curve (which plots hits against false alarm rates) *back* to the two distributions from which it was generated. As for $p \to 1$, $z(p) \to \infty$, and for $p \to 0$, $z(p) \to -\infty$, by convention p is usually set to a minimum value of $p = \frac{1}{n}$ and to a maximum of $p = \frac{n-1}{n}$ for n trials in the experiment. For 100 trials, d' therefore is bounded by

$$0.01 < d' < 4.66.$$

Note that d' is a shorthand for a whole ROC curve—the curve corresponding to $d' = 0$ is the diagonal, the curve corresponding to $d' = 1$ is $(0,0) - (0,1) - (1,1)$. Another way to find d' is then to compare ROC curves with standardized ROC curves of known d' values.

Testing whether two different groups of d' values (for example, for different experimental conditions) are different is usually done by a simple t-test or—for several groups—by a straightforward ANOVA. Confidence intervals for d'-values can also be determined, but usually require quite a large number of trials for accurate representation—see Miller (1996) for more information.

14.3.1.3 Guidelines

- *Normality of the underlying distribution.* For a larger number of trials and for most perceptual tasks assuming that the distribution is normal is probably fine. When there only few trials, however, estimating d' can become a problem and nonparametric measures should be used instead.

- *Negative d'.* The measure d' can become negative if the hit-rate is less than the false alarm rate. If the assumptions of d' are met, the simplest explanation for this problem is that the participant mislabeled the trials; that is, he or she answered "target present" when they actually meant to press "target absent," or vice versa. This is a problem that sometimes occurs in fast experiments in which two buttons need to be pressed quickly. Another possibility for mislabeling trials could be that the participant purposely answered incorrectly. In both cases, the solution is to simply flip d' and make it positive. However, one needs to check carefully that this was actually a valid explanation of the results—it could also be that the assumptions for d' have been violated, such as when noise and target+noise distributions have different variances.

14.3.1.4 Example

The d' values associated with the three participants from Figure 14.1 are $d' = 0.76$, $d' = 1.05$, and $d' = 0.76$, respectively. Note that only participants 1 and 3 actually have the same d', and thus also have the exact same sensitivity. Participant 2 has a slightly higher sensitivity, although with only 20 trials, the confidence intervals for each of the d' values would be fairly high indeed.

14.4 Psychometric Functions

A psychometric function is the mathematical description of how a percept relates to a specific stimulus dimension. It is one of the core mathematical and computational models of how lower-level sensory processes work and is an invaluable part of the tool chest of any perception scientist.

More specifically, a psychometric function is usually measured in a detection or discrimination task, which varies the input dimension of interest (contrast, brightness, color saturation, motion strength, motion coherence, tone pitch, smell intensity, face-morph level, speed of facial expression, etc.) and presents either one stimulus for detection, or a series of choices for discrimination. Either way the output is usually measured in terms of percentage correctly detected and the resulting data will almost always resemble a sigmoid (that is, an s-shaped) function.

The particular shape of the function comes about because—given a suitable range of tested input parameters—it is assumed that at the higher range of input parameters the detectability of the stimulus is very high, and hence participants will be correct most of the time, whereas at the lower range of input parameters, the stimulus becomes much harder to detect/discriminate and hence participants will be reduced to guessing, thus reaching chance level performance on the task. The interesting bit for the psychophysicist happens in the middle parameter region, as this represents the transition from guessing to certainty. The data usually bends upwards as the stimulus property increases and the slope or steepness of the curve determines the sensitivity of the participant(s). In addition, the point at which the curve inflects (that is, where the curvature changes) or point at which the performance reaches the average of chance level and perfect performance is defined as the *threshold level.*

14.4.1 Parametrizing Psychometric Functions

The most common parametric definition of a psychometric function is

$$\mathrm{PF}(\mathrm{x};\alpha,\beta,\gamma)=\gamma+(1-\gamma)f(x;\alpha,\beta),$$

where x is the stimulus parameter, α,β,γ are characteristics of the response curve, and $f(x;\alpha,\beta)$ is a sigmoid-shaped function.

The parameter γ is the lower bound of PF, or in other words, the chance level in absence of the target. In typical yes/no paradigms or detection tasks, this parameter technically should be $\gamma=0$, but it is often corrected to a non-zero value to account for answers due to guessing. In typical n-AFC paradigms, the parameter is fixed to the expected chance level for n alternatives, $\gamma=\frac{1}{n}$.

The function $f(x;\alpha,\beta)$ can be based on theoretical observations about expected response probabilities given a particular experimental paradigm. In signal-detection theory the ideal observer will be modeled with a Gaussian integral (which simply sums up the values over the Gaussian distribution), since target and target+noise distributions are modeled to be standard Gaussian functions. As this integral is rather cumbersome to evaluate and does not have an analytical solution, it is usually approximated by the *logistic function* (Green and Swets, 1974):

$$f_{\mathrm{log}}(x;\alpha,\beta)=\frac{1}{1+(\frac{x}{\alpha})^{\beta}}.$$

Other models for response probabilities are based on independent detection mechanisms which are summed up to provide the final detection performance and result in a *Weibull function* (Green and Swets, 1974)

$$f_{\mathrm{wei}}(x;\alpha,\beta)=1-\exp-(\frac{x}{\alpha})^{\beta}.$$

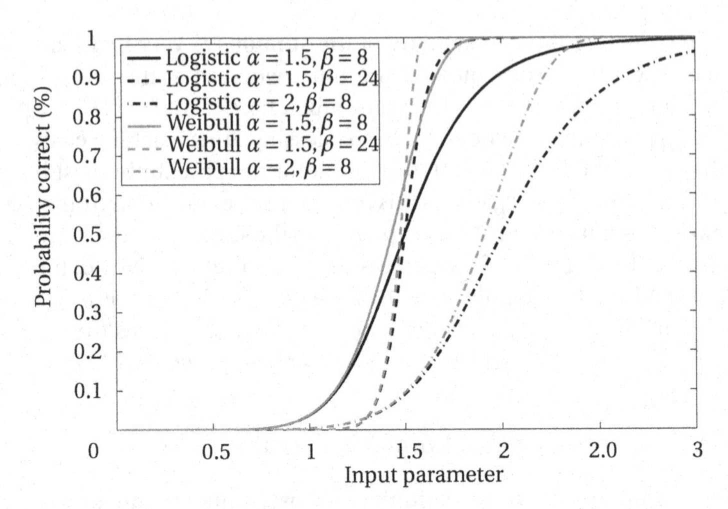

Figure 14.3. Two different, idealized psychometric functions $PF_{log,wei}$ based on logistic and Weibull distributions for different parameter combinations.

Note that both the logistic function as well as the Weibull function are derived from their corresponding probability distributions, but that they are not suitably normalized here, such that $\int_x f_{wei,log} \neq 1$.

The parameters α, β have specific meaning in terms of characterizing the shape of the curve, and with that the performance characteristics in the task: α is the input parameter at which PF has the largest slope $\frac{df}{dx}$, in other words, where the discriminability of the observer as characterized by the derivative is the largest. The parameter β is the steepness of the curve, that is it describes how small or large the transition from no detection to full detection is.

Figure 14.3 shows PF_{log} and PF_{wei} for $\gamma = 0$ and different parameter combinations of α, β. Note that the Weibull distribution is always a little steeper than the logistic function for the same parameter values. In general, it seems that Weibull distributions are able to model both detection and n-AFC tasks better, so that the community mostly uses PF_{wei} for modeling psychometric functions (Wichmann and Hill, 2001a,b).

As a final note, we should mention that the most general definition of the psychometric function actually includes another parameter called λ that specifies the lapses of a participant, in that $1 - \lambda$ defines the performance ceiling at full stimulus strength. These lapses come about because of errors in the percept, wrong response, etc. and usually should be only a very minor effect of a few percent. As shown by Wichmann and Hill (2001a), however, the exclusion of the parameter can sometimes dramatically bias the fitting results, with only a few points measured at high stimulus strength able to influence the whole psychometric function.

14.4.2 Fitting Psychometric Functions

Of course, the elegance of psychometric functions and the ability to model perception with only two (α, β) or three (α, β, λ) parameters can only be achieved when they are fitted to the behavioral data. There is a huge body of literature available on fitting psychometric functions, obtaining good confidence limits on thresholds from the psychometric functions, comparing two functions, etc. We refer the interested reader to the excellent companion papers by Wichmann and Hill (2001a,b) for an introduction into the problems associated with this topic. Here, we briefly review the most elementary principles of fitting psychometric functions.

14.4.2.1 Computation

First, in order to fit a function, we need to measure performance (perf) as a function of the stimulus strength s for n data points. That is, we get a set of value pairs (perf_i, s_i), $i = 1, \ldots, n$. These values were obtained through a task such as a detection (yes/no) task, or a n-AFC task. Given a particular task, the first step is to fix γ to the chance level of that task. The second step is to decide on a suitable function parametrization. As outlined above, the Weibull psychometric function seems to be a good choice; that is, we fix PF_{wei}.

Next, we need to define a criterion according to which we actually fit PF_{wei} to our data values (perf_i, s_i), so we will need to perform a regression. It might be tempting to return to our regression examples from earlier chapters and simply choose the parameters α, β, for which

$$\underset{\alpha,\beta}{\arg\min}\,(\text{SSE}) = \sum_{i}^{n} \left(\text{perf}_i - \text{PF}_{\text{wei}}(s_i; \alpha, \beta)\right)^2$$

is minimized using the sum of squared errors criterion—in other words, by using a least-squares optimization.

The problem in doing so is that the least-squares criterion requires the scale to be interval/ratio, so we would need to make sure that all differences in performance have equal meaning. In a detection or discrimination task, however, performance differences around the extreme values have different weighting associated with them than for other values. Hence, a difference of 5% around 95% can be more crucial than a difference of 5% around 50%. Instead, we need to look at the way the data was actually gathered, that is, as a series of statistical yes/no or present/absent decisions by the participants. The statistically correct way of looking at this data is by using probabilities and log-likelihoods as described in the following.

Note that the way the psychometric function is constructed describes the probability of a successful detection given a stimulus strength $p(s_i)$. Also note that the experimental paradigm usually consists of a series of yes/no answers that are generated at different stimulus strengths s_i using any of the classical psychophysical methods (methods of limits, methods of constant stimuli, etc.), or more sophisticated methods such as staircases or adaptive staircases. The likelihood P of observing the full sequence of the total of n trials including correct and incorrect responses perf_i is then given as a simple Bernoulli process

$$P = \prod_i^n \text{PF}(s_i;\alpha,\beta)^{c_i}(1-\text{PF}(s_i;\alpha,\beta))^{1-c_i},$$

where $c_i = 1$ for correctly identified trials, $c_i = 0$ for incorrectly identified trials, $\prod_i^n \text{PF}(s_i;\alpha,\beta)^{c_i}$ is the probability of observing the sequence of correct responses, and $\prod_i^n(1-\text{PF}(s_i;\alpha,\beta))^{1-c_i}$ the probability of observing the incorrect responses.

Obviously, we want to maximize P, that is, we want to find parameters α,β for which $\arg\max_{\alpha,\beta}(\text{PF})$. Because of numerical issues, we usually do not want to directly maximize P, but rather to maximize the log-likelihood

$$\log(P) = \sum_i^n c_i \log(\text{PF}(s_i;\alpha,\beta)) + (1-c_i)\log(1-\text{PF}(s_i;\alpha,\beta)),$$

as this avoids the sometimes excessively small values of P for large numbers of trials n.

One problem for this optimization process is when $\text{PF}(s_i;\alpha,\beta) = 1$ or $\text{PF}(s_i;\alpha,\beta) = 0$, as then one of the terms in the equation becomes undefined. It is usual practice in these cases to correct the function, such that

$$\text{PF}(s_i;\alpha,\beta) \rightarrow 0.99\text{PF}(s_i;\alpha,\beta) + 0.005,$$

which puts a lower bound of 0.005 and an upper bound of 0.995 on the function. In effect, this performs the correction for lapses at both high and low stimulus strengths. As mentioned above, these errors occur because the human is not a perfect system, so that errors creep in due to fatigue, attention lapses, motor noise, etc. It is important to note, however, that this correction does not lead to an additional free parameter. Rather, the function is constrained absolutely to these bounds.

In terms of finding the optimal parameter combination α,β that maximizes the log-likelihood of the data, we can resort to any numerical optimization scheme. The simplest would be a brute force combinatorial exploration of the two-dimensional search space, whereas more complex optimization procedures take only a few, carefully planned steps to home in on the optimal values. Almost all statistical software, as well as general mathematical software packages, deliver functions for optimization.

In general, once the psychometric function has been fully determined, one usually reads off the associated threshold for the task—which is provided by the average of the chance level and the highest performance level.

14.4.2.2 Example

Let us return to the data from Chapter 7 for characterizing the motion sensitivity for five facial expressions using a computer-animated avatar. This is a classic 2-AFC task, in which participants are asked to indicate in each trial which of the two presented stimuli is faster. The data was generated by a total of 900 trials using a method-of-constant-stimuli paradigm with the standard motion strength (100%) presented together with a comparison stimulus in each trial. We have 900 trials, as we have 9 different motion strengths × 5 expressions × 20 repetitions. Let us focus on the data from one participant for now, which is plotted in Figure 14.4.

The goal is to fit five different psychometric functions to the data for each facial expression and then to check whether the thresholds for detecting differences in motion strength might differ across the expressions.

Taking the first set of data for the disgust expression, we have 9 strengths × 20 repetitions = 180 trials for fitting the function. This data is plotted in Figure 14.5 (note, the y-axis now reads probability instead of percent) together with a guess at a Weibull psychometric function with a turning point of $\alpha = 100$, and a steepness of $\beta = 15$. This function already looks quite close to the data, but perhaps α could be adjusted a little further to the right.

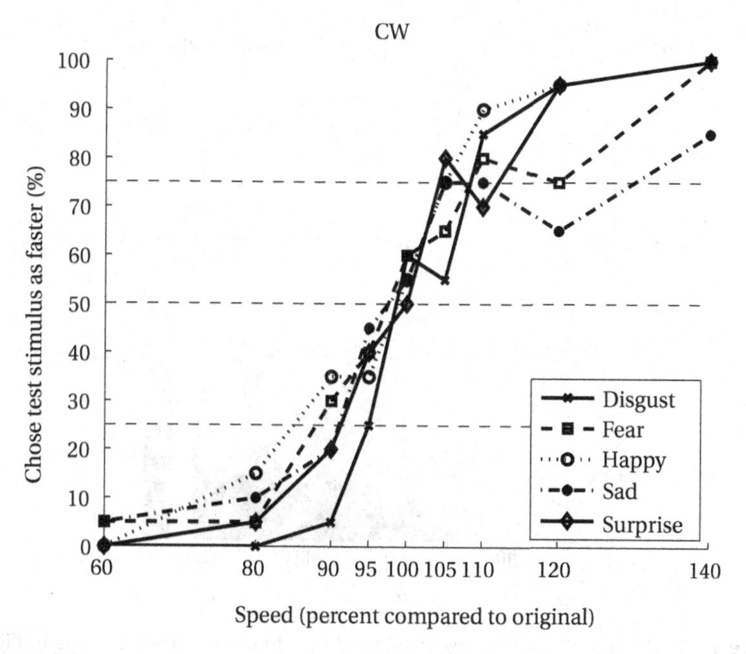

Figure 14.4. Data from one participant showing the detection of differences in motion strength for five facial expressions.

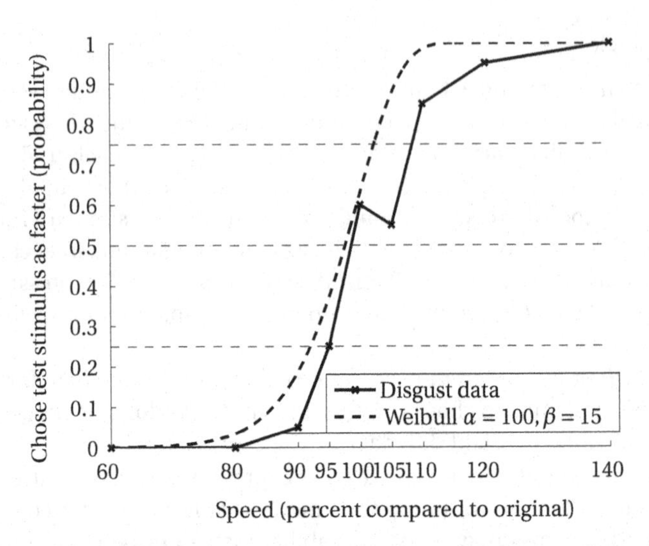

Figure 14.5. Data for the disgust expression together with a hypothetical Weibull fit.

Given our particular guess at the psychometric function, we can now evaluate the log-likelihood of this particular dataset

$$\log(P) = \sum_{i}^{n} c_i \log(\mathrm{PF}(s_i; \alpha, \beta)) + (1 - c_i)\log(1 - \mathrm{PF}(s_i; \alpha, \beta)),$$

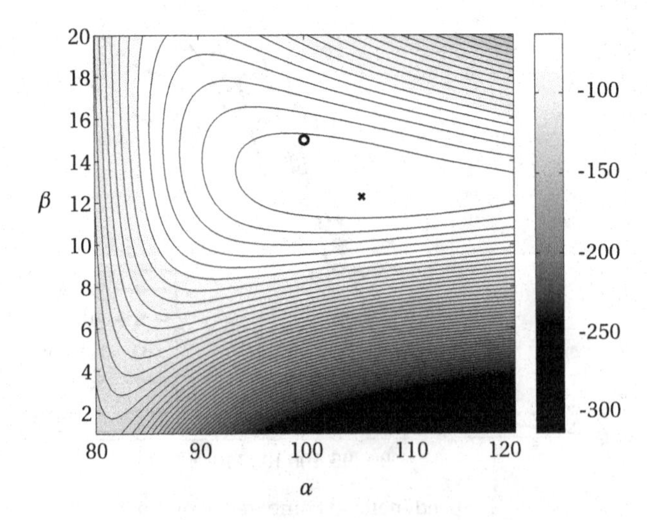

Figure 14.6. Contour plot of the log-likelihood $P(\alpha, \beta)$ for the observed data in Figure 14.5 for a parameter region. The hypothetical values for $\alpha = 100, \beta = 15$ are indicated with a circle, the optimal parameters $\alpha = 105, \beta = 12$ are indicated with a cross.

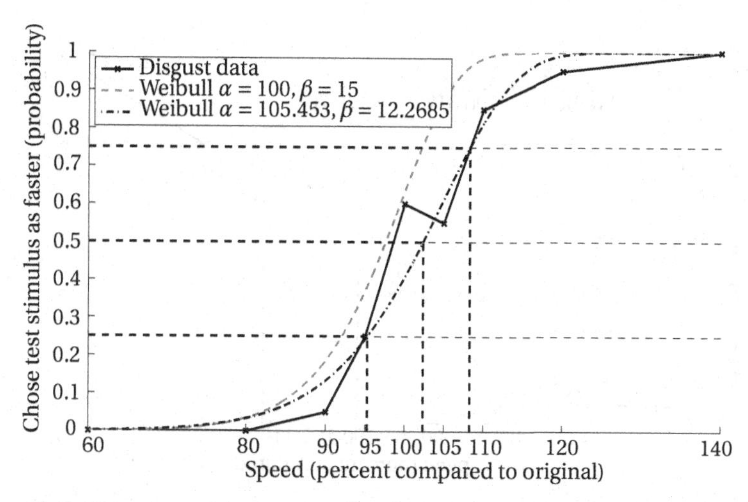

Figure 14.7. Data for the disgust expression together with the hypothetical Weibull fit and optimal fit.

where the c_i are simply our trial responses (0 if the participant chose the first stimulus and 1 if the participant chose the second stimulus) and the s_i are the nine levels of motion strengths we used.

Evaluating this using our hypothetical values of α, β gives a log-likelihood of $P = -74.9$. The next step is to optimize the function PF given the log-likelihood criterion. As we have already found quite a good guess for the parameters, we can actually visualize the log-likelihood in a region around the two parameter values. Doing so with $80 < \alpha < 120$ and $1 < \beta < 20$ yields a three-dimensional surface $P(\alpha, \beta)$, which is shown in Figure 14.6 as a contour plot. Our guess of $\alpha = 100, \beta = 15$ is shown as a circle—note that it does not quite hit the hilltop of the contour plot, but is already quite close.

Inserting our log-likelihood function into a gradient-free optimizer in Matlab, for example, will yield the following optimal parameters: $\alpha = 105.45$, $\beta = 12.27$, which are also plotted in Figure 14.6. The resulting, smoothly interpolated psychometric function for the disgust data is shown in Figure 14.7 as stippled lines together with the original guess and the experimental data—note how the optimal curve provides an excellent fit for our data.

In addition to the fit, the figure also shows the point of subjective equality (PSE), that is, the speed at which $PF(x; \alpha, \beta) = 0.5$, which in our case is PSE = 102.4. Hence, our data shows that point of subjective equality is very close to the normal speed of 100%. The just noticeable difference (JND) is given as the difference between the speeds at which $PF(x; \alpha, \beta) = 0.75$ and $PF(x; \alpha, \beta) = 0.25$, which here is JND = 108.8% − 95.3% = 13.5%: a speed difference of 13% between two facial expressions of disgust is therefore needed in order for them to be perceived as different.

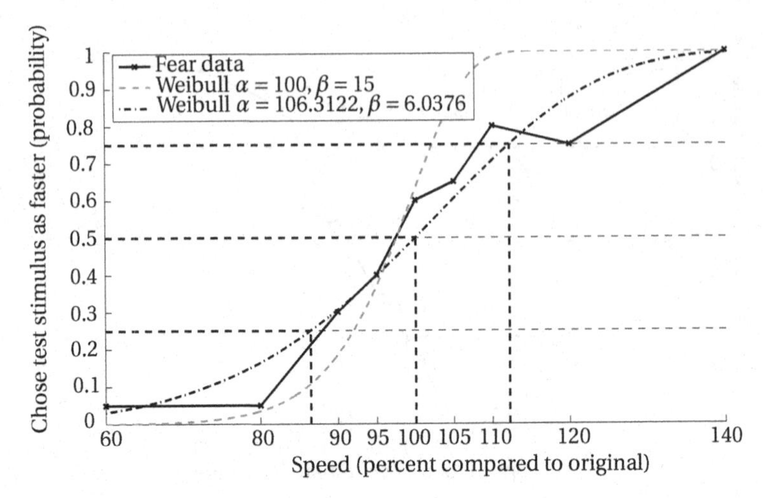

Figure 14.8. Data for the fear expression together with the hypothetical Weibull fit and optimal fit.

Figure 14.8 shows the results of the analysis for the fear expression. Notice that this curve does seem much shallower than the data for the disgust expression shown in Figure 14.7. Hence, our initial guess of $\alpha = 100, \beta = 15$ is definitely not fitting these results well. Accordingly, running the full optimization yields a much shallower β-value of $\beta = 6.04$, which results in a much higher JND of JND = 25.3. Note, however, that the PSE for this data is very similar, around PSE = 100. Whether or not the results for the fear and disgust expressions are actually different will be discussed in the next section.

14.4.3 Finding Confidence Intervals

After determining the psychometric function, the next problem that often occurs is how to calculate confidence intervals around, for example, the measured threshold, or PSE. As one might imagine, the confidence intervals are not normally distributed such that simple mean and standard error are not suitable. Whenever nonparametric confidence intervals need to be determined, one possible way is to resort to bootstrapping, or resampling of the data. The general idea behind bootstrapping is that one takes the optimal psychometric function and generates many different fictional datasets from this function that are randomly perturbed. For each of those datasets, the optimal psychometric function is determined and, with it, the corresponding threshold, or PSE. The resulting distribution of thresholds is a good, nonparametric representation of the real variation in the dataset and the associated uncertainty.

The only real trick in doing this is to create the random datasets for each bootstrapping iteration. For this, we create a series of random decisions for each trial in the series and *bias* it by the corresponding best-fit probability at that point, and then round down the result. By doing so we generate many different data points around the real curve that still follow the overall trend of the best-fitting function obtained for the real dataset.

It should be noted that this is not yet the full story, as the bootstrapping procedure needs to be corrected for biases, etc. However, the general idea for these methods is the same. For a full theoretical explanation, please refer to Wichmann and Hill (2001a,b). An excellent toolbox for Matlab and Python that can do many more sophisticated analyses is available from http://psignifit.sourceforge.net/.

14.4.3.1 Example I

Returning to the example above, we now wish to evaluate the confidence intervals around the PSE and the JND.

Figure 14.9 shows the results of such a bootstrap run for 1,000 repetitions, that is, for 1,000 different, randomized datasets for the the disgust data shown in Figure 14.7. It plots both datasets as a histogram with the optimal JND and PSE from the previous optimization together with a confidence interval of 75% around these values. The confidence interval for the JND values ranges from 12.2% to 13.4%, and that for the PSE from 101.2% to 104.9%. Note that the distribution of values in each case is non-Gaussian, confirming the need for this nonparametric estimate.

Figure 14.10 shows the results of the bootstrap run for 1,000 repetitions for the fear data. Again, both datasets are shown as a histogram with the optimal JND and PSE from the previous optimization with a 75% confidence interval around these values. The confidence interval for the JND values ranges from 24.0% to 27.1%, and for the PSE it ranges from 97.5% to 104.5%.

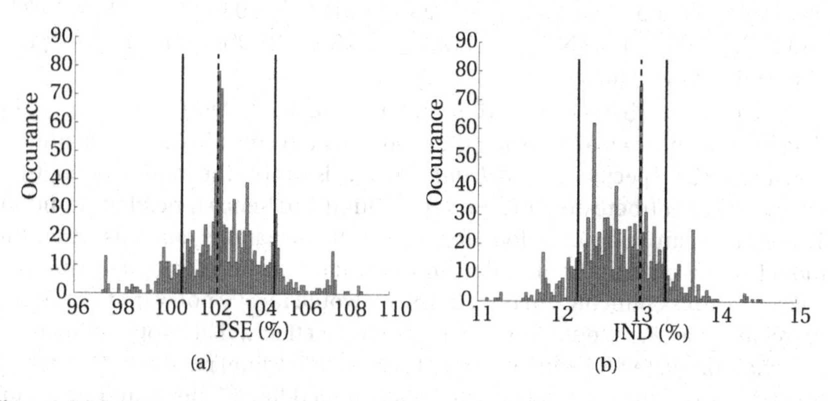

Figure 14.9. Histogram including bootstrapped confidence intervals for (a) PSE, and (b) JND for the disgust data shown in Figure 14.7.

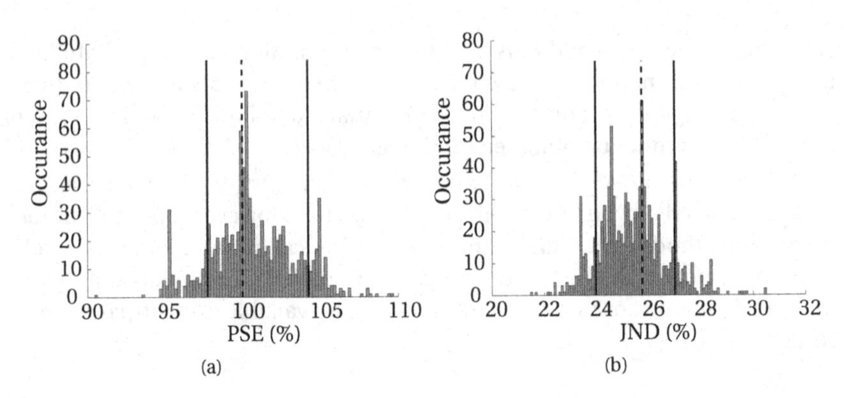

Figure 14.10. Histogram including bootstrapped confidence intervals for (a) PSE, and (b) JND for the fear data shown in Figure 14.8.

Comparing the two PSEs and JNDs for the disgust and fear expressions, we can conclude that whereas the PSEs for both expressions are similar with highly overlapping confidence intervals, the JNDs are not—that is, we need almost double the difference for the fear expression than for the disgust expression in order for the motion to become noticeable.

14.4.3.2 Example 2

Using the software toolbox psignifit, we obtain the two plot series for disgust and fear shown in Figure 14.11. In all cases, the data was fit with a Weibull psychometric function using maximum likelihood methods, and bootstrap methods were used to obtain confidence intervals. The maximum likelihood parameters for the disgust data are $\alpha = 106.0, \beta = 11.3$ and for the fear data $\alpha = 106.4, \beta = 6.0$, which is almost the same as the estimates obtained above. The PSE and JND values are $\text{PSE}_{\text{disg}} = 102.6\%$, $\text{JND}_{\text{disg}} = 109.1\% - 94.8\% = 14.3\%$, and $\text{PSE}_{\text{disg}} = 100.1\%$, $\text{JND}_{\text{disg}} = 112.4\% - 86.5\% = 25.9\%$ respectively—again, almost the same values.

The plot also shows the distribution of residuals (that is, the amount of the data left unexplained after fitting the model) with respect to the model prediction and with respect to the *block index*, which is simply the motion strength in our experiment (there are nine levels of motion and hence nine block indices). These plots can be used to look for systematic deviations (such as when the model always overshoots or undershoots the data), or for potential learning effects across task difficulty. In both cases, we would ideally not want to have any correlations as this would indicated systematic effects. The plots in the lower row show predicted (histogram) and observed (thick line) residuals. If the thick line falls out of the confidence intervals (stippled lines), this would be an indication of a bad fit, as then the predicted model would not be in accordance

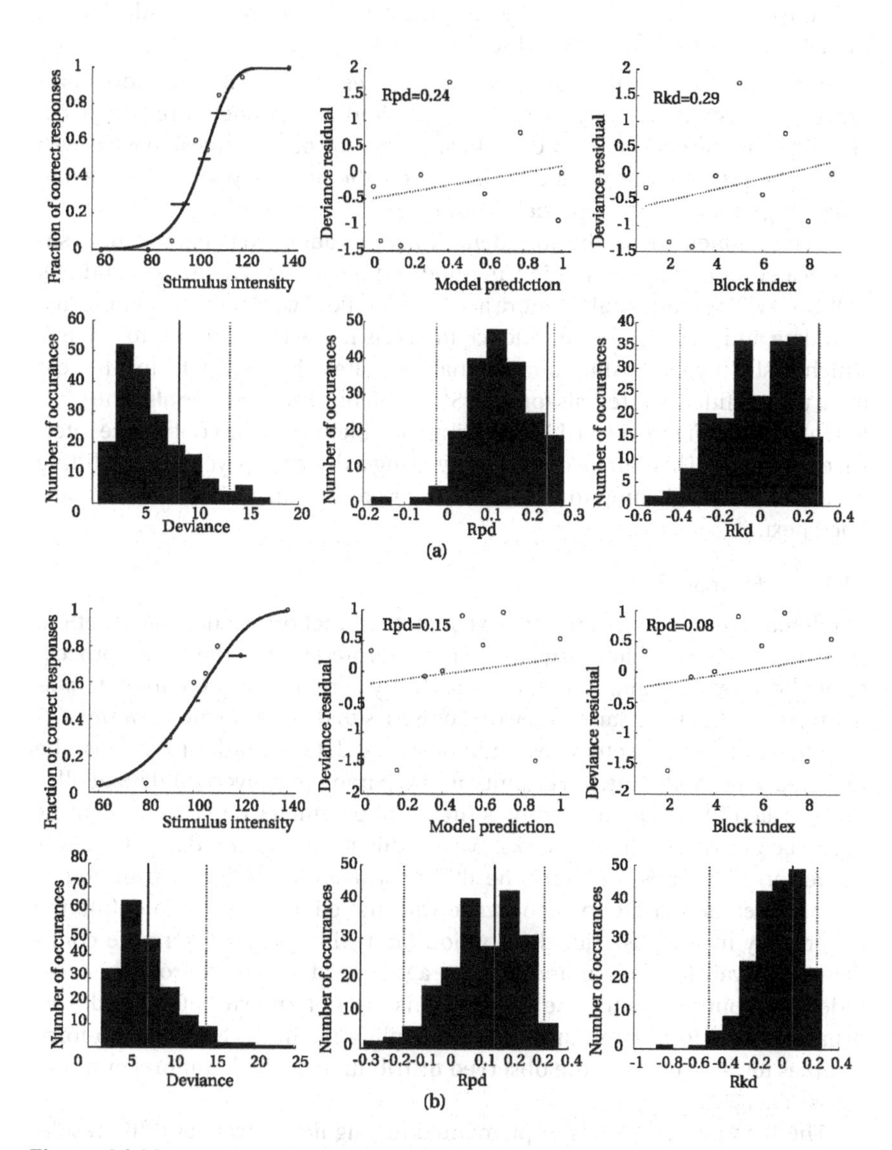

Figure 14.11. Series of analysis plots from the psignifit toolbox for the (a) disgust and (b) fear data. The plots show the fitted psychometric function including confidence intervals for the 0.25, 0.5, and 0.75 levels (top left plot), as well as various residual plots (see text for more details).

with the observed data. Again, the plots are available for residuals (deviance), model predictions, and block index.

In our case, we see that all correlations between residuals and model prediction or block indices are low (all $r < 0.2$) and we can therefore largely exclude systematic effects in our data. In addition, the observed residual lies well within the predicted residual distribution for the bootstrapped values, again indicating a good fit of the model to our data.

As mentioned above, the confidence intervals obtained using the more sophisticated bootstrap estimation methods employed by Wichmann and Hill (2001a,b) will usually result in more accurate confidence estimates. For the disgust data and the PSE the confidence intervals range from 100.1% to 105.8%, which is slightly larger than our original estimates. Interestingly, for the fear data, the confidence intervals for the PSE are highly distorted (see also the plotted intervals in Figure 14.11(b) indicating a problem with the confidence interval corrections. This can be corrected by using a different psychometric fitting function, or by switching to a different, more robust analysis method as outlined next.

14.4.3.3 Example 3

A different method for analyzing psychometric functions is the Bayes method (Kuss et al., 2005), which turns the usual interpretation of the fitting process on its head by regarding the data as perfectly certain and the parameters as unknown. In the Bayesian framework, one tries to determine the *posterior distribution* of the parameters given the observed data. In order to do this, the first step is to model the uncertainty in the parameters given no data at all—this is called the *prior* and captures the usual parameter range that might be observed across all possible tasks. Given this prior and the data, the Bayes framework now tries to determine the most likely posterior distribution of the parameters that are in accordance with the data. In order to define the uncertainty in the posterior distribution (and with that to determine confidence intervals for the parameters), we again need to sample from this posterior distribution. Since the distributions are not known beforehand, one usually resorts to Monte-Carlo type resampling methods, which try to force samples towards obeying the observed distribution as the iteration continues (Kuss et al., 2005).

The Bayes framework is implemented in psignifit as well, and the results for that estimation are shown in Figure 14.12. The resulting analysis plots are similar to the ones obtained in Figure 14.11, but now need to be interpreted slightly differently. First of all, the data is now modeled by a series of psychometric functions (remember that the parameters are the random variables in the Bayes sense), which are plotted in the top left panel. The average psychometric function is shown as the fully saturated curve with the whole family of curves sampled from the posterior distribution of parameters

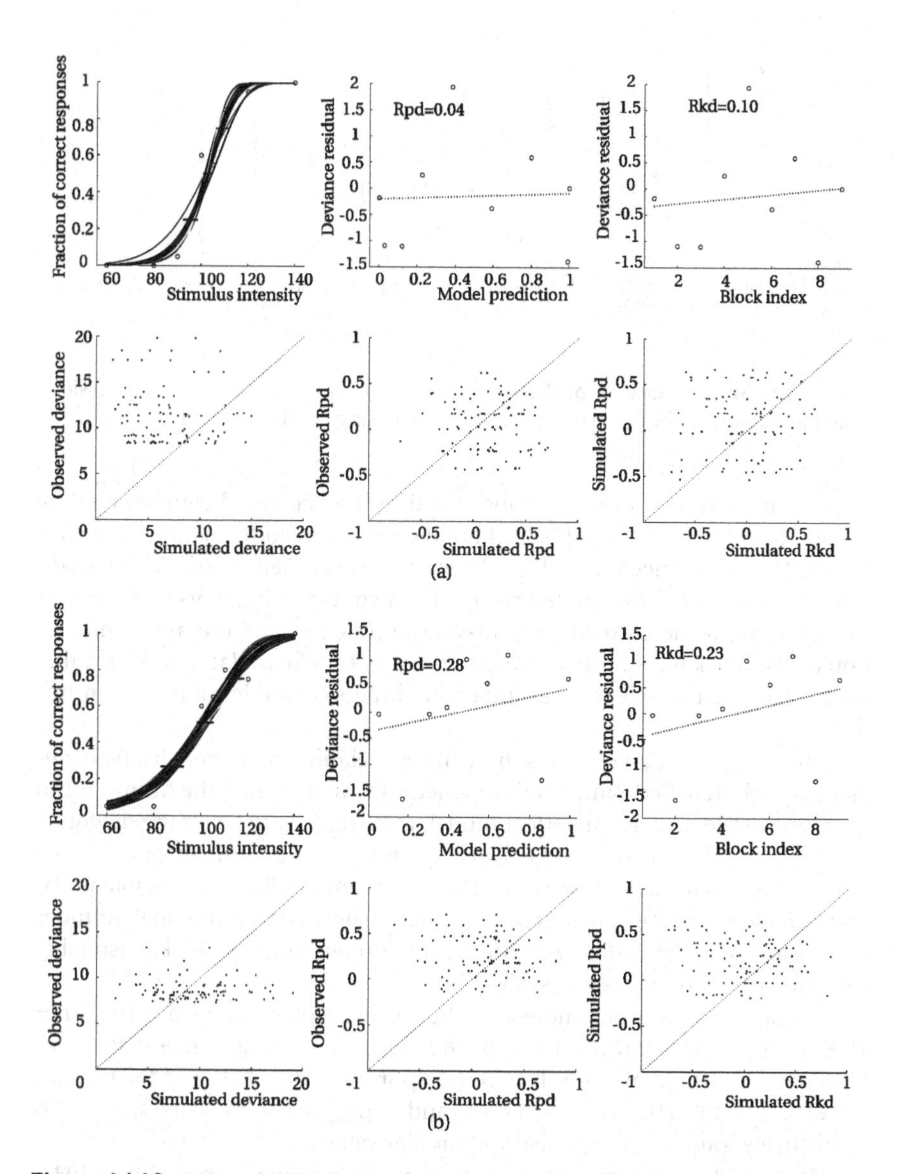

Figure 14.12. Series of analysis plots from the psignifit toolbox for the (a) disgust and (b) fear data based on Bayes fitting. The plots show the fitted psychometric function(s) including confidence intervals for the 0.25, 0.5, and 0.75 levels (top left plot), as well as various residual plots (see text for more details).

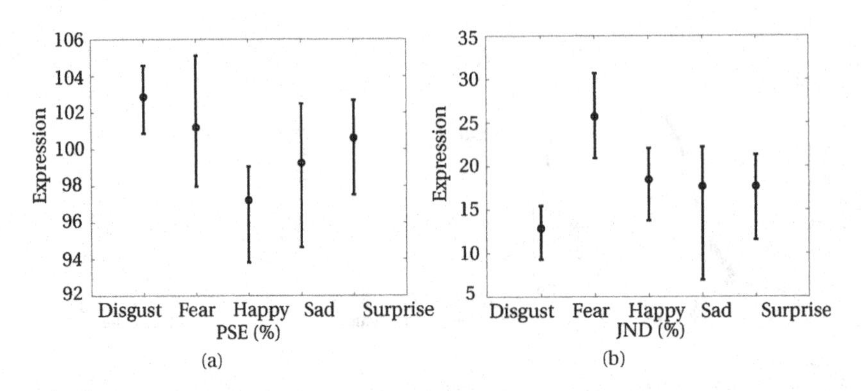

Figure 14.13. (a) PSEs and (b) JNDs with confidence intervals for all five expressions from participant 1. Confidence intervals obtained using the Bayesian method.

shown with saturation values proportional to the likelihood with which they explain the data (note that this likelihood does not include the prior distributions). The confidence intervals (in Bayesian terms called *credibility intervals*) are also indicated in the top left panel. The two remaining plots in the upper row again show the correlation between the residuals and the model prediction or the block index. Again, we observe very small correlations (all $r < 0.2$) for both disgust and fear data, indicating little systematic influence on our data.

The lower left panel now shows the correlation between observed deviances and predicted/simulated deviances obtained during the Monte-Carlo sampling procedure. For a perfect model, the data should lie on the diagonal; for well-fitted data it should at least be symmetric around the diagonal. Note that for the disgust data, there is a slight asymmetry indicating possible problems with the fit in the first panel, most likely associated with the small number of trials for each data point ($n = 20$). All other plots, however, look reasonably well-structured for both expressions.

The optimal Bayes parameters for the disgust expression are $\alpha = 105.9, \beta = 12.5$ and for the fear data $\alpha = 107.5, \beta = 6.1$; again very similar values to the ones obtained with the maximum likelihood approach. The PSE and JND values are $\text{PSE}_{\text{disg}} = 102.9\%$, $\text{JND}_{\text{disg}} = 12.8\%$, and $\text{PSE}_{\text{disg}} = 101.2\%$, $\text{JND}_{\text{disg}} = 25.7\%$ respectively—and again, we have very similar values.

The confidence intervals for the disgust PSE are now from 100.9 to 104.6 and for the fear expression from 98.0 to 105.1, respectively. Note that we now have a valid confidence interval for the fear data. Also note that both confidence intervals clearly overlap, showing that the PSEs for both expressions are likely not different. For the JND, however, we now have a much larger confidence interval from 20.3% to 30.1% for the fear expression than the one shown above. However, the confidence intervals for disgust (from 9.9% to 14.7%) and

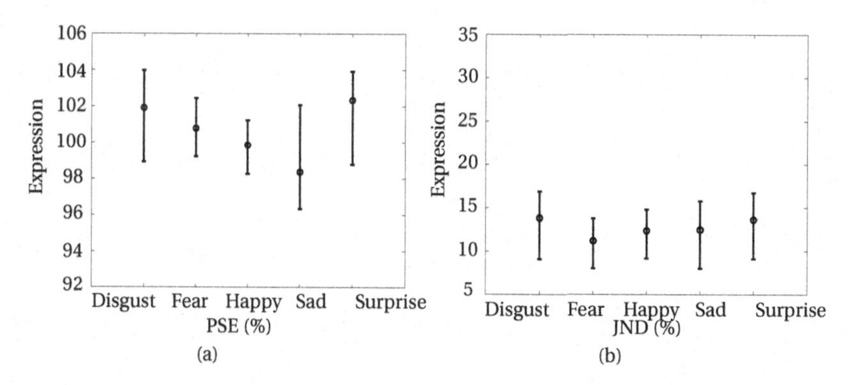

Figure 14.14. (a) PSEs and (b) JNDs with confidence intervals for all five expressions from participant 2. Confidence intervals obtained using the Bayesian method.

fear are very different, indicating that the JNDs for both expressions are likely different also in the Bayesian analysis framework.

Figure 14.13 plots the PSEs and JNDs together with the predicted Bayesian confidence intervals for all five expressions for participant 1. We can see that all PSEs hover around 100% with all confidence intervals overlapping, confirming the observation made earlier that all PSEs would be equal. For the JNDs, however, we see that the fear expression for this one participant does generate a much higher JND—at least when comparing disgust to fear.

Figure 14.14 plots the PSEs and JNDs together with the predicted Bayesian confidence intervals for all five expressions for participant 2. As for participant 1, all PSEs hover around 100% with all confidence intervals overlapping. In contrast to participant 1, however, all JNDs for this participant also seem to be equally placed (at around 13%) with all confidence intervals clearly overlapping. In order to conclude that expression thresholds for most expressions would be around 13%, we would need to run more participants as the one outlier data point for participant 1 does raise suspicions. However, the fits for the fear expression for that participant look reasonable, and thus without additional testing, it is hard to say whether the data was due to some erroneous decisions or whether this is a real effect.

Bibliography

D. C. Aboyoun and J. M. Dabbs. The Hess pupil dilation findings: Sex or novelty. *Social Behavior and Personality: an international journal*, 26(4):415–419, 1998.

R. Adolphs. Recognizing emotion from facial expressions: psychological and neurological mechanisms. *Behavioral and Cognitive Neuroscience Reviews*, 1:21–61, 2002.

K. Amaya, A. Bruderlin, and T. Calvert. Emotion from motion. In *Graphics Interface 1996*, pages 222–229, 1996.

K. Armel and V. Ramachandran. Projecting sensations to external objects: evidence from skin conductance response. *Proceedings of the Royal Society B: Biological Sciences*, 270(1523):1499, 2003.

S. E. Asch. Forming impressions of personality. *Journal of Abnormal Social Psychology*, 41:258–290, 1946.

American Psychological Association. *Publication Manual of the American Psychological Association, 6th edition.* American Psychological Association, Washington, DC, 2010.

J. S. Baily. Adaptation to prisms: Do proprioceptive changes mediate adapted behavior with ballistic arm movements? *Quarterly Journal of Experimental Psychology*, 24: 8–20, 1972.

A. Bartels and S. Zeki. The neural basis of romantic love. *Neuroreport*, 11(17):3829, 2000.

A. Battocchi and F. Pianesi. DaFEx: Un Database di Espressioni Facciali Dinamiche. In *SLI-GSCP Workshop "Comunicazione Parlata e Manifestazione delle Emozioni"*, pages 1–11. SLI-GSCP Workshop "Comunicazione Parlata e Manifestazione delle Emozioni," Padova (Italy) 30 Novembre–1 Dicembre, 2004.

R. M. Bauer. Autonomic recognition of names and faces in prosopagnosia: A neuropsychological application of the guilty knowledge test. *Neuropsychologia*, 22(4): 457–469, 1984.

W. M. Baum. *Understanding behaviorism: science, behavior, and culture.* HarperCollins College Publishers, New York, NY, 1994.

J. B. Bavelas, L. Coates, and T. Johnson. Listeners as co-narrators. *Journal of Personality and Social Psychology*, 79:941–952, 2000.

J. Beatty. Task-evoked pupillary responses, processing load, and the structure of processing resources. *Psychological Bulletin*, 91(2):276–292, 1982.

F. L. Bedford. Constraints on learning new mappings between perceptual dimensions. *Journal of Experimental Psychology: Human Perception and Performance*, 15:232–248, 1989.

F. L. Bedford. Perceptual and cognitive spatial learning. *Journal of Experimental Psychology: Human Perception and Performance*, 19:517–530, 1993a.

F. L. Bedford. Perceptual learning. *The Psychology of Learning and Motivation*, 30:1–60, 1993b.

H. Berger. Über das Elektrenkephalogramm des Menschen. *European Archives of Psychiatry and Clinical Neuroscience*, 87(1):527–570, 1929.

D. Bimler and G. Paramei. Facial-expression affective attributes and their configural correlates: Components and categories. *The Spanish Journal of Psychology*, 9:19–31, 2006.

V. Blanz and T. Vetter. A morphable model for the synthesis of 3d faces. In *SIGGRAPH'99 Conference Proceedings*, pages 187–194, 1999.

S. A. Book. Why $n-1$ in the formula for the sample standard deviation? *The Two-Year College Mathematics Journal*, 10(5):330–333, 1979.

I. Borg and P. Groenen. *Modern multidimensional scaling*. Springer, New York, 2nd edition, 2005.

J. Bosten, J. Robinson, G. Jordan, and J. Mollon. Multidimensional scaling reveals a color dimension unique to "color-deficient" observers. *Current Biology*, 15:950–952, 2005.

J. Botella, M. L. Garcia, and M. Barriopedro. Intrusion patterns in rapid serial visual presentation tasks with two response dimensions. *Perception & Psychophysics*, 52: 547–552, 1992.

R. Bracewell. *The Fourier Transform and Its Applications*. McGraw-Hill, Boston, 2000.

A. Bradley, B. C. Skottun, I. Ohzawa, G. Sclar, and R. D. Freeman. Visual orientation and spatial frequency discrimination: a comparison of single neurons and behavior. *Journal of Neurophysiology*, 57:755–772, 1987.

D. H. Brainard. The Psychophysics Toolbox. *Spatial Vision*, 10:433–436, 1997.

M. Breidt, C. Wallraven, D. W. Cunningham, and H. H. Bülthoff. Facial Animation Based on 3D Scans and Motion Capture. In N. Campbell, editor, *SIGGRAPH '03 Sketches & Applications*, New York, 2003. ACM Press.

D. E. Broadbent. *Perception and communication*. Pergamon, Oxford, 1958.

D. E. Broadbent and M. H. Broadbent. From detection to identification: response to multiple targets in rapid serial visual presentation. *Perception & Psychophysics*, 42: 105–113, 1987.

M. Brodeur, E. Dionne-Dostie, T. Montreuil, and M. Lepage. The Bank of Standardized Stimuli (BOSS), a New Set of 480 Normative Photos of Objects to Be Used as Visual Stimuli in Cognitive Research. *PLoS ONE*, 5(5):e10773, 05 2010.

M. W. Browne. A comparison of factor analytic techniques. *Psychometrika*, 33:267–334, 1968.

G. T. Buswell. *How people look at pictures*. University of Chicago Press, Chicago, 1935.

J. T. Cacioppo, L. G. Tassinary, and G. G. Berntson. *Handbook of psychophysiology*. Cambridge Univ Pr, 2007.

A. Caclin, S. McAdams, B. Smith, and S. Winsberg. Acoustic correlates of timbre space dimensions: A confirmatory study using synthetic tones. *Journal of the Acoustical Society of America*, 118:471–482, 2005.

P. Carrera-Levillain and J. Fernandez-Dols. Neutral faces in context: Their emotional meaning and their function. *Journal of Nonverbal Behavior*, 18:281–299, 1994.

E.C. Carter and R.C. Carter. Color and conspicuousness. *Journal of the Optical Society of America*, 71:723–729, 1981.

R.C. Carter. Visual search with color. *Journal of Experimental Psychology: Human Perception and Perception*, 8:127–136, 1982.

K. R. Castleman. *Digital Image Processing*. Prentice Hall, Englewood Cliffs, NJ, 1996.

J. M. Cattell. The inertia of the eye and brain. *Brain*, 8:295–312, 1886.

R. B. Cattell. The scree test for the number of factors. *Multivariate Behavioral Research*, 1:245–276, 1966.

R. B. Cattell and J. A. Jaspers. A general plasmode for factor analytic exercises and research. *Multivariate Behavioral Research Monographs*, pages 1–212, 1967.

C. R. Chapman, S. Oka, D. H. Bradshaw, R. C. Jacobson, and G. W. Donaldson. Phasic pupil dilation response to noxious stimulation in normal volunteers: Relationship to brain evoked potentials and pain report. *Psychophysiology*, 36(01):44–52, 1999.

C. Chiarello, S. Nuding, and A. Pollock. Lexical decision and naming asymmetries: Influence of response selection and response bias. *Brain and Language*, 34:302–314, 1988.

S. L. Chow. Iconic store and partial report. *Memory & Cognition*, 13:256–264, 1985.

N. Cliff. Adverbs as multipliers. *Psychological Review*, 66:27–44, 1959.

H. B. Cohen. Some critical factors in prism adaptation. *American Journal of Psychology*, 79:285–290, 1966.

J. Cohen. Things I Have Learned (So Far). *American Psychologist*, 45(12):1304–1312, 1990.

J. Cohen. A power primer. *Psychological bulletin*, 112(1):155, 1992.

J. Cohen, P. Cohen, S. West, and L. Aiken. *Applied multiple regression/correlation analysis for the behavioral sciences*, volume 1. Lawrence Erlbaum, Mahwah, NJ, 2003.

M. Cohen, S. Kosslyn, H. Breiter, G. DiGirolamo, W. Thompsonk, A. Anderson, S. Bookheimer, B. Rosen, and J. Belliveau. Changes in cortical activity during mental rotation A mapping study using functional MRI. *Brain*, 119(1):89–100, 1996.

T. Cooke, F. Jäkel, C. Wallraven, and H. Bülthoff. Multimodal similarity and categorization of novel, three-dimensional objects. *Neuropsychologia*, 45:484–495, 2007.

E. P. Cox, III. The optimal number of response alternatives for a scale: A review. *Journal of Marketing Research*, 17:407–422, 1980.

L. Coyne and P. S. Holzman. Three equivalent forms of a semantic differential inventory. *Educational and Psychological Measurement*, 21:665–674, 1966.

J. W. Creswell. *Research Design: Qualitative, Quantitative, and Mixed Methods Approaches, 3rd edition*. Sage Publications, Thousand Oaks, CA, 2009.

W. W. Cumming and R. Berryman. Some data on matching behavior in the pigeon. *Journal of the Experimental Analysis of Behavior*, 4:281–284, 1961.

D. W. Cunningham, M. Breidt, M. Kleiner, C. Wallraven, and H. H. Bülthoff. How believable are real faces?: Towards a perceptual basis for conversational animation. In *Computer Animation and Social Agents 2003*, pages 23–29, 2003a.

D. W. Cunningham, M. Breidt, M. Kleiner, C. Wallraven, and H. H. Bülthoff. The inaccuracy and insincerity of real faces. In *Proceedings of Visualization, Imaging, and Image Processing 2003*, 2003b.

D. W. Cunningham, M. Kleiner, C. Wallraven, and H. H. Bülthoff. Manipulating video sequences to determine the components of conversational facial expressions. *ACM Transactions on Applied Perception*, 2(3):251–269, 2005.

D. W. Cunningham, M. Nusseck, C. Wallraven, and H. H. Bülthoff. The role of image size in the recognition of conversational facial expressions. *Computer Animation & Virtual Worlds*, 15(3-4):305–310, July 2004a.

D. W. Cunningham, T. F. Shipley, and P. J. Kellman. The dynamic specification of surfaces and boundaries. *Perception*, 27:403–416, 1998.

D. W. Cunningham and C. Wallraven. Dynamic information for the recognition of conversational expressions. *Journal of Vision*, 9:1–17, 2009.

D. W. Cunningham, C. Wallraven, and H. H. Bülthoff. The semantic space of conversational expressions. 2012.

D. W. Cunningham, C. Wallraven, M. Kleiner, and H. H. Bülthoff. The components of conversational facial expressions. In *APGV 2004–Symposium on Applied Perception in Graphics and Visualization*, pages 143–149. ACM Press, 2004b.

A. R. Damasio, D. Tranel, and H. Damasio. Somatic markers and the guidance of behavior. *Human emotions: A reader*, pages 122–135, 1998.

F. C. Donders. On the speed of mental processes. *Acta Psychologica*, 30:412–431, 1969/1868.

J. Duncan. The locus of interference in the perception of simultaneous stimuli. *Psychological Review*, 87:272–300, 1980.

M. P. Eckstein. The lower efficiency for conjunctions is due to noise and not serial attentional processing. *Psychological Science*, 9:111–118, 1998.

M. P. Eckstein, J. P. Thomas, J. Palmer, and S. S. Shimozaki. A signal detection model predicts the effects of set size on visual search accuracy for feature, conjunction, triple conjunction, and disjunction displays. *Perception and Psychophysics*, 62:425–451, 2000.

S. Edelman. *Representation and recognition in vision*. MIT Press, 1999.

W. Einhäuser and P. König. Does luminance-contrast contribute to a saliency map for overt visual attention? *European Journal of Neuroscience*, 17(5):1089–1097, 2003.

N. I. Eisenberger, M. D. Lieberman, and K. D. Williams. Does rejection hurt? An fMRI study of social exclusion. *Science*, 302(5643):290, 2003.

P. Ekman. Universal and cultural differences in facial expressions of emotion. In J. R. Cole, editor, *Nebraska Symposium on Motoivation 1971*, pages 207–283. University of Nebraska Press, Lincoln, NE, 1972.

P. Ekman and W. Friesen. *Pictures of facial affect*. Consulting Psychologists Press, Palo Alto, CA, 1976.

P. Ekman and W. V. Friesen. *Facial Action Coding System*. Consulting Psychologists Press, Inc., Palo Alto, CA, 1978.

C. W. Eriksen and J.F. Collins. Visual perceptual rate under two conditions of search. *Journal of Experimental Psychology*, 80:489–492, 1969.

C. W. Eriksen and T. Spencer. Rate of information processing in visual perception: Some results and methodological considerations. *Journal ofExperimental Psychology*, 79:1–16, 1969.

J-C. Falmagne. *Elements of Psychophysical Theory*. Oxford University Press, 1985.

G. T. Fechner. *Elemente der Psychophysik*. Breitkopf und Hartel, Leipzig, 1860.

J. Fernandez-Dols, H. Wallbott, and F. Sanchez. Emotion category accessibility and the decoding of emotion from facial expression and context. *Journal of Nonverbal Behavior*, 15:107–124, 1991.

C. B. Ferster. Intermittent reinforcement of matching to sample in the pigeon. *Journal of the Experimental Analysis of Behavior*, 3:259–272, 1960.

D. H. Ffytche, R. J. Howard, M. J. Brammer, A. David, P. Woodruff, and S. Williams. The anatomy of conscious vision: An fMRI study of visual hallucinations. *Nature Neuroscience*, 1(8):738–742, 1998.

A. P. Field. *Discovering statistics using SPSS.* Sage Publications, 2009.

D. Field, T. F. Shipley, and D. W. Cunningham. Prism adaptation to dynamic events. *Perception and Psychophysics*, 61:161–176, 1999.

L. Fields and J. A. Nevin. Stimulus equivalence: A special issue of the psychological record. *The Psychological Record*, 43:541–844, 1993.

J. Fischer, D. Bartz, and W. Straßer. Artistic reality: Fast brush stroke stylization for augmented reality. In *Proc. of ACM Symposium on Virtual Reality Software and Technology (VRST)*, pages 155–158, November 2005a.

J. Fischer, D. Bartz, and W. Straßer. Illustrative Display of Hidden Iso-Surface Structures. In *Proc. of IEEE Visualization*, pages 663–670, October 2005b.

J. Fischer, M. Eichler, D. Bartz, and W. Straßer. Model-based Hybrid Tracking for Medical Augmented Reality. In *Eurographics Symposium on Virtual Environments (EGVE)*, 2006.

J. Fontaine, K. Scherer, E. Roesch, and P. Ellsworth. The world of emotions is not two-dimensional. *Psychological Science*, 18(12):1050–1057, 2007.

M. G. Frank and J. Stennett. The forced-choice paradigm and the perception of facial expressions of emotion. *Journal of Personality and Social Psychology*, 80:75–85, 2001.

M. Fredrikson and A. Öhman. Cardiovascular and electrodermal responses conditioned to fear-relevant stimuli. *Psychophysiology*, 16(1):1–7, 1979.

K. Friston, J. Ashburner, S. Kiebel, T. Nichols, and W. Penny, editors. *Statistical parametric mapping: The analysis of functional brain images.* Academic Press, 2006.

K. Friston, L. Harrison, and W. Penny. Dynamic causal modelling. *Neuroimage*, 19(4): 1273–1302, 2003.

K. Friston, A. Holmes, K. Worsley, J. Poline, C. Frith, and R. Frackowiak. Statistical parametric maps in functional imaging: a general linear approach. *Human Brain Mapping*, 2(4):189–210, 1995.

K. Friston, P. Jezzard, and R. Turner. Analysis of functional MRI time-series. *Human Brain Mapping*, 1(2):153–171, 2004.

F. Galton. Co-relations and their measurement. *Proceedings of the Royal Society of London*, 45:135–145, 1888.

X. M. Gao and J-Y. Lee. A factor analysis approach to measure the biased effects of retail fruit juice advertising. *Empirical Economics*, 20:93–107, 1995.

H. Gardner. *Frames of mind: The theory of multiple intelligences.* Basic Books, 1983.

I. Gauthier and M. J. Tarr. Becoming a "Greeble" expert: exploring mechanisms for face recognition. *Vision Research,* 37(12):1673–1682, 1997.

G. Gescheider. *Psychophysics: the fundamentals (3rd edition).* Lawrence Erlbaum Associates, 1997.

F. Gibbs, H. Davis, and W. Lennox. The electro-encephalogram in epilepsy and in conditions of impaired consciousness. *Archives of Neurology & Psychiatry (Chicago),* 34:1133–1148, 1935.

J. J. Gibson. *The ecological approach to visual perception.* Lawrence Erlbaum, Hillsdale, NJ, 1979.

G. Gillund and R. M. Shiffrin. A retrieval model for both recognition and recall. *Psycholical Review,* 91(1):1–67, 1984.

R. Goldstone and J. Son. *Similarity,* pages 13–36. New York: Cambridge University Press, 2005.

P. Gomez, R. Ratcliff, and M. Perea. A model of the go/no-go task. *Journal of experimental psychology: General,* 136(3):389–413, 2007.

R. C. Gonzalez and R. E. Woods. *Digital Image Processing, 3rd ed.* Prentice-Hall, Upper Saddle River, NJ., 2008.

M. Goto, H. Hashiguchi, T. Nishimura, and R. Oka. RWC music database: Music genre database and musical instrument sound database. In *Proceedings of ISMIR,* volume 3, pages 229–230. Citeseer, 2003.

N. J. Grahame, S. C. Hallam, L. Geier, and R. R. Miller. Context as an occasion setter following either cs acquisition and extinction or cs acquisition alone. *Learning and Motivation,* 21(3):237–265, 1990.

D. M. Green and J. A. Swets. Signal detection theory and psychophysics. 1974.

J. D. Greene, L. E. Nystrom, A. D. Engell, J. M. Darley, and J. D. Cohen. The neural bases of cognitive conflict and control in moral judgment. *Neuron,* 44(2):389–400, 2004.

R. T. Griesser, D. W. Cunningham, C. Wallraven, and H. H. Bülthoff. Psychophysical investigation of facial expressions using computer animated faces. In *APGV'07, 4th Symposium on Applied Perception in Graphics and Visualization,* pages 11–18. ACM Press, 2007.

L. Gugerty. Evidence from a partial report task for forgetting in dynamic spatial memory. *Human Factors,* 40:498–508, 1998.

K. L. Gwet. Computing inter-rater reliability and its variance in the presence of high agreement. *British Journal of Mathematical and Statistical Psychology,* 61(1):29–48, 2008.

K. L. Gwet. *Handbook of Inter-Rater Reliability, Second edition.* StatAxis Publishing, 2010.

A. R. Hakstian, W. D. Rogers, and R. B. Cattell. The behavior of numbers of factors rules with simulated data. *Multivariate Behavioral Research*, 17:193–219, 1982.

C. S. Harris. Perceptual adaptation to inverted, reversed, and displaced vision. *Psychological Review*, 72:419–444, 1965.

C. S. Harris. Insight or out of sight? two examoles of perecptual plasticity in the human adult. In C. S. Harris, editor, *Visual Coding and adaptability*, pages 95–149. Erlbaum, Hillsdale, NJ, 1980.

J. V. Haxby, M. I. Gobbini, M. L. Furey, A. Ishai, J. L. Schouten, and P. Pietrini. Distributed and overlapping representations of faces and objects in ventral temporal cortex. *Science*, 293(5539):2425, 2001.

W. L. Hayes. *Statistics, fifth edition.* Wadsworth Publishing, 1994.

J. D. Haynes and G. Rees. Decoding mental states from brain activity in humans. *Nature Reviews Neuroscience*, 7(7):523–534, 2006.

Z. J. He and K. Nakayama. Visual attention to surfaces in three-dimensional space. *Proceedings of the National academy of sciences*, 92:11155–11159, 1995.

R. Held and A. Hein. Adaptsation to disarranged hand-eye coordination contingent upon re-afferent stimulation. *Perceptual and Motor Skills*, 8:87–90, 1958.

H. von Helmholtz. *Handbook of Physiological Optics (3rd edit, J.P.C. Southall trans.).* Dover, New York, 1962/1867.

J. Henrich, S. J. Heine, and A. Norenzayan. The weirdest people in the world? *Behavioral and Brain Sciences*, 1(-1):1–23, 2006.

J. Henrich, S. J. Heine, and A. Norenzayan. Most people are not WEIRD. *Nature*, 466 (7302):29, 2010a.

J. Henrich, S. J. Heine, and A. Norenzayan. The weirdest people in the world? *Behavioral and Brain Sciences*, 33:61–83, 2010b.

R. Henson. What can functional neuroimaging tell the experimental psychologist? *The Quarterly Journal of Experimental Psychology Section A*, 58(2):193–233, 2005.

E. H. Hess. *Pupillometrics: A method of studying mental, emotional and sensory processes*, pages 491–531. 1972.

E. H. Hess and J. M. Polt. Pupil size as related to interest value of visual stimuli. *Science*, 132(3423):349–350, 1960.

D. R. Hiese. The semantic differential and attitude research. In Gene F. Summers, editor, *Attitude Measurement*, chapter 14. Chicago: Rand Mcnally, 1970.

S. A. Hillyard, R. F. Hink, V. L. Schwent, and T. W. Picton. Electrical signs of selective attention in the human brain. *Science*, 182(4108):177, 1973.

J. E. Hoffman. A two-stage model of visual search. *Perception and Psychophysics*, 25: 319–327, 1979.

P. C. Holland. Occasion setting in pavlovian feature positive discriminations. In M. L. Commons, R. J. Herrnstein, and A. R. Wagner, editors, *Quantitative analyses of behavior: Discrimination processes*, volume 4, pages 183–206. Ballinger, New York, 1983.

M. Hollins, R. Faldowski, S. Rao, and F. Young. Perceptual dimensions of tactile surface texture: A multidimensional scaling analysis. *Perception and Psychophysics*, 54(6): 687–705, 1993.

E. S. Howe. Probabilistic adverbial qualifications of adjectives. *Journal of Verbal Learning and Verbal Behavior*, 1:225–242, 1962.

E. S. Howe. Associative structure of quantifiers. *Journal of Verbal Learning and Verbal Behavior*, 5:156–162, 1966a.

E. S. Howe. Verb tense, negatives and other determinants of the intensity of evaluative meaning. *Journal of Verbal Learning and Verbal Behavior*, 5:147–155, 1966b.

D. C. Howell. *Statistical methods for psychology*. Wadsworth, Belmont, CA, 2009.

D. C. Howell. *Fundamental statistics for the behavioral sciences*. Wadsworth, Belmont, CA, 2010.

G. B. Huang, M. Ramesh, T. Berg, and E. Learned-Miller. Labeled faces in the wild: A database for studying face recognition in unconstrained environments. *University of Massachusetts, Amherst, Technical Report 07*, 49:1, 2007.

L. Huang and H. Pashler. Attention capacity and task difficulty in visual search. *Cognition*, 94:101–111, 2005.

E. B. Huey. *The psychology and pedagogy of reading*. Macmillan, 1908.

H. Intraub. Rapid conceptual identification of sequentially presented pictures. *Journal of Experimental Psychology: Human Perception and Performance*, 7:604–610, 1981.

L. Itti. Quantifying the contribution of low-level saliency to human eye movements in dynamic scenes. *Visual Cognition*, 12(6):1093–1123, 2005.

L. Itti and C. Koch. Computational modelling of visual attention. *Nature Reviews Neuroscience*, 2(3):194–203, March 2001.

D. Jack, R. Boian, A. Merians, S. V. Adamovich, M. Tremaine, M. Recce, G. C. Burdea, and H. Poizner. A virtual reality-based exercise program for stroke rehabilitation. In *Assets '00: Proceedings of the fourth international ACM conference on Assistive technologies*, pages 56–63, New York, 2000. ACM Press.

M. Janisse. Pupil size and affect: A critical review of the literature since 1960. *Canadian Psychologist/Psychologie canadienne*, 14:311–329, 1973.

Joint Commission Resources. *Approaches to pain management: an essential guide for clinical leaders*. The Joint Commission, Oakbrook Terrace, IL, 2003.

A. D. Jones, R. Cho, L. E. Nystrom, J. D. Cohen, and T. S. Braver. A computational model of anterior cingulate function in speeded response tasks: Effects of frequency, sequence and conflict. *Journal of Cognitive, Affective and Behavioral Neuroscience*, 2: 300–317, 2002.

F. Jones, K. Roberts, and E. Holman. Similarity judgments and recognition memory for common spices. *Perception and Psychophysics*, 24:2–6, 1978.

D. Kahneman and J. Beatty. Pupil diameter and load on memory. *Science*, 154(3756): 1583–1585, 1966.

H. F. Kaiser. The application of electronic computers to factor analysis. *Educational and Psychological Measurement*, 20:141–151, 1960.

T. Kanade, J.F. Cohn, and Y. Tian. Comprehensive database for facial expression analysis. In *Fourth IEEE International Conference on Automatic Face and Gesture Recognition*, pages 46–53, 2000.

G. K. Kanji. *100 statistical tests*. Sage Publications, 2006.

N. Kanwisher, J. McDermott, and M. M. Chun. The fusiform face area: a module in human extrastriate cortex specialized for face perception. *Journal of Neuroscience*, 17(11):4302, 1997.

K. Kaulard, D. W. Cunningham, H. H. Bülthoff, and C. Wallraven. The mpi facial expression database: A validated database of emotional and conversational facial expressions. *submitted*, 2012.

K. Kaulard, C. Wallraven, D. W. Cunningham, and H. H. Bülthoff. Laying the foundations for an in-depth investigation of the whole space of facial expressions. In *Journal of Vision*, volume 10, page 606, 2010.

K. N. Kay, T. Naselaris, R. J. Prenger, and J. L. Gallant. Identifying natural images from human brain activity. *Nature*, 452(7185):352–355, 2008.

D. Kehneman. *Attention and Effort*. Prentice-Hall, 1973.

H. H. Kelley. The warm-cold variable in first impressions of persons. In Arie W. Kruglanski and E. Tory Higgins, editors, *Social Psychology: A General Reader*. Social Psychology: A General Reader, 2003.

S. Khalfa, P. Isabelle, B. Jean-Pierre, and R. Manon. Event-related skin conductance responses to musical emotions in humans. *Neuroscience letters*, 328(2):145–149, 2002.

W. Kienzle, M. O. Franz, B. Schölkopf, and F. A. Wichmann. Center-surround patterns emerge as optimal predictors for human saccade targets. *Journal of Vision*, 9(5): 1–15, 2009.

H. Kirchner and S. J. Thorpe. Ultra-rapid object detection with saccadic eye movements: Visual processing speed revisited. *Vision Research*, 46(11):1762–1776, 2006.

M. Kleiner, D. Brainard, D. Pelli, A. Ingling, R. Murray, and C. Broussard. What's new in Psychtoolbox-3? *Perception (ECVP Abstract Supplement)*, 14, 2007.

M. Kleiner, C. Wallraven, and H. H. Bülthoff. The MPI Videolab–a system for high quality synchronous recording of video and audio from multiple viewpoints. Technical Report 123, Max Planck Institute for Biological Cybernetics, Tübingen, Germany, 2004.

W. Klimesch. EEG alpha and theta oscillations reflect cognitive and memory performance: a review and analysis. *Brain Research Reviews*, 29(2–3):169–195, 1999.

B. B. Koltuv. Some characteristics of intrajudge trait intercorrelations. *Journal of Abnormal and Social Psychology*, 552:all, 1962.

Z. Kourtzi and N. Kanwisher. Cortical regions involved in perceiving object shape. *Journal of Neuroscience*, 20(9):3310, 2000.

D. H. Kranz. A theory of magnitude estimation and crossmodality. *Journal of Mathematical Psychology*, 9:168–199, 1972.

M. Kuss, F. Jäkel, and F. A. Wichmann. Bayesian inference for psychometric functions. *Journal of Vision*, 5(5), 2005.

K. Kwong, J. Belliveau, D. Chesler, I. Goldberg, R. Weisskoff, B. Poncelet, D. Kennedy, B. Hoppel, M. Cohen, and R. Turner. Dynamic magnetic resonance imaging of human brain activity during primary sensory stimulation. *Proceedings of the National Academy of Sciences*, 89(12):5675–5679, 1992.

M. Land, N. Mennie, and J. Rusted. The roles of vision and eye movements in the control of activities of daily living. *Perception*, 28(11):1311–1328, 1999.

C. Landis. Studies of emotional reactions: General behavior and facial expressions. *Journal of Comparative Psychology*, 4:447–509, 1924.

J. R. Landis and G. G. Koch. The measurement of observer agreement for categorical data. *Biometrics*, 33(1):159–174, 1977.

K. E. Lasch. Culture, pain, and culturally sensitive pain care. *Pain Management Nursing*, 1:16–22, 2000.

D. H. Lawrence. Two studies of visual search for word targets with controlled rates of presentation. *Perception & Psychophysics*, 10:85–89, 1971.

S. Z. Li and A. K. Jain. *Handbook of face recognition*. Springer Verlag, 2005.

R. Likert. A technique for the measurement of attitudes. *Archives of Psychology*, 140: 1–55, 1932.

J. Lim and H. Lawless. Qualitative Differences of Divalent Salts: Multidimensional Scaling and Cluster Analysis. *Chemical Senses*, 30(9):719–726, 2005.

R. L. Linn. A monte carlo approach to the number of factors problem. *Psychometrika*, 33:37–71, 1968.

E. F. Loftus and J. C. Palmer. Reconstruction of auto-mobile destruction: An example of the interaction between language and memory. *Journal of Verbal Learning and Verbal Behaviour*, 13:585–589, 1974.

N. K. Logothetis. What we can do and what we cannot do with fMRI. *Nature*, 453(7197): 869–878, 2008.

N. K. Logothetis, J. Pauls, M. Augath, T. Trinath, and A. Oeltermann. Neurophysiological investigation of the basis of the fMRI signal. *Nature*, 412(6843):150–157, 2001.

R. D. Luce. *Response times: Their role in inferring elementary mental organization.* Oxford University Press, 1986.

R. D. Luce. "On the possible psychophysical laws" revisited: Remarks on cross-modal matching. *Psychological Review*, 97:66–77, 1990.

M. J. Lyons, S. Akamatsu, M. Kamachi, and J. Gyoba. Coding Facial Expressions with Gabor Wavelets. In *Third IEEE International Conference on Automatic Face and Gesture Recognition*, pages 200–205. IEEE Computer Society, 1998.

N. A. Macmillan. Signal detection theory. In J. Wixted, editor, *Stevens' Handbook of Experimental Psychology*, pages 43–90. John Wiley & Sons Inc., 2002.

K. Mania, H. H. Bülthoff, D. W. Cunningham, B. D. Adelstein, N. Mourkoussis, and J. Edward Swan II. Human-centered fidelity metrics for virtual environment simulations. In *VR '05: Proceedings of the IEEE Virtual Reality Conference 2005 (VR'05)*, page 308, Washington, DC, 2005. IEEE Computer Society.

S. Mannan, K. Ruddock, and D. Wooding. The relationship between the locations of spatial features and those of fixations made during visual examination of briefly presented images. *Spatial Vision*, 10(3):165–188, 1996.

A. A. J. Marley. Internal state models for magnitude estimation and related experiments. *Journal of Mathematical Psychology*, 9:306–319, 1972.

D. Marr. *Vision.* W.H. Freeman, San Francisco, 1982.

A. M. Martínez and R. Benavente. The ar face database. Technical Report 24, Computer Vision Center (CVC), June 1998.

E. Matin. Saccadic suppression: A review and an analysis. *Psychological Bulletin*, 81 (12):899–917, 1974.

D. Matsumoto and P. Ekman. *Japanese and Caucasian Facial Expression of Emotion (JACFEE) and Neutral Faces (JACNeuF).* SM Francisco. CA: Intercultural and Emotion Research Laboratory, Department of Psychology, San Francisco, 1988.

S. E. Maxwell and H. D. Delaney. *Designing Experiments and Analyzing Data.* Wadsworth Publishing, 1990.

G. Measso and E. Zaidel. Effect of response programming on hemispheric differences in lexical decision. *Neuropsychologia*, 28:635–646, 1990.

L. Mega and D. W. Cunningham. Completely non-verbal conversations. 2012.

A. Mehrabian and S. Ferris. Inference of attitudes from nonverbal communication in two channels. *Journal of Consulting Psychology*, 31:248–252, 1967.

R. Melzack and J. Katz. The mcgill pain questionnaire: appraisal and current status. In D. C. Turk and R. Melzack, editors, *Handbook of pain assessment*, pages 35–52. Guilford Press, New York, 2001.

J. Miller. The sampling distribution of d'. *Attention, Perception, & Psychophysics*, 58(1): 65–72, 1996.

M. S. Miron. The influence of instruction modification upon test-retest reliabilities of the semantic differential. *Educational and Psychological Measurement*, 21:883–893, 1961.

S. B. Mitsos. Personal constructs and the semantic differential. *Journal of Abnormal and Social Psychology*, 62:433–434, 1961.

Y. Miyawaki, H. Uchida, O. Yamashita, M. Sato, Y. Morito, H. C. Tanabe, N. Sadato, and Y. Kamitani. Visual image reconstruction from human brain activity using a combination of multiscale local image decoders. *Neuron*, 60(5):915–929, 2008.

F. Molnar. About the role of visual exploration in aesthetics. *Advances in intrinsic motivation and aesthetics*, pages 385–413, 1981.

B. Moulden. Adaptation to displaced vision: Reafference is a special case of the cue-discrepance hypothesis. *Quarterly Journal of Experimental Psychology*, 23:113–117, 1971.

G. Müller, J. Meseth, M. Sattler, R. Sarlette, and R. Klein. Acquisition, Synthesis and Rendering of Bidirectional Texture Functions. *Computer Graphics Forum*, 24(1): 83–109, 2005.

F. C. Müller-Lyer. Optische Urteilstäuschungen. *Archiv für Anatomie und Physiologie, Physiologische Abteilung*, 2:263–270, 1889.

A. L. Nagy and R. R. Sanchez. Critical color differences determined with a visual search task. *Journal of the Optical Society of America–A*, 7:1209–1217, 1990.

K. Nakayama and G. H. Silverman. Serial and parallel processing of visual feature conjunctions. *Nature*, 320:264–265, 1986.

G. Naor-Raz, M. J. Tarr, and D. Kersten. Is color an intrinsic property of object representation? *Perception*, 32(6):667–680, 2003.

M. C. Narayan. Culture's effects on pain assessment and management. *American Journal of Nursing*, pages 38–47, 2010.

L. Narens. A theory of ratio magnitude estimation. *Journal of Mathematical Psychology*, 40:109–129, 1996.

U. Neisser. *Cognitive Psychology*. Appleton-Century-Crofts, 1967.

W. T. Newsome, K. R. Britten, J. A. Movshon, and M. Shadlen. Single neurons and the perception of visual motion. In *Neural Mechanisms of Visual Perception*. Portfolio Publishing, 1989.

H. Nguyen. *GPU Gems 3*. Addison-Wesley Professional, 2007.

R. Nosofsky. Tests of an exemplar model for relating perceptual classification and recognition memory. *Journal of Experimental Psychology: Human Perception and Performance*, 17(1):3–27, 1991.

M. Nusseck, D. W. Cunningham, C. Wallraven, and H. H. Bülthoff. The contribution of different facial regions to the recognition of conversational expressions. *Journal of Vision*, 8:1–23, 2008.

K. O'Craven, P. Downing, and N. Kanwisher. fmri evidence for objects as the units of attentional selection. *Nature*, 401:584–587, 1999.

K. O'Craven and N. Kanwisher. Mental imagery of faces and places activates corresponding stimulus-specific brain regions. *Journal of Cognitive Neuroscience*, 12(6): 1013–1023, 2000.

S. Ogawa, T. Lee, A. Kay, and D. Tank. Brain magnetic resonance imaging with contrast dependent on blood oxygenation. *Proceedings of the National Academy of Sciences of the United States of America*, 87(24):9868–9872, 1990.

A. N. Oppenheim. *Questionnaire Design, Interviewing, and Attitude Measurement*. Continuum, 1992.

S. J. Orfanidis. *Introduction to signal processing*. Prentice-Hall, Upper Saddle River, NJ, 1995.

C. Osgood. *The Measurement of Meaning*. University of Illinois Press, 1957.

C. Osgood. Semantic differential technique in the comparative study of cultures. *American Anthropologist, New Series*, 66:171–200, 1964.

A. J. O'Toole, P. J. Phillips, F. Jiang, J. Ayyad, N. Pénard, et al. Face recognition algorithms surpass humans matching faces over changes in illumination. *IEEE Transactions on Pattern Analysis and Machine Intelligence*, pages 1642–1646, 2007.

D. A. Overton. State-dependent or "dissociated" learning produced with pentobarbital. *Journal of Comparative and Physiological Psychology*, 57:3–12, 1964.

D. A. Overton. Historical context of state dependent learning and discriminative drug effects. *Behavioral Pharmacology*, 2:253–264, 1991.

M. Pantic, M. F. Valstar, R. Rademaker, and L. Maat. Web-based database for facial expression analysis. In *IEEE Int'l Conf. Multimedia and Expo (ICME)*, Amsterdam, The Netherlands, July 2005.

D. J. Parkhurst and E. Niebur. Texture contrast attracts overt visual attention in natural scenes. *European Journal of Neuroscience*, 19(3):783–789, 2004.

T. Partala and V. Surakka. Pupil size variation as an indication of affective processing. *International Journal of Human-Computer Studies*, 59(1–2):185–198, 2003.

P. Paysan, R. Knothe, B. Amberg, S. Romdhani, and T. Vetter. A 3d face model for pose and illumination invariant face recognition. *2009 Advanced Video and Signal Based Surveillance*, pages 296–301, 2009.

W. Peavler. Pupil size, information overload, and performance differences. *Psychophysiology*, 11(5):559, 1974.

D. G. Pelli. The VideoToolbox software for visual psychophysics: Transforming numbers into movies. *Spatial Vision*, 10:437–442, 1997.

P. Phillips, W. Scruggs, A. O'Toole, P. Flynn, K. Bowyer, C. Schott, and M. Sharpe. FRVT 2006 and ICE 2006 Large-Scale Experimental Results. *IEEE Transactions on Pattern Analysis and Machine Intelligence*, 32(5):831–846, Jan 2010.

P. J. Phillips, P. J. Flynn, T. Scruggs, K. W. Bowyer, J. Chang, K. Hoffman, J. Marques, J. Min, and W. Worek. Overview of the face recognition grand challenge. 2005.

P. J. Phillips, H. Wechsler, J. Huang, and P. J. Rauss. The FERET database and evaluation procedure for face-recognition algorithms. *Image and Vision Computing*, 16(5): 295–306, 1998.

K. S. Pilz, I. M. Thornton, and H. H. Bülthoff. A search advantage for faces learned in motion. *Experimental Brain Research*, 171:436–437, 2006.

E. T. Pivik. *Sleep and Dreaming*, pages 633–662. Cambridge Univ Pr, New York, 2007.

D. Premack. *Intelligence in ape and man*. L. Erlbaum Associates, 1976.

G. M. Redding and B. Wallace. *Adaptive spatial alignment*. Erlbaum, Mahwah, NJ, 1997.

J. L. Renault et al. Brain potentials reveal covert facial recognition in prosopagnosia. *Neuropsychologia*, 27(7):905–912, 1989.

M. Richardson. Multidimensional psychophysics. *Psychological Bulletin*, 35:659–660, 1938. (Abstract cited by Torgerson, 1952).

B. Riecke, D. W. Cunningham, and H. H. Bülthoff. Spatial updating in virtual reality: the sufficiency of visual information. *Psychological Research*, 71:298–313, 2006.

B. Riecke, M. von der Heyde, and H. H. Bülthoff. Visual cues can be sufficient for triggering automatic, reflex-like spatial updating. *ACM Transactions on Applied Perception*, 2:183–215, 2005.

B. Rossion and G. Pourtois. Revisiting Snodgrass and Vanderwart's object pictorial set: The role of surface detail in basic-level object recognition. *Perception*, 33(2):217–236, 2004.

M. Rugg, A. Milner, C. Lines, and R. Phalp. Modulation of visual event-related potentials by spatial and non-spatial visual selective attention. *Neuropsychologia*, 25(1): 85–96, 1987.

J. A. Russell and M. Bullock. Fuzzy concepts and the perception of emotion in facial expressions. *Social Cognition*, 4:309–341, 1986.

J. A. Russell and B. Fehr. Relativity in the perception of emotion in facial expression. *Journal of Experimental Psychology: General*, 116:223–237, 1987.

A. Santi and W. A. Roberts. Prospective representation: The effects of varied mapping of sample stimuli to comparison stimuli and differential trial outcomes on pigeons' working memory. *Animal Learning and Behavior*, 13:103–108, 1985.

H. Schlosberg. Three dimensions of emotion. *Psychological review*, 61(2):81–88, 1954.

E. Schubert and D. Fabian. The dimensions of baroque music performance: a semantic differential study. *Psychology of Music*, 34:573–587, 2006.

A. Schwaninger, C. Wallraven, D. W. Cunningham, and S. D. Chiller-Glaus. Processing of facial identity and expression: A psychophysical, physiological and computational perspective. *Progress in Brain Research*, pages 325–348, 2006.

M. H. Segall, D. T. Campbell, and M. J. Herskovits. *The Influence of Culture on Visual Perception: An Advanced Study in Psychology and Anthropology*. Bobbs-Merrill, 1966.

R. Shepard. The analysis of proximities: multidimensional scaling with an unknown distance function. I. *Psychometrika*, 27:125–140, 1962.

R. N. Shepard. On the status of "direct" psychological measurement. In C. W. Savage, editor, *Minnesota studies in the philosophy of science (Vol 9)*. Minneapolis, MN: University of Minnesota Press, 1978.

R. N. Shepard. Psychological relations and psychophysical scales: On the status of "direct" psychophysical measurement. *Journal of Mathematical Psychology*, 24:21–57, 1981.

R. N. Shepard and J. Metzler. Mental rotation of three-dimensional objects. *Science*, 171(3972):701–703, 1971.

P. Shilane, P. Min, M. Kazhdan, and T. Funkhouser. The Princeton Shape Benchmark. In *Proceedings of Shape Modeling Applications, 2004*, pages 167–178. IEEE, 2004.

P. E. Shrout and J. L. Fleiss. Intraclass correlations: uses in assessing rater reliability. *Psychological Bulletin*, 86(2):420–428, 1979.

T. Sim, S. Baker, and M. Bsat. The CMU Pose, Illumination, and Expression Database. *IEEE Transactions on Pattern Analysis and Machine Intelligence*, 25(12):1615–1618, December 2003.

C. Spearman. "General intelligence" objectively determined and measured. *American Journal of Psychology*, 15:201–293, 1904.

R. Spence. Rapid, serial and visual: a presentation technique with potential. *Information Visualization*, 1:13–19, 2002.

G. Sperling. The information available in brief visual presentations. *Psychological Monographs*, 74:1–29, 1960.

G. Sperling and J. Weichshelgartner. Episodic theory of the dynamics of spatial attention. *Psychological Review*, 102:503– 532, 1995.

R. J. Sternberg. *Beyond IQ: A Triarchic Theory of Intelligence.* Cambridge University Press, 1985.

S. S. Stevens. On the theory of scales of measurement. *Science*, 103:677–680, 1946.

S. S. Stevens. Mathematics, measurement and psychophysics. In S. S. Stevens, editor, *Handbook of experimental psychology*, page 149. New York, Wiley, 1951.

G. Stratton. Vision without inversion of the retinal image. *Psychological Review*, 4: 341–360, 1897.

S. Sutton, M. Braren, J. Zubin, and E. R. John. Evoked-potential correlates of stimulus uncertainty. *Science*, 150(3700):1187, 1965.

B. W. Tatler, R. J. Baddeley, and I. D. Gilchrist. Visual correlates of fixation selection: Effects of scale and time. *Vision Research*, 45(5):643–659, 2005.

W. Thompson, R. Fleming, S. Creem-Regehr, and J. K. Stefanucci. *Visual Perception from a Computer Graphics Perspective.* AK Peters, 2011.

I. M. Thornton and Z. Kourtzi. A matching advantage for dynamic faces. *Perception*, 31:113 –132, 2002.

S. Thorpe, D. Fize, and C. Marlot. Speed of processing in the human visual system. *nature*, 381(6582):520–522, 1996.

R. H. Thouless. Phenomenal regression to the "real" object I. *British Journal of Psychology*, 21:339–359, 1931.

L. L. Thurstone. Multiple factor analysis. *Psychological Review*, 38:406–427, 1931.

L. L. Thurstone. *Primary mental abilities.* University of Chicago Press, 1938.

W. Torgerson. Multidimensional scaling: I. theory and method. *Psychometrika*, 17: 401–419, 1952.

J. T. Townsend. Serial and within-stage independent parallel model equivalence on the minimum completion time. *Journal of Mathematical Psychology*, 14:219–238, 1976.

J. T. Townsend. Serial vs. parallel processing: Sometimes they look like tweedledum and tweedledee but they can (and should) be distinguished. *Psychological Science*, 1:46–54, 1990.

D. Tranel and A. R. Damasio. Knowledge without awareness: An autonomic index of facial recognition by prosopagnosics. *Science*, 228(4706):1453, 1985.

D. Tranel, H. Damasio, and A. R. Damasio. Double dissociation between overt and covert face recognition. *Journal of Cognitive Neuroscience*, 7(4):425–432, 1995.

D. Tranel, D. Fowles, and A. R. Damasio. Electrodermal discrimination of familiar and unfamiliar faces: A methodology. *Psychophysiology*, 22(4):403–408, 1985.

A. Treisman and G. Gelade. A feature integration theory of attention. *Cognitive Psychology*, 12:97–136, 1980.

A. M. Treisman and S. Sato. Conjunction search revisited. *Journal of Experimental Psychology: Human Perception and Performance*, 16:459–478, 1990.

H. C. Triandis. Differential perception of certain jobs and people by managers, clerks, and workers in industry. *Journal of Applied Psychology*, 43:221–225, 1959.

N. F. Troje and H. H. Bülthoff. Face recognition under varying poses: The role of texture and shape. *Vision Research*, 36(12):1761–1771, 1996.

R. Ulrich, S. Mattes, and J. O. Miller. Donders's assumption of pure insertion: An evaluation on the basis of response dynamics. *Acta Psychologica*, 102:43–75, 1999.

T. Valentine. A unified account of the effects of distinctiveness, inversion and race in face recognition. *Quarterly Journal of Experimental Psychology*, 43A:161–204, 1991.

M. L. Vo, A. M. Jacobs, L. Kuchinke, M. Hofmann, M. Conrad, A. Schacht, and F. Hutzler. The coupling of emotion and cognition in the eye: Introducing the pupil old/new effect. *Psychophysiology*, 45(1):130–140, 2008.

E. Vul, C. Harris, P. Winkielman, and H. Pashler. Puzzlingly high correlations in fmri studies of emotion, personality, and social cognition. *Perspectives on Psychological Science*, 4(3):274, 2009.

N. Wade and B. W. Tatler. *The moving tablet of the eye: The origins of modern eye movement research.* Oxford University Press, 2005.

H. Wallbott and K. Scherer. Cues and Channels in Emotional Recognition. *Journal of Personality and Social Psychology*, 51(4):690–699, October 1986.

G. Wallis, A. Chatziastros, and H. H. Bülthoff. An unexpected role for visual feedback in vehicle steering control. *Current Biology*, 12:295–299, 2002.

G. Wallis, A. Chatziastros, J. Tresilian, and N. Tomasevic. The role of visual and non-visual feedback in a vehicle steering task. *Journal of Experimental Psychology: Human Perception and Performance*, 33:1127–1144, 2007.

C. Wallraven, M. Breidt, D. W. Cunningham, and H. H. Bülthoff. Psychophysical evaluation of animated facial expressions. In *APGV '05: Proceedings of the 2nd symposium on Appied perception in graphics and visualization*, pages 17–24, New York, 2005. ACM Press.

C. Wallraven, M. Breidt, D. W. Cunningham, and H. H. Bülthoff. Evaluating the perceptual realism of animated facial expressions. *ACM Transactions on Applied Perception*, 2006.

C. Wallraven, H. H. Bülthoff, J. Fischer, D. W. Cunningham, and D. Bartz. Evaluation of real-world and computer-generated stylized facial expressions. *ACM Transactions on Applied Perception*, 4(3):1–24, 11 2007.

C. Wallraven, D. W. Cunningham, M. Breidt, and H. H. Bülthoff. View dependence of complex versus simple facial motions. In H. H. Bülthoff and H. Rushmeier, editors, *Proceedings of the First Symposium on Applied Perception in Graphics and Visualization*, page 181. ACM SIGGRAPH, 2004.

C. Wallraven, D. W. Cunningham, J. Rigau, M. Feixas, and M. Sbert. Aesthetic appraisal of art: from eye movements to computers. In *Fifth International Symposium on Computational Aesthetics in Graphics, Visualization, and Imaging*, pages 137–144, 05 2009.

C. Wallraven, M. Schultze, B. Mohler, A. Vatakis, and K. Pastra. The POETICON enacted scenario corpus–a tool for human and computational experiments on action understanding. In *Proceedings of IEEE Conference on Automatic Face and Gesture Recognition (FG 2011)*, 2011.

D. Walther and C. Koch. Modeling attention to salient proto-objects. *Neural Networks*, 19(9):1395–1407, Jan 2006.

J. P. Wann, S. K. Rushton, M. Smyth, and D. Jones. Rehabilitative environments for attention and movement disorders. *Commun. ACM*, 40(8):49–52, 1997.

R. B. Welch. *Perceptual modification: Adapting to altered sensory environments*. Academic Press, New York, 1978.

G. L. Wells and E. A. Olson. Eyewitness testimony. *Annual Review of Psychology*, 54: 277–295, 2003.

W. D. Wells and G. Smith. Four semantic rating scales compared. *Journal of Applied Psychology*, 44:393–397, 1960.

F. A. Wichmann and N. J. Hill. The psychometric function: I. fitting, sampling, and goodness of fit. *Perception and Psychophysics*, 63(8):1293–1313, 2001a.

F. A. Wichmann and N. J. Hill. The psychometric function: II. bootstrap-based confidence intervals and sampling. *Perception and Psychophysics*, 63(8):1314–1329, 2001b.

L. M. Williams, M. L. Phillips, M. J. Brammer, D. Skerrett, J. Lagopoulos, C. Rennie, H. Bahramali, G. Olivieri, A. S. David, A. Peduto, et al. Arousal dissociates amygdala and hippocampal fear responses: evidence from simultaneous fMRI and skin conductance recording. *Neuroimage*, 14(5):1070–1079, 2001.

J. M. Wolfe. Visual search. In H. Pashler, editor, *Attention*. University College London Press, 1998.

J. M. Wolfe, K. R. Cave, and S. L. Franzel. Guided search: an alternative to the feature integration model for visual search. *Journal of Experimental Psychology: Human Perception and Performance*, 3:419–433, 1989.

J. R. Wolpaw and D. J. McFarland. Control of a two-dimensional movement signal by a noninvasive brain-computer interface in humans. *Proceedings of the National Academy of Sciences of the United States of America*, 101(51):17849, 2004.

J. R. Wolpaw, D. J. McFarland, G. W. Neat, and C. A. Forneris. An EEG-based brain-computer interface for cursor control. *Electroencephalography and clinical neurophysiology*, 78(3):252–259, 1991.

W. Wundt. *Grundzuge der physiologischen Psychologie [Principles of physiological psychology]*. Verlag W. Engimann, 1880.

A. L. Yarbus. *Eye movements and vision*. Plenum, New York, 1967.

L. Yin, X. Chen, Y. Sun, T. Worm, and M. Reale. A high-resolution 3D dynamic facial expression database. In *Automatic Face & Gesture Recognition, 2008. FG'08. 8th IEEE International Conference on*, pages 1–6. IEEE, 2008.

L. Yin, X. Wei, Y. Sun, J. Wang, and M. J. Rosato. A 3D Facial Expression Database For Facial Behavior Research. *Proceedings of the 7th International Conference on Automatic Face and Gesture Recognition*, pages 211–216, 2006.

V. H. Yngve. On getting a word in edgewise. In *Papers from the Sixth Regional Meeting of the Chicago Linguistic Society*, pages 567–578. Chicago Linguistic Society, Chicago, 1970.

Index

Printed in the United States
by Baker & Taylor Publisher Services

Printed in the United States
by Baker & Taylor Publisher Services